Springer Series in **Materials Science** 17

Springer Series in **Materials Science**

Advisors: M. S. Dresselhaus · H. Kamimura · K. A. Müller
Editors: U. Gonser · A. Mooradian · R. M. Osgood · M. B. Panish · H. Sakaki
Managing Editor: H. K. V. Lotsch

1 **Chemical Processing with Lasers**
By D. Bäuerle

2 **Laser-Beam Interactions with Materials**
Physical Principles and Applications
By M. von Allmen

3 **Laser Processing of Thin Films and Microstructures**
Oxidation, Deposition and Etching of Insulators
By I. W. Boyd

4 **Microclusters**
Editors: S. Sugano, Y. Nishina, and S. Ohnishi

5 **Graphite Fibers and Filaments**
By M. S. Dresselhaus, G. Dresselhaus, K. Sugihara, I. L. Spain, and H. A. Goldberg

6 **Elemental and Molecular Clusters**
Editors: G. Benedek, T. P. Martin, and G. Pacchioni

7 **Molecular Beam Epitaxy**
Fundamentals and Current Status
By M. A. Herman and H. Sitter

8 **Physical Chemistry of, in and on Silicon**
By G. F. Cerofolini and L. Meda

9 **Tritium and Helium-3 in Metals**
By R. Lässer

10 **Computer Simulation of Ion–Solid Interactions**
By W. Eckstein

11 **Mechanisms of High Temperature Superconductivity**
Editors: H. Kamimura and A. Oshiyama

12 **Dislocation Dynamics and Plasticity**
By T. Suzuki, S. Takeuchi, and H. Yoshinaga

13 **Semiconductor Silicon**
Materials Science and Technology
Editors: G. Harbeke and M. J. Schulz

14 **Graphite Intercalation Compounds I**
Structure and Dynamics
Editors: H. Zabel and S. A. Solin

15 **Crystal Chemistry of High T_c Superconducting Copper Oxides**
By B. Raveau, C. Michel, M. Hervieu, and D. Groult

16 **Hydrogen in Semiconductors**
By S. J. Pearton, M. Stavola, and J. W. Corbett

17 **Ordering at Surfaces and Interfaces**
Editors: A. Yoshimori, T. Shinjo, and H. Watanabe

18 **Graphite Intercalation Compounds II**
Editors: S. A. Solin and H. Zabel

19 **Laser-Assisted Microtechnology**
By S. M. Metev and V. P. Veiko

20 **Microcluster Physics**
By S. Sugano

A. Yoshimori T. Shinjo
H. Watanabe (Eds.)

Ordering at Surfaces and Interfaces

Proceedings of the Third NEC Symposium
Hakone, Japan, October 7–11, 1990

With 215 Figures

Springer-Verlag
Berlin Heidelberg New York
London Paris Tokyo
Hong Kong Barcelona
Budapest

Professor Akio Yoshimori
Department of Material Physics,
Osaka University,
Toyonaka, Osaka 560, Japan

Dr. Hisatsune Watanabe
Fundamental Research Laboratories,
NEC Corporation, 34 Miyukigaoka,
Tsukuba, Ibaraki 305, Japan

Professor Teruya Shinjo
Institute for Chemical Research,
Kyoto University,
Uji, Kyoto-fu 611, Japan

ISBN-13:978-3-642-84484-3 e-ISBN-13:978-3-642-84482-9
DOI: 10.1007/978-3-642-84482-9

Library of Congress Cataloging-in-Publication Data NEC Symposium (3rd : 1990 : Hakone-machi, Japan) Ordering at surface and interfaces : proceedings of the Third NEC Symposium, Hakone, Japan, October 7–11, 1990 / A. Yoshimori, T. Shinjo, H. Watanabe, Eds. p. cm – (Springer series in materials sciences ; v. 17) Includes bibliographical references and index.ISBN-13:978-3-642-84484-31.Semiconductors–Surfaces–Congresses.2.Order-disorder models–Congresses. 3. Surfaces (Physics)–Congresses. I. Yoshimori, A. (Akio), 1927– . II. Shinjo, Teruya, 1938–. III. Watanabe, H. Hisatsune), 1943– . IV. Title. V. Series. QC611.6.S9N43 1990 530.4'17–dc20 91-33140

Softcover reprint of the hardcover 1st edition 1992

Typesetting: Camera ready by authors
54/3140-543210 – Printed on acid-free paper

Preface

This volume contains the proceedings of the third in a series of biennial NEC Symposia on Fundamental Approaches to New Material Phases sponsored by the NEC Corporation, Tokyo, Japan. The symposium was held from October 7 to 11, 1990, at the Hakone Kanko Hotel in Hakone. About 40 invited participants stayed together, became involved in intense discussions, and freely exchanged ideas both in and out of the conference room, which faced Mt. Fuji, the beautiful lake Ashinoko, and the quiet landscape in the old crater.

The title of this volume, *Ordering at Surfaces and Interfaces*, which was also the title of the third symposium, describes the aim of the symposium: to discuss ordering properties and their underlying mechanisms at surfaces and interfaces. The topics treated include the reconstruction of surfaces of semiconductors and metals, atomic and magnetic ordering at interfaces, theoretical tools to study ordering mechanisms at surfaces and interfaces, ordering in adsorbate–surface systems, such as alkali-adsorbed silicon surfaces, electric current effects on semiconductor surfaces and many related STM (scanning tunneling microscopy) results.

Surface and interface physics is in a dramatic period of rapid development at present. Not only STM but many other powerful tools, both experimental and theoretical, are now available. Our understanding of the fundamentals of surfaces and interfaces is advancing very rapidly, and phenomena, concepts, and even principles unique to surfaces and interfaces are beginning to show up to such an extent that it is tempting to consider surfaces and interfaces as new material phases. As was expected, it proved most fruitful to discuss the present status of these fields to obtain an overview of them for further developments.

On this occasion, we would like to express our sincere thanks to all the participants for their contributions. We are also indebted to the following members of the Steering Committee of the NEC Symposium for choosing "ordering on surfaces and interfaces" as the subject of the third symposium: Professor S. Sugano, Himeji Institute of Technology, Chair of the Steering Committee and Advisor to the Organizing Committee; Dr.M. Uenohara, Executive Advisor, NEC Corporation; Dr.D. Shinoda, Executive Vice President, NEC Research Institute, Inc. in USA; Dr.F. Saito, Vice President, NEC Corporation; Professor H. Kamimura, University of Tokyo; Professor Y. Nishina, Tohoku University; and Dr. S. Onishi (Secretary), NEC Corporation, who is also a member of the

Organizing Committee. We thank the members of the Organizing Committee for the Third Symposium: Professor A. Ichimiya, Professor O. Nishikawa, Professor K. Takayanagi, Professor K. Terakura, Professor M. Tsukada and Dr.T. Mizutani (Secretary).

Osaka, Kyoto, Ibaraki
November 1990

A. Yoshimori
T. Shinjo
H. Watanabe

Contents

Part I Surface Analysis

Part II Surface Ordering

Part IV **Interface Ordering**

Part I

Surface Analysis

Ion and X-Ray Scattering Studies of Surface Disorder

J.F. van der Veen, R.G. van Silfhout, and A.W. Denier van der Gon

FOM-Institute for Atomic and Molecular Physics,
Kruislaan 407, 1098 SJ Amsterdam, The Netherlands

Abstract. We discuss two different types of surface disorder: atomic-scale roughness and disorder induced by surface melting. The roughening of a surface during, e.g., a crystal growth process can be monitored in situ by X-ray scattering. As an example the deposition of Ge atoms on Ge(111) is followed in time. For the detection of disorder induced by surface melting the technique of medium-energy ion scattering can be used to advantage. We will discuss recent ion scattering studies of the thermal disordering of the Al(110) and Ge(111) surfaces.

1. Introduction

The surface of a single crystal rarely presents itself as an abrupt, ideally flat, termination of the bulk lattice. Often the surface is to some extent disordered. The disorder may involve local variations of the height at which the crystal lattice terminates (roughness due to steps and kinks) or irregular displacements of atoms from lattice sites (positional disorder). In this paper we identify the different types of disorder by X-ray scattering and medium-energy ion sacattering (MEIS). We demonstrate, by giving some examples, that X-ray scattering is particularly suited for investigating atomic-scale surface roughness and that MEIS is the technique of choice for determining the amount of positionally disordered atoms on the surface.

A crystal surface may be *rough* for quite different reasons. Atomic-scale roughness develops during deposition of atoms on the surface or during a crystal growth process such as molecular beam epitaxy (MBE) or CVD. These processes occur at conditions away from thermodynamic equilibrium, at temperatures where the surface morphology is largely determined by kinetic effects. On the other hand, a surface may also be rough when it is in perfect equilibrium with the vapor. This type of roughness is purely thermally induced. It occurs at the temperature, T_r, at which the free energy for the formation of a step vanishes [1]. Above T_r, steps proliferate on the surface, causing the height-height correlation function to diverge logarithmically with increasing distance over the surface. By contrast, the roughness resulting from atom depositon on the surface is generally restricted to a finite number of height levels.

A *positional* disordering of the surface may be thermally induced or caused by, e.g., ion bombardment or a chemical reaction with deposited atoms. The latter processes will not be further discussed. Here we are interested in the disorder that may spontaneously develop on a crystal face as the temperature is raised towards the bulk melting point T_m. More specifically, close to T_m a liquidlike ('quasiliquid') layer may form on the surface as a precursor to melting of the bulk solid [2,3]. This phenomenon is termed 'surface melting' (or 'surface-induced melting') if the quasiliquid layer thickness diverges as $T \rightarrow T_m$. Surface melting can be regarded as a case of complete wetting, namely a complete wetting of the solid by its own melt. Other possibilities are that the surface remains dry ("surface non-melting') or that it becomes covered by a quasiliquid layer whose thickness does not diverge but instead remains constant ('incomplete surface melting', in analogy with incomplete wetting). Experimental evidence for the existence of surface melting has been found on a variety of crystal faces of Pb, Al, ice and noble gas crystals. These experiments are reviewed in refs. [2] and [3], where also the thermodynamic conditions for the occurrence of surface melting or non-melting are discussed. It is generally found that the metal surfaces with open atom packing (e.g., fcc(110)) exhibit surface melting, whereas the close-packed surfaces (fcc(111)) do not.

We note that surface melting is a phenomenon very different from thermal roughening. In the case of roughening the atomic registry in the surface region is preserved whereas in the case of surface melting the atoms are displaced from lattice sites.

2. Experimental methods

2.1 X-ray scattering

X-ray diffraction is widely used as a crystallographic tool, allowing bulk crystal structures to be determined with great precision. It can also be applied to surfaces [4], thanks to the availability of intense X-ray beams from synchrotron radiation sources. For a determination of the structure of an ordered crystal surface one typically measures the scattered intensities along fractional and integer-order diffraction rods perpendicular to the surface plane, yielding a set of structure factors. The structure is then found by searching for the best fit between the experimentally determined structure factors and those predicted by various structure models.

The intensity profile along an integer-order rod (also called crystal truncation rod or CTR [4,5]) is the result of interference between scattering from the surface and from the bulk, the relative phase depending on the value of momentum transfer perpendicular to the surface. This makes these CTR profiles not only sensitive to the structure of the surface unit cell and its registry with the bulk, but also to the atomic-scale surface roughness. In particular, it proves possible to deduce from the CTR profiles the statistical distribution of heights at which the crystal terminates [6]. The technique is well-suited for an in-situ monitoring of the surface roughness developing during, e.g., MBE crystal growth. But one may also study thermal roughening [7,8]. To prove in the latter case the existence of a roughening transition, one has to measure the diffraction line shape *across* the rod, preferably at an intersection point along the rod for which there is maximum destructive interference. Above the roughening temperature the line shape should then follow a power law, reflecting the logarithmic divergence of the height fluctuations [7]. In the case of roughness resulting from a deposition process, the line shape is often found to be Lorentzian, reflecting an exponential correlation function [6].

X-ray scattering is potentially a powerful probe of surface melting [9, 10]. In principle, one can derive the pair correlation function in the disordered layer by measuring the momentum-transfer dependence of the elastic diffuse scattering intensity at very grazing angles [10]. A grazing scattering geometry, below the critical angle for total external reflection, is required in order to achieve surface sensitivity and to reduce thermal diffuse scattering from the bulk. Such measurements, however, impose extreme demands on the surface flatness and are very difficult to perform. Similar difficulties are encountered with determining the temperature dependence of surface order parameters close to T_m, where CTR intensities also have to be measured under very grazing angles. Reliable surface melting studies of the latter type have yet to be carried out.

2.2 Ion scattering

MEIS is a direct real-space probe of both ordered surface structures [11] and of positional disorder at surfaces [2]. Using geometric shadowing and blocking effects one is able to determine the depth over which the short-range crystalline order is lost or diminished.

The principle of the method, as applied to surface melting, is illustrated in fig. 1. First consider the surface at a temperature at which it is still *ordered* (fig. 1a). If a parallel beam of energetic ions, typically ~100 keV protons, is aligned with a low-index crystal direction and the backscattered ions are detected along another low-index direction, then the resulting shadowing and blocking effects give rise to a distinct surface peak in the backscattering energy spectrum (fig. 1a). The area of the surface peak (SPA) is a direct, absolute, measure of the number of atoms per unit area that are 'visible' to ion beam and detector [11]. Next consider the same surface, but now at a temperature at which it is covered with a thin *disordered* melt layer (fig. 1b). The atoms in that layer, being displaced from lattice sites, do not contribute to shadowing and blocking and hence will be fully visible to beam and detector. This causes an increase in the SPA with respect to the one expected for an ordered surface at that temperature. This increase yields directly the number of disordered atoms per unit area on the surface.

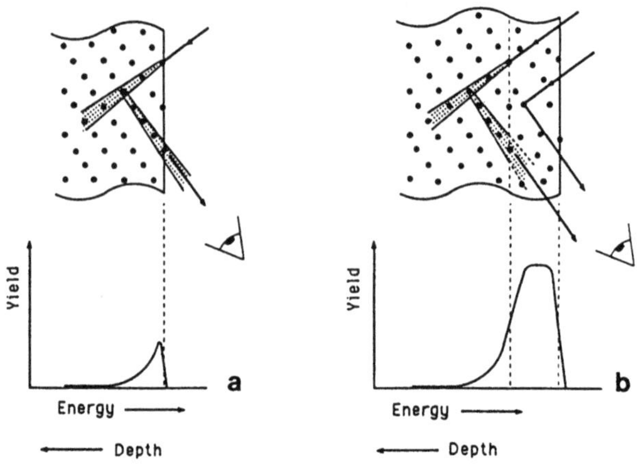

Fig. 1. Schematic representation of surface peaks in MEIS energy spectra, obtained in shadowing/blocking geometry for (a) a well-ordered crystal surface and (b) a crystal with a disordered melt layer on top. The shadowing and blocking effects are indicated as shaded cones.

Note that roughening, which usually precedes surface melting, has hardly any effect on the MEIS measurements, because the atoms on a rough surface are still orderly positioned along the atomic rows in the shadowing and blocking directions; the rows just terminate at different heights on the surface.

3. Surface morphology during crystal growth

X-ray scattering has a large sensitivity to the nucleation of islands on the surface [6,12]. Here we briefly discuss, as an example, a study of the surface roughness which builds up during deposition of Ge on Ge(111) [13]. The substrate on which the Ge atoms are deposited is the c(2x8) reconstructed Ge(111) surface. Its structure, as determined by X-ray diffraction [14, 15], is shown in fig. 2. Its main feature is the presence of a quarter of a monolayer of Ge adatoms occupying T_4 sites and inducing elastic distortions in the substrate (one monolayer is defined as 0.7×10^{15} atoms cm^{-2}). CTR profiles, measured on thoroughly sputtered and annealed samples, showed extreme surface flatness [15].

For monitoring the nucleation of two-dimensional (2D) islands on this surface, a specular reflection geometry was selected. The angle of incidence was chosen such that the wave amplitudes scattered from two terrace levels one bilayer distance apart are 180^0 out of phase (see inset of fig. 3, where for clarity the adatoms are left out). The resulting destructive interference between the scattered waves yields a reflected intensity much lower than for the flat starting surface. If half of the substrate surface is covered by islands of bilayer

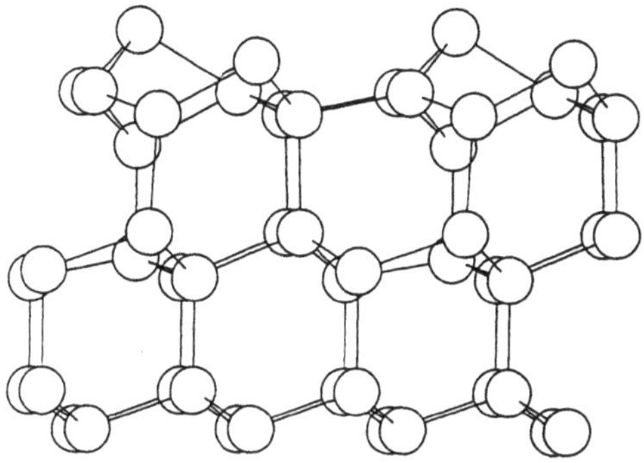

Fig. 2. Ball-and-stick model of the reconstructed Ge(111) surface, in side view. The model yields the best fit to the X-ray diffraction data of ref. [15]. The adatoms occupy the T_4 sites. Distances and atom displacements are drawn to scale.

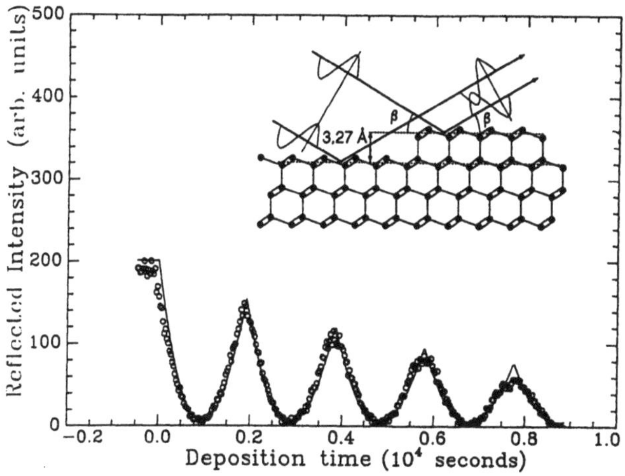

Fig. 3. Specularly reflected X-ray intensity from Ge(111) during Ge deposition, measured in the 180° out-of phase scattering geometry shown in the inset. The full curve shows the fit made to the data using a multi-level model of growth.

height (θ = 1/2), then there is exact cancellation of the scattered wave amplitudes, giving zero specularly reflected Bragg intensity. In fact, this is precisely what is observed in the case of Ge deposition on Ge(111). Figure 3 shows the reflected X-ray intensity as a function of deposition time. The substrate was held at a temperature of 470 K and the deposition rate was ~5.8×10^{13} atoms cm^{-2} min^{-1}. The intensity is seen to go through a sequence of oscillations, where each oscillation corresponds to the advancement of the growth front by about one bilayer. These oscillations have the same origin as those seen in reflection high-energy electron diffraction (RHEED) studies of growth [16]; they are a signature of layer-by-layer growth, proceeding by cycles of 2D nucleation and coalescence of islands.

The advantage of X-ray scattering over RHEED is that the observed intensity changes can be readily interpreted on the basis of kinematic scattering theory. This theory, when applied to a two-level model of the growth (i.e., 2D nucleation of islands followed by *complete* coalescence), predicts an identical, parabolic, coverage dependence $I_s = (1 - 2\theta)^2$ of the scattered intensity for each successive cycle [6,13]. However, only the first oscillation period is seen to be strictly parabolic and, in addition, the intensity maxima damp out. This can only be explained by a growth model which includes more than two levels [13]. The solid curve in fig. 3 shows the fit of a multi-level model to the first four periods. After these four periods the occupancies of the three top levels in this particular example are about 80, 16, and 4 %. Clearly, the system has not yet reached a steady-state growth mode.

The intensity oscillations are observed not only in the specularly reflected beam but also in the diffracted beams. Hence, the growth is epitaxial at a temperature of ~470 K [12]. It appears that the growing surface retains the adatoms, though probably in somewhat lower concentration than on the starting surface. We note that growth at temperatures above ~510 K proceeds by 'step flow' rather than by 2D nucleation, as oscillations are no longer observed [12]. Growth at room temperature, on the other hand, involves in its initial stage the occupancy of non-regular sites, as is evident from a distinctly different oscillatory behaviour of the different diffracted beams [17].

4. Melting-induced disorder

Evident cases of disordering due to surface melting are found on the (110) crystal faces of Pb [18] and Al [19]. Figure 4 shows the temperature dependence of the SPA for Al(110), measured by MEIS. In this experiment a beam of 100 keV protons was aligned with the [$\bar{1}$01] direction and backscattered protons were detected which emerge from the crystal along the [011] direction (see inset). Up to a temperature of ~800 K the number of visible atoms per unit area increases slowly as expected for a well-ordered crystal with increasing thermal vibrations. The solid curve, with which the data are seen to be in agreement, represents a Monte Carlo simulation of the experiment for a thermally vibrating crystal *without* melt layer. Above ~800 K, however, the SPA increases much faster than the simulation does. The number of disordered atoms per unit area, as derived from this additional increase, is seen to diverge as T approaches $T_m = 933.5$ K. This disordering process therefore is a case of complete wetting.

Within the temperature range of $880 < T < T_m - 0.5$ K the areal density of disordered atoms is found to diverge logarithmically as $\sim\ln[T_m/(T_m-T)]$. This is precisely the temperature dependence expected for the melted layer thickness in a system governed by short-range atomic interactions.

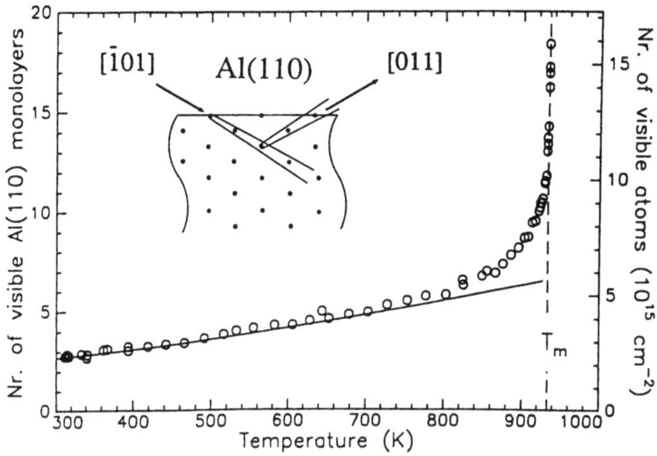

Fig. 4. Surface peak area as function of temperature for the Al(110) surface, as obtained in the MEIS geometry shown in the inset. The solid curve represents the yield expected for an ordered Al(110) surface, as calculated in a Monte Carlo simulation of the experiment. The difference between the experimental data and the simulated yield equals the number of disordered atoms per unit area due to surface melting. From ref. [19].

The same type of experiment, when performed on the (111) surfaces of Al [19] and Pb [20], shows no positional disordering at all; the measured SPA closely follows the Monte Carlo simulations for the ordered surface up to T_m. These surfaces are therefore non-melting. This marked difference in the thermal behavior of the (111) and (110) surfaces relates to their different atom packing densities [2].

On the (110) surfaces of Al and Pb we find no evidence of layering effects in their melting behavior; the number of disordered atoms rises smoothly with temperature (though for Pb(110) a weak bump is seen at ~570 K [18]). But on the Ge(111) surface, which is also known to disorder at high temperature [21], a strong layering effect might be expected since the diamond structure of the Ge lattice provides a pronounced layered stacking of atoms in the [111] direction (see fig. 2). MEIS measurements indeed show a distinctly different thermal behavior [22]. The temperature dependence of the SPA, measured in the [00$\bar{1}$]/[11$\bar{1}$] shadowing/blocking geometry, is shown in fig. 5 for proton beam energies of 60, 100 and 175 keV. The dashed and dash-dotted curves represent Monte Carlo simulations for a bulklike crystal termination and for the relaxed adatom structure shown in fig. 2. Between 1020 K and 1120 K the experimental SPA rises much faster with temperature than either of the simulations but above ~1120 K it levels off. The disorder remains finite as $T \to T_m$. We believe this to be a case of incomplete surface melting.
Because the disorder saturates at a level equivalent to only a single monolayer, it is likely that a substantial amount of lattice order is present in the layer. This

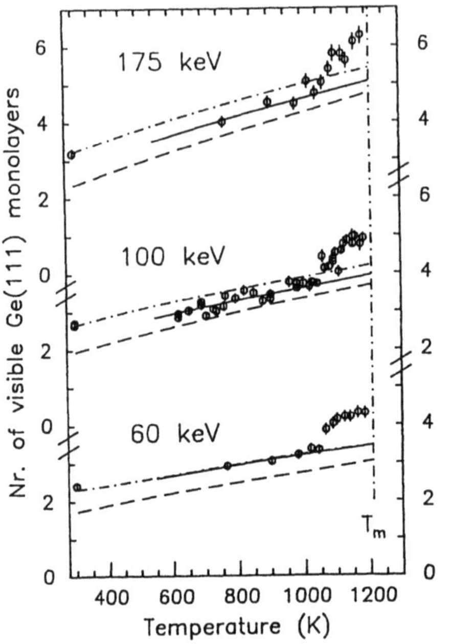

Fig. 5. Surface peak area as a function of temperature for the Ge(111) surface, measured for three different energies of the proton beam. The incoming beam was aligned with the $[00\bar{1}]$ crystal axis, the detection direction of backscattered protons was aligned with the $[11\bar{1}]$ axis. The open circles show the experimental results, the dashed curves represent simulations on a bulklike surface and the dash-dotted curves represent simulations for the adatom model shown in fig. 2. The solid curves are fits to the data between 600 K and 1000 K and are given the same temperature dependence as the simulations for the corresponding beam energy. From ref. [22].

is supported by our observation of a larger amount of disorder for higher proton energy, i.e., for smaller shadow and blocking cone radii. If the atoms were randomly positioned, as would be the case when the presence of the substrate is not felt, then there would have been no energy dependence.

We believe that the incomplete melting of the Ge(111) surface is caused by a layering effect which confines the disorder largely to the outermost bilayer [22]. But a deeper layerwise melting does not occur, at least not up to 25 K below Tm. Apparently, it is too costly in energy to create, below Tm, an interface with a thicker, more liquidlike layer. For a detailed consideration of the interfacial free energies involved, see ref. [22].

It is finally noted that the Ge(111) surface exhibits an additional order-disorder transition from c(2x8) to (1x1) at 573 K [23]. This transition involves a ~40 % reduction of the adatom concentration on the surface [22,24]. In fig. 5 this shows up as a slight decrease, above ~600 K, of the SPA with respect to the Monte Carlo simulation for the fully adatom-covered surface.

5. Outlook

Two different types of surface disorder have been discussed: surface roughness, in which case the crystal terminates at different heights but the atoms remain on lattice positions, and positional disorder, with atoms irregularly displaced from lattice sites. X-ray scattering and medium-energy ion scattering as probes of disorder have the advantage over other techniques that the interpretation of the experimental results is straightforward.

Elastic diffuse X-ray scattering under very grazing incidence is a promising technique for the structural characterization of ultrathin surface melt layers and of macroscopic solid-liquid interfaces. But one needs very flat sample surfaces and a high-intensity short-wavelength X-ray source. The advent of a new generation of high-brilliance synchrotron radiation sources (ESRF in Europe, SPRING-8 in Japan and APS in the U.S.A.) will undoubtedly open new possibilities in this research area.

Acknowledgements

The X-ray scattering work on Ge surfaces was performed at the Synchrotron Radiation Source (SRS), Daresbury, U.K. We enjoyed stimulating discussions with J.W.M. Frenken and E. Vlieg. This work is part of the research program of the Stichting voor Fundamenteel Onderzoek der Materie (FOM) and is made possible by financial support from the Nederlandse Organisatie voor Wetenschappelijk Onderzoek (NWO).

References

1. See. e.g. M. Wortis, in: Chemistry and Physics of Solid Surfaces VII, Springer Series in Surface Sciences, Vol. 10, Eds. R. Vanselow and R.F. Howe (Springer, Heidelberg, 1988) p.367.
2. J.F. van der Veen, B. Pluis and A.W. Denier van der Gon, in: Chemistry and Physics of Solid Surfaces VII, Springer Series in Surface Sciences, Vol. 10, Eds. R. Vanselow and R.F. Howe (Springer, Heidelberg, 1988) p.455.
3. J.G. Dash, Contemp. Phys. 30, 89 (1989).
4. R. Feidenhans'l, Surface Sci. Rept. 10, 105 (1980) and references therein.
5. I.K.Robinson, Phys. Rev. B33, 3830 (1986).
6. E. Vlieg, J.F. van der Veen, S.J. Gurman, C. Norris and J.E. Macdonald, Surface Sci. 210, 301 (1989).
7. J. Lapujoulade and B. Salanon, in: Phase Transitions in Surface Films, NATO ASI series, Ed. H. Taub (Plenum, New York), in press.
8. I.K. Robinson, E.H. Conrad and D.S. Reed, J. Phys. France 51, 103 (1990).
9. B. Pluis, J.-M. Gay, J.W.M. Frenken, S. Gierlotka, J.F. van der Veen, J.E. Macdonald, A.A. Williams, N. Piggins and J. Als-Nielsen, Surface Sci. 222, L845 (1989).

10. P.H. Fuoss, L.J. Norton and S. Brennan, Phys. Rev. Letters 60, 2046 (1988).
11. J.F. van der Veen, Surface Sci. Rept. 5, 199 (1985).
12. E. Vlieg, A.W. Denier van der Gon, J.F. van der Veen, J.E. Macdonald and C. Norris, Phys. Rev. Lett. 61, 2241 (1988).
13. R.G. van Silfhout, J.W.M. Frenken, J.F. van der Veen, S. Ferrer, A. Johnson, H. Derbyshire, C. Norris and J.E. Macdonald, J. Phys.France 50, C7, 295 (1989).
14. R. Feidenhans'l, J.S. Pedersen, J. Bohr, M. Nielsen, F. Grey and R.L. Johnson, Phys. Rev. B38, 9715 (1988).
15. R.G. van Silfhout, J.F. van der Veen, C. Norris and J.E. Macdonald, J. Chem. Soc. Faraday Discussion 89 (1990), in press.
16. See, e.g.: High-Energy Electron Diffraction and Reflection Electron Imaging of Surfaces, NATO ASI series, Eds. P.K. Larsen and P.J. Dobson (Plenum, New York, 1988).
17. R.G. van Silfhout et al., to be published.
18. J.W.M. Frenken, P.M.J. Marée, and J.F. van der Veen, Phys. Rev. B34, 7506 (1986).
19. A.W. Denier van der Gon, R.J. Smith, J.-M. Gay, D.J. O'Connor and J.F. van der Veen, Surface Sci. 227, 143 (1990).
20. B. Pluis, A.W. Denier van der Gon, J.W.M. Frenken and J.F. van der Veen, Phys. Rev. Lett. 59, 2678 (1987).
21. E.G. McRae and R.A. Malic, Phys. Rev. B38, 13163 (1988).
22. A.W. Denier van der Gon, J.M. Gay, J.W.M. Frenken and J.F. van der Veen, Surface Sci., in press.
23. P.W. Palmberg, Surface Sci. 11, 153 (1968).
24. R.G. van Silfhout et al., to be published.

Properties of Graphite and Monolayer Graphite Studied by Electron Energy Loss Spectroscopy and Scanning Tunneling Microscopy

C. Oshima[1], *H. Itoh*[1], *T. Ichinokawa*[1], *T. Aizawa*[2], *R. Souda*[2], *S. Otani*[2], *and Y. Ishizawa*[2]

[1]Department of Applied Physics, Waseda University, Okubo, 3-4-1, Shinjuku, Tokyo 169, Japan
[2]National Institute for Research in Inorganic Materials, Namiki 1-1, Tsukuba 305, Japan

Abstract. The dispersion curves of surface phonons localized at a topmost layer of graphite and monolayer graphite have been investigated by using electron energy loss spectroscopy. The observed curves of the surface phonons of graphite are quite similar to those of the graphite crystal, while the dispersion curves of the monolayer graphite are very different from those in the bulk; some vibrational modes soften strongly. Scanning tunneling microscopy has shown that the electronic structure of the monolayer graphite is modified largely with the periodicity of the substrate.

1. Introduction

Graphite is a typical layered-structure material, in which intralayer bonds have a strong covalent character and the interlayer bond is a weak van der Waals interaction. Since no dangling bond exists on the (0001) surface, the surface is extremely inert for reaction with various gases. This is the main reason why atomic-scale scanning tunneling microscopy (STM) images of graphite are obtained even in air /1/. Therefore, we expect that the properties of the graphite surface are not greatly changed from those inside the crystal. In this paper, we show that the expectation is partially true for the graphite surface, but not for the monolayer graphite owing to the interaction between overlayer and substrate.

Graphite is a standard substrate in low temperature physics. In addition, graphite intercalation compounds (GIC) have been attracting attention as a new type of compound. Therefore, a lot of experiments have been done so far for the graphite crystal and its related materials. Concerning the dynamical properties, many techniques have been applied to those materials /2/-/5/, but none of them clarified not only the dispersion curves of the bulk phonons, but also the surface phonons of graphite and its related compounds. The purpose of this paper is to clarify the dynamical properties of the topmost layer of graphite and the monolayer graphite by using electron energy loss spectroscopy (ELS). In addition, the modified electronic structure of the monolayer graphite has been studied by STM /6/-/8/.

2. Experimental

Two specimens were used in this experiment. One is single crystal graphite of high quality, and the other is an overlayer of graphite on TiC(111), which was prepared as follows: first, the TiC(111) surface was cleaned by flash-heating in an ultra-high vacuum, and then exposed to ethylene gas at the sample temperature of 1100°C for 200L. Owing to the reaction with the gas, the LEED pattern changed from a 1 x 1 structure to a complex structure with three kinds of diffraction spots. They are due to the diffraction from 1 x 1-TiC, 1 x 1-graphite and double diffraction. The overlayer growth stops spontaneously, when the TiC(111) is exposed for a much longer exposure time, for example 2000L, which is one order of magnitude larger than the exposure time required for the formation of the initial overlayer. Namely, after such a large exposure, no changes of overlayer were detected in AES spectra, ELS spectra and LEED patterns. This strongly suggests that the overlayer has a certain thickness.

The apparatus for ELS is composed of a tandem monochromator and a tandem analyzer. An energy resolution of 3-4 meV (FWHM of the elastic peak) was obtained at a sample current of 10^{-10} A /9/. The ELS equipment was fixed in a vacuum chamber that was evacuated to 1 x 10^{-8} Pa. In the same chamber, we have a cylindrical mirror analyzer for Auger analysis, a LEED screen, a gas inlet and an ion gun for surface cleaning. The STM experiment was done in another chamber, of which the operation pressure is of the order of 10^{-4} Pa. The ELS experiment was done in the chamber in which the specimen of the overlayer was prepared and then the specimen was moved to the other chamber for the STM measurement /10/.

3. Results

Figure 1(a) shows the dispersion curves of the surface phonons of graphite crystal. We observed 6 branches corresponding to the 6 degrees of freedom of 2 atoms in a unit cell of the topmost layer of graphite. They are an acoustic longitudinal mode (LA), an acoustic transverse one (ZA), an optical longitudinal one (LO), an optical transverse one (ZO) and two shear horizontal modes (SH). The bulk data measured by using infrared absorption /12/ and neutron inelastic scattering /13/ are indicated by the filled circles in Fig.1(a); they gave the dispersion curves only in the limited region. On the other hand, the whole dispersion curves were obtained in the entire Brillouin zone by ELS, which are shown by open circles. Owing to the strong electronic excitation, the phonon signals at the Γ point were not measured by ELS. The extrapolation of those surface branches to the Γ point provides good agreement with the bulk data, which means that the vibrational properties of the topmost layer of graphite are quite similar to those in the bulk. It is very different from the surface phonons of normal three-dimensional materials.

The phonon dispersion curves of the graphite overlayer on TiC(111) along ΓM are shown in Fig.1(b). Five branches were observed, and two SH branches were not detected due to the selection rule for the electron scattering /11/. The observed dispersion curves are similar to the surface phonons of the graphite crystal in Fig.1(a) except for energy shifts, which

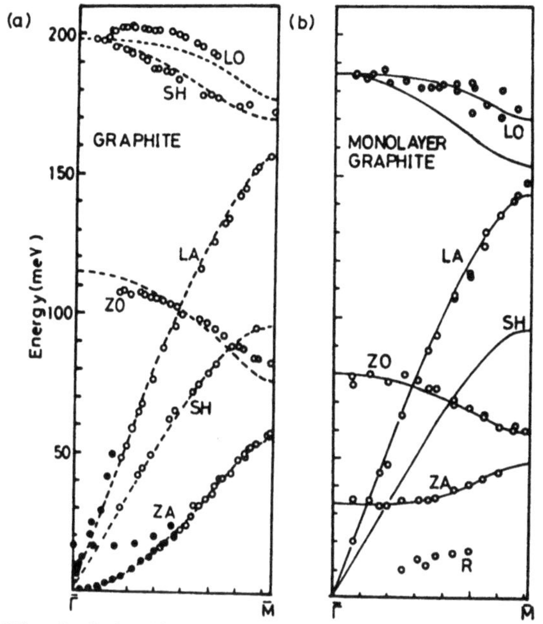

Fig.1.(a) The surface-phonon dispersion curves of the graphite crystal and (b) the monolayer graphite on the TiC(111). The dotted and solid curves are the theoretical ones calculated on the basis of a force constant model.

indicate that the overlayer has the same atomic structure as bulk graphite. The highest branch (LO) located at 170-185 meV is the optical longitudinal phonon, and the LA branch with a large dispersion from 0-150 meV is due to the acoustic longitudinal one. Such large dispersion branches correspond to the existence of the graphite overlayer, which is consistent with the other observations; instead of the "carbidic" carbon peak, the AES spectra shows the "graphitic" carbon peak after overlayer formation /9/. In the LEED pattern, as we mentioned before, we observed the bright diffraction spots from graphite and satellite spots due to double diffraction from the graphite overlayer on the TiC(111) substrate. The lattice constant of the overlayer graphite is expanded by 3 % from the bulk value.

The vibrational energies of the overlayer in Fig.1 (b) are very different from those in the graphite crystal of Fig.1(a). Three modes (LO, ZA and LA) soften. Especially, the vibrational energy of the ZO mode, which is the transverse optical mode is largely reduced; two carbon atoms in the unit cell are displaced out of phase perpendicular to the layer. On the other hand, the ZA mode, in which two carbon atoms vibrate in phase perpendicular to the layer, stiffens. This corresponds to the enhancement of the bond strength between the overlayer and the TiC(111) substrate.

Figure 2(a) is a typical STM image of the graphite crystal. The bright spots align with the periodicity of the graphite, although the unit cell of the graphite layer has two carbon atoms. The origin of the strong contrast between the two atoms in the

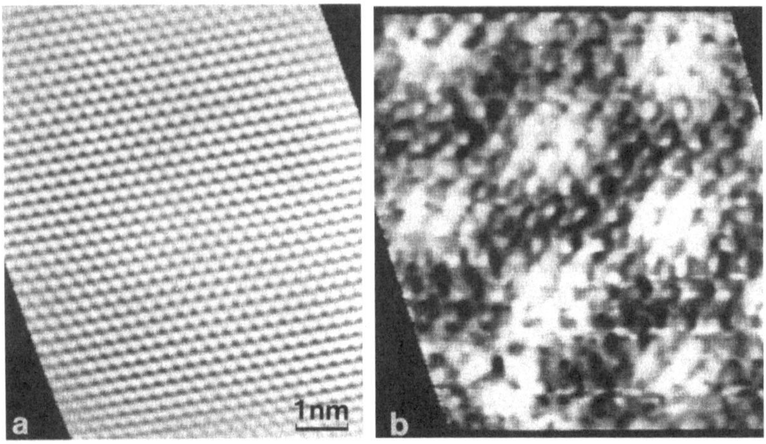

Fig.2 STM images of graphite (a) and monolayer graphite on a TiC(111) substrate (b). There are two kinds of periodicity in the STM image of (b).

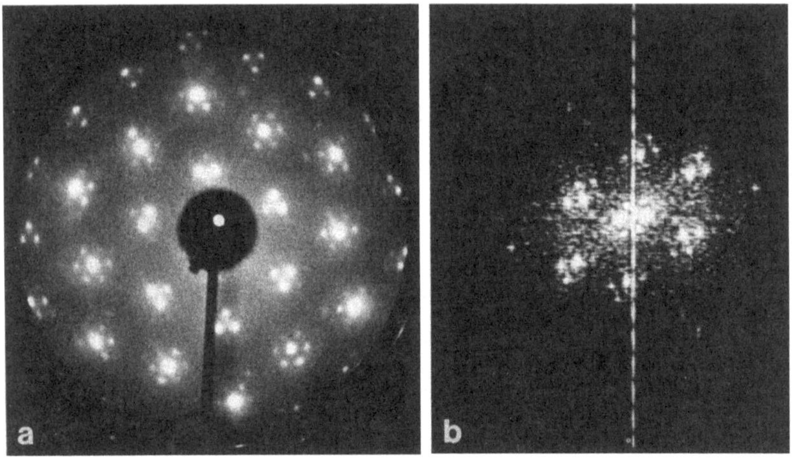

Fig.3 The LEED pattern of the monolayer graphite, E=98eV (a) and the Fourier transform (FT) pattern of the monolayer graphite (b) on TiC(111).

STM image was initially interpreted as a rippled structure /1/; one atom is considered to be located at 0.1nm above the other atom. But theoretical calculation suggested that the different electronic states of the two atoms result in the strong contrast /12/. A similar conclusion was also indicated by the dispersion curves of surface phonons in Fig.1(a); the agreement between the two dispersion curves in the bulk and the surface refutes the large rippled structure with a width of 0.1nm. Therefore, the electronic origin seems to be possible.

Figure 2 (b) is the STM image of the overlayer graphite at the bias condition of 700 mV. The tunneling electrons passed from the graphite to a tip. We observed the modulated image with two kinds of periodicity: a large distance of 2.7nm and short distance of 0.6nm, which are much larger than the nearest-neighbour atomic distance (0.14nm). It should be noted that the observed contrast is not the atomic image, which is also supported by the following fact. When we changed the polarity of the tunneling bias, the modulated structure with the large periodicity of 2.7nm almost disappeared. Therefore, it is concluded that the observed structure of the STM image is ascribed to the modulated electronic structure.

Figure 3 (a) and (b) show the LEED pattern of the graphite overlayer on TiC(111) and the two-dimensional Fourier transform (FT) pattern of the STM image, respectively. Spots appearing in the FT pattern correspond to the satellite LEED spots; namely, two modulations in the STM image are constructed by the interference between two kinds of periodicity : the graphite overlayer and the TiC(111) substrate. The origin of the modulated electronic structure is not yet clear, but it is clear evidence for the strong interaction between the overlayer graphite and the substrate.

4. Discussions

4.1 Dynamical properties of the topmost layer of graphite

As shown in Fig. 1 (a), the dispersion curves of the topmost layer are quite similar to those of bulk graphite. Does it mean that the topmost layer does not soften, being different from the normal materials with three-dimensional chemical bonds? The answer is no. The average vibrational amplitude of surface atoms has been measured with the LEED intensity by Wu /13/, and this indicated that the vibrational amplitude of the surface atoms is large in comparison with those inside the graphite crystal, which appears to contradict the dispersion curves in Fig. 1 (a). Additionally, the theoretical calculation also suggested similar softening /14/. The ZA mode, of which the dispersion curves near the Γ point were not determined in this experiment, presumably softens near the Γ point.

The graphite crystal is a kind of molecular crystal, which possesses a strong intralayer interaction and a weak interlayer interaction. At the topmost layer, one of the weak van der Waals bonds is broken. Therefore, the topmost layer vibrates as a whole very easily against the second layer and the third layer with a low vibrational frequency. This is presumably the origin of the large atomic displacement at the topmost layer of the graphite observed by the LEED intensity. Concerning the vibrational mode with a finite wave vector, the dynamical properties of the topmost layer are quite similar to the bulk ones as shown in Fig. 1 (a).

4.2 Identification of " one monolayer" thickness

A lot of authors have discussed the thickness of the graphite overlayer on different substrates by using AES, LEED and UPS /14/. But no one showed direct experimental evidence for a thickness of one monolayer, although most of the authors have speculated it. In this experiment, we have shown the following direct evidence for one monolayer thickness, together with indirect support.

(1) As shown in Fig. 1(b), we detected only one set of the modified phonon dispersion curves. If the overlayer consists of more than two atomic layers, we can expect at least two sets of dispersion curves; since the interaction between the first layer and the second layer is possibly weak, as in bulk graphite, the non-modified curves like the ones shown in Fig.1 (a) should appear, and the strong interface interaction may provide the additional modified ones as shown in Fig.1(b). Therefore, Both the dispersion curves in Fig.1 (a) and (b) should be measured at the same time. But we observed only one set of curves in this experiment, which indicates that the overlayer is one monolayer in thickness.

(2) A similar discussion is possible for the STM image. In Fig.2(b), we showed the STM image strongly modified by the periodicity of the substrate. This is only possible for one monolayer in thickness: if this is not the case, since the substrate does not influence the electronic structure of the topmost layer because of the weak van der Waals interaction between two outermost layers, the modified STM image cannot be expected. Hence, the modified electronic structure in the STM image also gives strong evidence for a thickness of one monolayer.

The overlayer formation provides other indirect evidence. As we have already mentioned, the overlayer formation spontaneously stops. If the surface is covered with a layer more than one monolayer thick, it is not easy to find a reason for the spontaneous stop. If it is monolayer, the spontaneous stop is easily explained by the change in activity of reaction from the substrate to the graphite. All these experimental facts indicate that the overlayer has a thickness of one monolayer.

4.3 The properties of the monolayer graphite

We have analyzed the observed dispersion curves on the basis of the force constant (FC) model. In this calculation, 6 parameters are introduced to describe the dispersion curves of the monolayer graphite and the topmost layer graphite crystal; two of them are the stretching FC for the nearest-neighbour interaction (α_1) and the second nearest-neighbour interaction (α_2). The γ_1 and γ_2 represent the three-body in-plane angle-bending FC and the out-of-plane angle bending FC, respectively. The 5th parameter, δ ,is the twisting FC, and the last one, α_s is the stretching FC between the monolayer graphite and the TiC(111) substrate /7/. The results of the analysis are the following. The FC changes of the monolayer graphite are expressed as the ratio to the values of the graphite surface; $\Delta\alpha_1$: -5%, $\Delta\alpha_2$:-32%, $\Delta\gamma_1$:-10%, $\Delta\gamma_2$:-50%, $\Delta\delta$:-55%. The last parameter of α_s was 3.12×10^4 dyne/cm.

The FC values due to the many-body interaction of γ_2 and δ decrease greatly, which means that the restoring force for the bond angle bending is greatly reduced. A similar tendency was reported for donor-type intercalation compounds (GIC); the electrons transfer from the substrate to the anti-π-bonding state of the graphite layer from the intercalent, which inevitably weakens the nearest-neighbour covalent bonds, resulting in the reduction of the vibrational energies and the expansion of lattice constant /15/.

When we compare the monolayer graphite with GIC, we notice three differences. One is a variety of substrates. We can select many materials as a substrate in comparison with GIC. Secondly, large changes in physical quantities of the monolayer graphite were observed; the changes in vibrational energies and lattice constant are 2-7 times as large as those of GIC. Therefore, monolayer graphite is much more largely modified from the graphite as compared with GIC. The third difference is the stability of the material. No change in STM images was detected for the monolayer graphite on TiC(111), after we exposed it to an air atmosphere for about one month. On the other hand, GIC has to be maintained in an inert gas atmosphere in order to keep the GIC crystal.

In principle, GIC is the material in which the interaction between graphite and intercalent is dominant. The interaction between intercalents seems to be relatively weak as compared with that interaction between a graphite layer and an intercalent. In the case of the monolayer graphite, on the other hand, the graphite layer interacts with the two-dimensional substrate. The substrate atoms interact with each other similarly to the interaction between the monolayer and substrate. It possibly provides the stable character of the monolayer graphite.

5. Summary

The graphite surface and monolayer graphite on TiC(111) have been studied by high resolution ELS and STM. The dynamical properties of the topmost layer of graphite are very similar to those of bulk graphite except for vibrations at the Γ point, while the monolayer graphite shows dynamical properties strongly modified by interaction with the TiC(111) substrate. The STM study suggested a modulated electronic structure with the periodicity of the substrate. Monolayer graphite is a promising material, for which we expect a variety of physical properties as a new graphite-related compound at the surface.

References

[1]S.Park and C.F.Quate, Appl. Phys. Lett. **48**, 112 (1986).
[2]R.J.Nemanich, G.Lucovsky and S.A. Solin, Solid State Commun. **23**, 117 (1977).
[3]R.Niklow, N.Wakabayashi and H.G.Smith, Phys. Rev. **B5**, 4951 (1972).
[4]D.P.E.Smith, G.Binnig and C.F.Quate, Appl.Phys.Lett. **49**, 1641 (1986).
[5]M.Maeda, Y.Kuramoto, and C.Horie, J. Phys. Soc.Japan, **47**, 337 (1979).
[6]C. Oshima, T.Aizawa, R.Souda, Y.Ishizawa and Y.Sumiyoshi, Solid State Commun. **65**, 1601 (1988).
[7]T. Aizawa, R. Souda, S.Otani, Y.Ishizawa and C. Oshima, Phys. Rev. Lett. **64**, 768 (1990).
[8]T. Aizawa, R.Souda, S. Otani, Y. Ishizawa and C. Oshima, unpublished.
[9]C. Oshima, R.Souda, M.Aono and Y.Ishizawa, Rev. Sci. Instrum. **56**, 227 (1985).

[10]T.Ichinokawa, T.Ichinose, M.Tohyama and H.Itoh, J.Vac.Sci. Technol. **A8**, 500 (1990).
[11]H. Ibach and D.L.Mills, **Surface Vibration and Electron Energy Loss Spectroscopy**, Academic Press, New York (1982).
[12]K. Kobayashi and M.Tukada, J.Vac. Sci. Technol. **A8**, 170 (1990).
[13]N.J. Wu, Surface Sci. **163**, 51 (1985).
[14]E.de Rouffignac, G.P.Alldredge and F.W. DeWette, Phys. Rev. **B23**, 4208 (1981).
[15]D.E. Nixon and G.S. Parry; J.Phys. **C3**, 1732 (1969).

Surface Order and Disorder Studied by Scanning Tunneling Microscopy

A.A. Baski, C.A. Lang, J. Nogami, and C.F. Quate

E.L. Ginzton Laboratory, Stanford University,
Stanford, CA 94305, USA

Abstract. Epitaxial growth of sub-monolayer coverages of an adsorbate on a substrate produces a surface that simultaneously exhibits order and disorder on various length scales below 1000 Å. We present examples of different ordering phenomena from scanning tunneling microscope (STM) studies of the growth of metals on semiconductor and metal surfaces. These results illustrate the local ordering of surface reconstructions and the effect of substrate surface reconstruction on nucleation and growth. The STM provides information about these surfaces that is largely inaccessible to other analytical techniques.

1. Introduction

Many techniques used to determine the atomic structure of surfaces involve beams of photons, electrons, or ions incident on the surface. All of these probes give information that has been averaged over a lateral distance of at least several hundred angstroms, resulting in a probed area that includes thousands of surface atoms. As a result, these techniques are most precise when the sample surface is uniform and it is appropriate to consider spatially averaged properties during data analysis.

Numerous surfaces, however, are intrinsically non-uniform. For example, this is typically the case when one material is grown on top of another. Even in the case of perfect layer-by-layer growth, the surface is not spatially uniform during the growth of the first atomic layer. In this regime many phenomena of interest can occur, such as nucleation, adsorbate-induced surface reconstruction, and changes in growth behavior near defects or step edges.

Most surface science techniques are limited in their study of these non-uniform systems by insufficient spatial resolution (e.g. scanning electron microscopy or photo electron microscopy) or by reliance on diffraction processes that can only detect periodic surface structure (e.g. low energy electron diffraction or glancing angle x-ray diffraction). The scanning tunneling microscope (STM) probes a surface directly in real space with atomic resolution. It reveals the atomic structure of even disordered regions on length scales down to 5 Å, beyond the capabilities of any other surface analytic instrument. Hence, it is an ideal tool to study non-periodic or localized structures on a non-uniform surface. Often, it is these deviations from perfect order on the surface that provide valuable insight into the growth process.

In this article, we present some examples of STM studies involving the growth of metals on either metal or semiconductor surfaces. In all of these cases, the growth is epitaxial and the resulting surfaces are ordered in the sense that the adsorbate atoms are bonded in a well-defined way with the substrate. Beyond the local bonding environment of the surface atoms, however, we see with the STM that the surface structure can be complex and that many interesting phenomena are visible on length scales from 5 Å to 5000 Å.

2. Disorder in Ordered Surface Reconstructions

In this section we present two examples of surface reconstructions, one demonstrating long range order but less short-range order, and the other showing short range order but imperfect long range order.

Figure 1 shows a Si(100) surface with 0.7 mono-layers (ML) of tin after a 550°C anneal. An extensive STM study [1] of the Sn/Si(100) system has shown

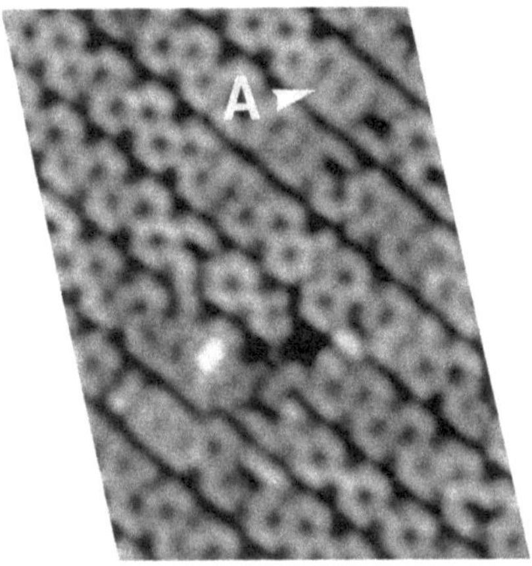

Fig. 1: Si(100) surface with 0.7 ML Sn deposited and then annealed. This (85 × 110 Å) image was taken with tip bias $V_t = 1.8$ V and tunneling current I = 7 nA. Buckled and unbuckled dimers form chain-like structures between the dark trenches.

that in this coverage range most of the surface atoms are bonded as dimers, just as for the clean Si(100) surface. The STM topograph is taken with a positive tip voltage and therefore represents the filled electronic states of the sample. A few rows of unbuckled dimers are visible (arrow A), but most dimers are buckled and, consequently, only one of the two atoms in each dimer is visible as a bright spot in the image. This buckling can occur in or out of phase for adjacent rows, resulting in various chain-like structures. These structures are periodically disrupted by black trenches oriented perpendicular to the dimer rows and typically spaced five unit cells apart. The periodicity of these trenches is sufficiently strong that 5× satellite dots are visible in the low energy electron diffraction (LEED) pattern. The complex structure of the dimer rows between the trenches, however, is insufficiently ordered to contribute to the diffraction pattern. Different combinations of the buckled dimers and the trench separations can result in a variety of local orderings, including 6×4, 5×4, and c(8×4).

In this system the local bonding environment of the surface atoms is largely uniform in a structural sense, since most of the surface consists of dimers. Small compositional disorder effects are possible, though, because the surface layer does not consist entirely of silicon or tin atoms at this particular metal coverage. On a slightly longer length scale, the complex arrangement of buckled and unbuckled dimers between the dark trenches forms areas of higher order surface reconstructions, but these areas are too small (< 30Å) for LEED measurements to provide information about this phenomenon. LEED detects only the relatively strong long range ordering due to the dark trenches, as well as some intensity in half-order dots due to the dimerization in the perpendicular direction. Only the atomic resolution images provided by STM reveal the complex nature of the surface structure on all of these length scales.

Figure 2 shows an 1200 Å × 1200 Å image of the Au(111) surface. This close packed surface is known to reconstruct by a slight compression of the surface layer (approximately 5%) along a $\langle 1\bar{1}0 \rangle$ direction [2,3,4]. This contraction causes a periodic modulation of the registry between gold atoms of the topmost and the second layer, leading to a 22×1 reconstruction. This reconstruction is visible in STM topographs as faint parallel white lines along $\langle 11\bar{2} \rangle$ directions with a height corrugation of 0.16 Å and a periodicity of approximately 65 Å [5,6,7]. The white lines mark the elevated transition regions between areas of normal fcc (ABC) stacking and areas of hcp (ABA) stacking. The three possible contraction axes correspond to three possible reconstruction domain orientations. As illustrated by the image, in some areas the individual domains of coherent reconstruction can be quite small.

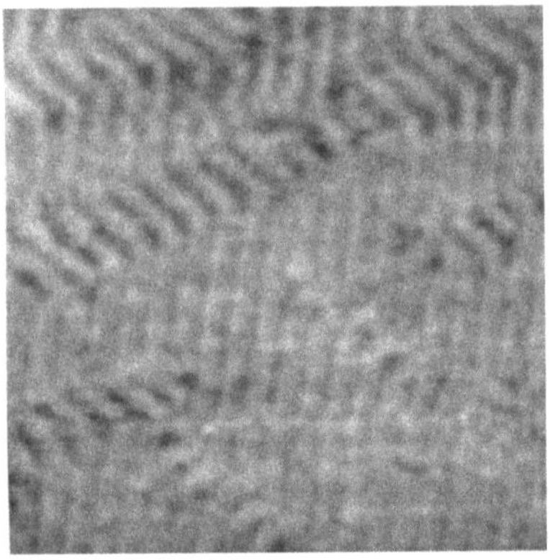

Fig. 2: A (1200 Å)^2 image of the domain structure of the 22×1 reconstructed Au(111) surface, taken with V_t = - 0.2 V and I = 0.1 nA.

The kink-to-kink spacing varies significantly over the sample surface, but it is usually between 200 Å to 800 Å. This value is consistent with recently reported x-ray diffraction measurements that indicate an average kink spacing of 325 Å [8]. The details of the domain structure appear to depend on the surface preparation.

This complex domain structure of the surface reconstruction illustrates another type of disorder that is difficult to study by other techniques. As will be shown in the next section, it is not merely a curiosity of the clean Au(111) surface, but also plays an important role in nucleation and growth on this substrate.

3. Order and Disorder in Nucleation and Growth

Nucleation phenomena provide ideal examples of surface structures that are both localized and non-uniform. Figure 3 illustrates the initial stages of indium growth on Si(100) [9]. In this particular image, 0.05 ML of indium has been deposited onto Si(100) at room temperature. A single height atomic step runs diagonally across the image, and the unperturbed rows of silicon dimers on both sides rotate by 90° across the step. The deposited indium is mobile enough to form bright rows aligned perpendicular to the underlying silicon dimer rows on both terraces. These bright rows are relatively long, indicating anisotropic growth along the row direction. Smaller scale images show that the atoms in each row pair up to form indium dimers. This anisotropic, dimer row growth mode has also been observed for gallium and tin on Si(100), suggesting that it reflects some characteristic of the Si(100) surface, rather than specific interactions between the particular metal adsorbate and the substrate.

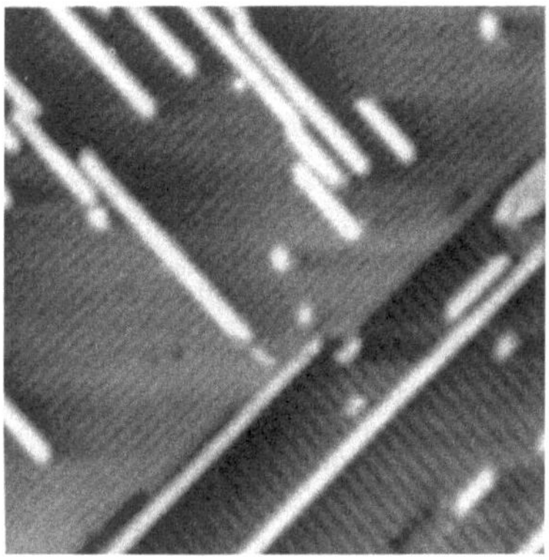

Fig. 3

Fig. 4

Fig. 3: Si(100) surface with 0.05 ML In deposited at room temperature. This (320 Å)^2 image was taken with V_t = - 1.8 V and I = 0.2 nA. Bright In rows are seen aligned perpendicular to the underlying Si dimer rows.

Fig. 4: Au growth on the Au(111) surface, with clusters nucleating along 22×1 domain boundaries. This (1100 Å)^2 image was taken with V_t = -0.2 V and I = 0.1 nA.

Figure 4 shows a Au(111) surface after deposition of 0.1 ML of gold at room temperature. An extensive STM study of gold autoepitaxy has shown that at very low coverages the gold grows as small clusters, each tens of angstroms in diameter and only one atomic layer high [10]. Remarkably, these clusters align preferentially along $<11\bar{2}>$ directions. The reason for this ordering is apparent in the STM image, in which the reconstruction is resolved. Above and below the horizontal rows of islands, the 22×1 reconstruction changes direction by 120°. This indicates that the gold islands are lying along a domain boundary, suggesting that these boundaries play an important role in nucleation. Nucleation at local reconstruction defects is also occasionally observed, sometimes obscuring nucleation along domain boundaries in areas with a more complicated domain structure. Similar examples of nucleation along domain boundaries have also been observed for the growth of nickel on Au(111) [11] and silicon on Si(111) [12].

In both of these examples, the STM provides information about the growth shapes of small clusters and their spatial distribution. This information is valuable in determining mechanisms for the nucleation and growth of the overlayers.

4. Other Forms of Disorder on Ordered Surfaces

Figure 5 shows a Si(111) surface after deposition and annealing of 0.8 ML of gold [13]. At this coverage, Au is known to form a $\sqrt{3}\times\sqrt{3}$ reconstruction. This structure appears in STM images as a centered hexagonal array of bright dots with one maximum per unit cell. The most striking feature of this image, however, is the irregular network of white domain walls that interrupt the periodicity of the $\sqrt{3}\times\sqrt{3}$ structure. These domain walls are associated with dislocations in the surface reconstruction, and break up the long range surface order by forming sub-100 Å domains. As the gold coverage is increased, the density of the domain walls also increases. This suggests that each domain wall is associated with a certain increase in the Au density per unit length.

Previous electron diffraction and ion scattering data had suggested that this Au/Si(111) $\sqrt{3}\times\sqrt{3}$ phase was not perfectly ordered. However, these earlier measurements did not elucidate the nature of the disorder in the surface structure. The STM images clearly show that the disorder is caused by the division of the $\sqrt{3}\times\sqrt{3}$ phase into small domains. In fact, the contraction of the domain size with increasing Au coverage is part of an interesting phase transition between the $\sqrt{3}\times\sqrt{3}$ phase and a higher coverage 6×6 reconstruction.

Figure 6a shows a Au(111) surface after exposure to 3 keV neon ions at a total dose of 3.2×10^{13} ions/cm^2. This surface appears quite disordered, with many single atomic layer deep pits having been created. Holes of various sizes are visible, in contrast to the more uniform crater sizes observed for

Fig. 5: Si(111) surface with 0.8 ML Au deposited and then annealed. This (160 Å)^2 image was taken with V_t = - 2.4 V and I = 2.2 nA. A network of white domain walls interrupt the $\sqrt{3}\times\sqrt{3}$ structure.

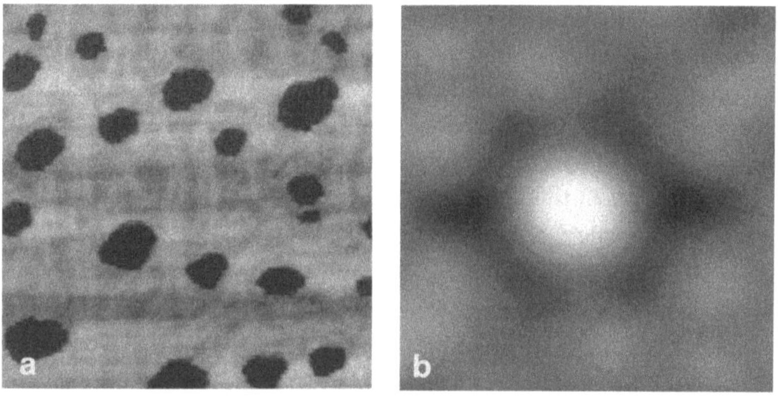

Fig. 6: a) Au(111) surface after exposure to 3 keV Ne atoms at a total dose of 3.2×10^{13} ions/cm^2. This (600 Å)^2 image was taken with $V_t = 0.2$ V and I = 0.1 nA.
b) The autocorrelation function of the topography in Fig. 6a.

lower ion doses. The larger holes most likely were created relatively early and experienced a larger ion exposure than smaller craters.

In addition, the bright lines showing the surface reconstruction are seen to be wiggly, rather than straight as for well ordered surfaces. Unsputtered surfaces often have complicated domain structures exhibiting numerous 120° kinks. In contrast, sputtered surfaces show significantly more forked reconstruction lines, located at craters. This observation is consistent with a possible impact-induced reordering of the reconstruction.

The ion impacts should be randomly distributed and, indeed, the distribution of the pits appears to be quite irregular. An autocorrelation of the STM topograph, however, shows some unexpected order. It is clear from the autocorrelation in Figure 6b that the spatial distribution of the pits shows both a well-defined inter-pit distance and an overall six-fold symmetry that is not apparent in the real-space data. Thus, quantifying the spatial distribution of features in an STM image by using autocorrelation can reveal hidden order in the surface structure. Similar analyses have been useful in studying autoepitaxy on Au(111) and for other sputtering studies [10,14].

5. Conclusion

We have described a number of systems that exhibit various ordering phenomena on length scales which are ideally suited to STM studies. One facet of these studies involved observing the initial stages of surface reconstruction induced by sub-monolayer metal depositions on semiconductor surfaces. For example, the Sn/Si(100) system involved a complex arrangement of buckled and unbuckled dimers which formed localized areas of higher order reconstructions. In a very low coverage regime, the In/Si(100) system illustrated the growth of isolated metal adatom rows which turn out to be a structural element of the 2×2 surface reconstruction at 0.5 ML. Another facet of non-uniform systems involved the examination of complex domain structures. STM images of the 22×1 Au(111) surface reconstruction revealed small domains with a variety of kinked and forked structures. The Au/Si(111) $\sqrt{3}\times\sqrt{3}$ surface also exhibited domains, but they were unique in that the domain size contracted as a function of coverage. Lastly, we presented the Au(111) system, examining both the nucleation of deposited gold clusters and the removal of gold by sputtering. These above examples should provide convincing evidence for the enormous potential of STM work in the study of non-uniform surfaces.

Acknowledgments

This research was supported by the U.S. Office of Naval Research and the Joint Services Electronics Program. One of the authors (A.A.B.) acknowledges partial support from AT&T.

References

1. A. A. Baski, J. Nogami, and C. F. Quate, (in preparation).
2. M. A. van Hove, R. J. Koestner, P. C. Stair, J. P. Biberian, L. L. Kesmodel, I. Bartos, and G. A. Somorjai, Surf. Sci. **103**, 189 (1981).
3. Y. Tanishiro, H. Kanamori, K. Takayanagi, K. Yagi, and G. Honjo, Surf. Sci. **111**, 395 (1981).
4. U. Harten, A. M. Lahee, J. P. Toennies, and Ch. Wöll, Phys. Rev. Lett. **54**, 2619 (1985).
5. C. A. Lang: Ph.D. thesis, Stanford University (1990).
6. Ch. Wöll, S. Chiang, R. J. Wilson, and P. H. Lippel, Phys. Rev. B **39**, 7988 (1989).
7. R. J. Behm in "Scanning Tunneling Microscopy and Related Methods", ed. R. J. Behm, N. Garcia, and H. Rohrer, NATO ASI Series, Series E: Applied Sciences, p. 173, vol.184, (1990) and J. V. Barth, H. Brune, G. Ertl, and R. J. Behm, (submitted for publication).
8. K. G. Huang, D. Gibbs, D. M. Zehner, A. R. Sandy, and S. G. J. Mochrie, (submitted for publication).
9. A. A. Baski, J. Nogami, and C. F. Quate, presented at the 37th American Vacuum Society Symposium, Toronto, October 1990.

10. C. A. Lang, M. M. Dovek, J. Nogami, and C. F. Quate, Surf. Sci. Lett. **224**, L947 (1989).
11. D. D. Chambliss, R. J. Wilson, and S. Chiang, presented at STM '90/NANO 1, Baltimore, July 1990.
12. U. Köhler, J. E. Demuth, and R. J. Hamers, J. Vac. Sci. Technol. A**7**(4), 2860 (1989).
13. J. Nogami, A. A. Baski, and C. F. Quate, Phys. Rev. Lett. **65**, 1611 (1990).
14. C. A. Lang, J. Nogami, and C. F. Quate, (submitted for publication).

Electron Holography and Its Applications to Surface Observation

A. Tonomura

Advanced Research Laboratory, Hitachi, Ltd., Hatoyama, Saitama 350-03, Japan

Abstract. The phase distribution of an electron beam is now measurable to a precision of 1/100 of the wavelength by using both electron holography and a "coherent" field-emission electron beam. Holographic interference micrographs obtained in this way can display the thickness distribution of a thin film or the topography of a single-crystalline surface in atomic dimensions.

1. Introduction

Electron microscopy has made a great contribution to the microscopic observation of solid surfaces. In electron microscopy types, such as transmission[1], reflection[2] and scanning[3], the intensity of an electron beam has been detected. The other independent information of an electron beam, the phase distribution, has not been employed here.

Recent developments of a "coherent" field-emission electron beam[4] have made electron holography[5] more practical, and improved the accuracy of phase measurements up to an order of 1/100 of the wavelength.[6] This opens the way for use of electron holography as high-precision electron interference microscopy[7] for quantitatively observing solid surfaces.

2. Electron Holography

Electron holography[5] is a two-step imaging method, as schematically shown in Fig.1. First, a scattered electron wave from an object is recorded on film as an interference pattern (hologram), resulting from the overlap of the scattered wave with a reference wave. Secondly, the incidence of a light wave onto this hologram reconstructs wavefronts identical in shape to the original electron wavefronts.

This imaging technique does not require lenses and only uses the most fundamental wave properties, interference and diffraction. Therefore, imaging is possible even between completely different waves. However, well-defined wavefronts are indispensable for holography. This is why the intrinsic value of holography had not been recognized until 1962 when a "coherent" laser beam could be used for optical holography.[8] The situation is the same for electron holography. A "coherent" field-emission electron beam developed in 1979 brought about remarkable progress in electron holography.

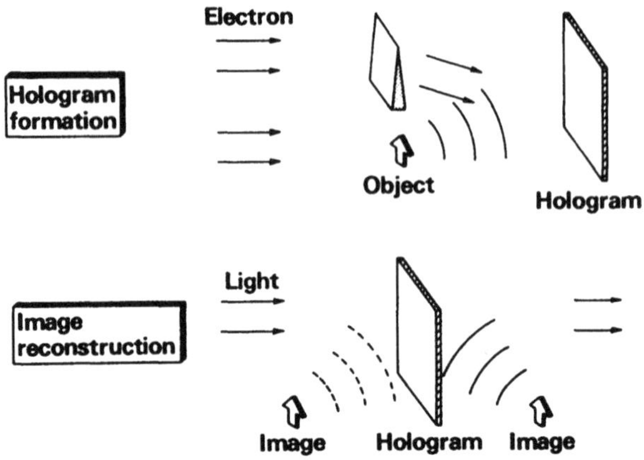

Fig.1. Principle behind electron holography.

3. Interference Electron Microscopy

Once the electron wavefronts are faithfully reconstructed as light wavefronts, versatile optical techniques can be applied to electron optics. The most effective application is interference electron microscopy, where the contour map of the wavefront can be obtained with a phase sensitivity of 1/100 of the wavelength. This is simply done by superimposing optical plane waves on the optically reconstructed image (Fig.2). Interference micrographs, though they are obtained with the help of optical techniques, can be regarded purely as electron micrographs.

By using a technique peculiar to holography, the amplified phase distribution of the electron beam can be observed. As shown in Fig.1, in holography in addition to the original image, a conjugate image is produced having the amplitude of the complex conjugate. In contrast to the method in the interference microscope which produces an interference pattern of plane waves and a

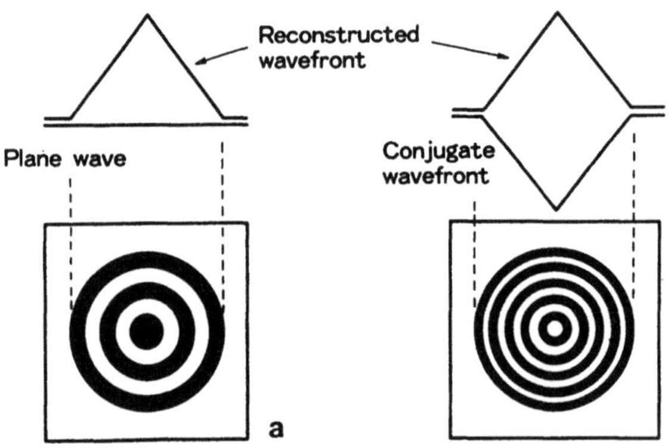

Fig.2. Principle behind phase amplification.
(a) Conventional interference micrograph
(b) Interference micrograph phase-amplified twice

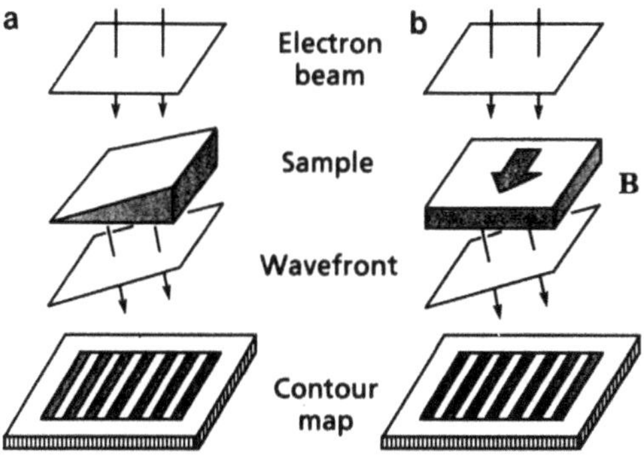

Fig.3. Principle behind electromagnetic-field observation.
(a) Thickness distribution
(b) Magnetic lines of force

reconstructed image, this type of holography (Fig.2) superimposes the conjugate image instead of plane waves. Therefore, the phase difference is exactly doubled. As a result, the phase is amplified. By repeating this process the amplification can be increased. Currently, the phase of an electron beam that is one hundredth of a wavelength can be detected [6].

The effects of both the electric and the magnetic field can be observed with interference electron microscopy. This method leads to intuitive understanding, especially in cases of both non-magnetic samples with thickness variations and magnetic samples.

When an electron penetrates a non-magnetic sample (see Fig.3(a)), it travels faster than in a vacuum, since it is accelerated by an inner potential Vo (=10-30 V). When the sample has a uniform inner potential Vo and consequently has a uniform refractive index given by n=1+Vo/2E (E:accelerating voltage of an electron beam), the transmitted electron wavefront is retarded in proportion to the thickness distribution. Therefore, the resultant contour map of the wavefront indicates the thickness contour map. Contour lines appear every thickness change of (2 E/V) λ, where λ is the electron wavelength. When E=100(kV) and Vo=10-30(V), the thickness change for a 2 π phase shift is 200-700 Å.

The measurement sensitivity seems very low as compared with the extremely short wavelength of an electron beam. This comes from the fact that the refractive index is very near to that of a vacuum. This is the reason why the phase-amplification technique is indispensable in this field.

Quite recently, reflection electron holography has become feasible [9]. Since an electron phase shift due to surface topography is measured by an extremely short electron wavelength, measurement sensitivity becomes very high.

Similar considerations can be applied to a magnetic sample (see Fig.3(b)). An electron is deflected by the magnetic field in passing through the sample. Incident parallel electrons correspond to a plane wave, and the transmitted electrons to an inclined plane wave. Detailed calculations show that contour lines of the wavefront indicate in-plane magnetic lines of force in h/e flux units.

4. Experimental Procedure

4.1 Electron hologram formation

Off-axis electron holograms are formed in an electron microscope, where an electron biprism[10] is installed. A field emission gun is required, due to the necessity of large spatial coherence of an electron beam for hologram formation.

A schematic diagram for hologram formation is shown in Fig.4. Electrons at a divergent angle of less than 1×10^{-7} rad illuminate the specimen. The specimen is located in one half of the specimen plane. The other half is for a reference electron beam whose wavefront is not disturbed by the specimen. An image of the specimen is formed through the objective lens. The electron biprism is installed below the objective lens. It is composed of a central 0.3 µm diameter wire and two ground potential electrodes on both sides. Reference and object waves overlap by applying a positive potential of up to 200 V on the central wire. The interference pattern is magnified by enlarging lenses and recorded on film as a hologram.

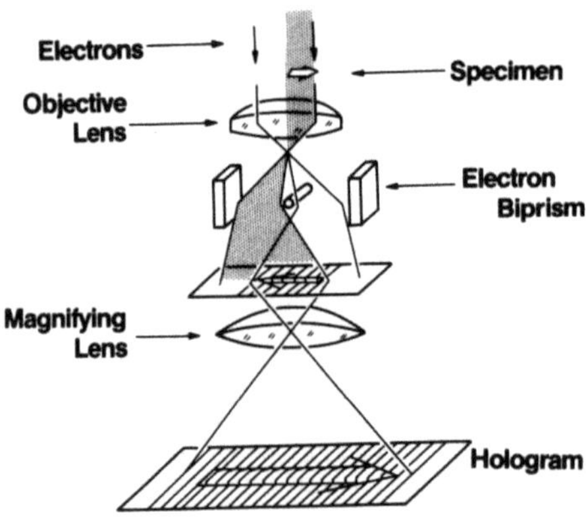

Fig.4. Schematic diagram for electron hologram formation.

4.2 Optical reconstruction -- interference microscopy

When a laser beam illuminates an electron hologram, the electron wavefront is reconstructed with a laser beam. The optical image reconstructed is essentially different from the photographed image and contains all the information on phase as well as on intensity. Therefore, the phase distribution can be displayed on the electron microscopic image in an optical reconstruction stage.

A reconstruction system for interference microscopy is shown in Fig.5. A collimated laser beam from a He-Ne laser (λ = 6328 Å) is split into two beams traveling in different directions by a Mach-Zehnder interferometer. These two beams, A and B, illuminate the hologram and produce two sets of a reconstructed image and conjugate, whose phase distributions are opposite in sign. An interference image is observed when the reconstructed image of beam A, and the transmitted beam of beam B, or vice versa, pass through a slit and overlap.

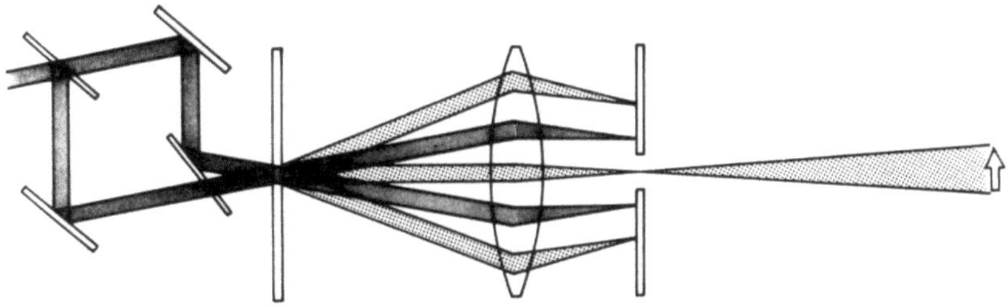

Fig.5. Optical reconstruction system for interference microscopy.

A contour map of the phase distribution in the electron microscopic image is obtained when the interferometer is adjusted so that two interfering beams may travel in the same direction. Contour mapping conditions are ascertained by observing the even background, rather than through a reference system of regular fringes in the specimen free region of an interference image.

This reconstruction system can be used for phase-amplified interference microscopy merely by changing the directions of the two incident beams. Twice the phase amplification can be done by selecting the reconstructed image of beam A and the conjugate image of beam B, then overlapping them on the observation plane. Repetition of this amplification produces a higher amplification rate.

5. Experimental Results

An example of an interference micrograph[11] is shown in Fig.6. The specimen is a Beryllium fine particle. In reconstructed image(a), the electron intensity is recorded, and consequently only the outer shape can be observed. However, in interference micrograph(b), the thickness distribution can be seen at a glance. Thickness contour lines appear every 440 Å change. The micrograph shown in Fig.6(c) is phase-amplified 32 times, and therefore, contour lines of only 13 Å units can be observed. One might doubt whether these fringes represent the thickness distribution precisely, or whether they are only interpolations between broad fringes in micrograph(b) lacking any great degree of precision.

This question was solved through observation of surface steps[6] where the thickness change was definitely known.

The experimental results are shown in Fig.7. The specimen shown here is a cleaved Molybdenite thin film. The surface is almost atomically flat, but has some surface steps. In interferogram(a), where the phase amplification rate is one, no deviation from regular fringes seems able to be observed. However, as the amplification increases, surface steps gradually become observable. Micrograph(c) is phase-amplified 24 times. Fringe steps can be recognized along lines designated as A, B, and C in the micrograph. These steps correspond to one, three and five atomic steps in cleaved surfaces. The smallest thickness change due to an atomic step is 6.2 Å. This is one-half of the c-axis lattice spacing and produces a phase shift of 1/50 of the wavelength. In

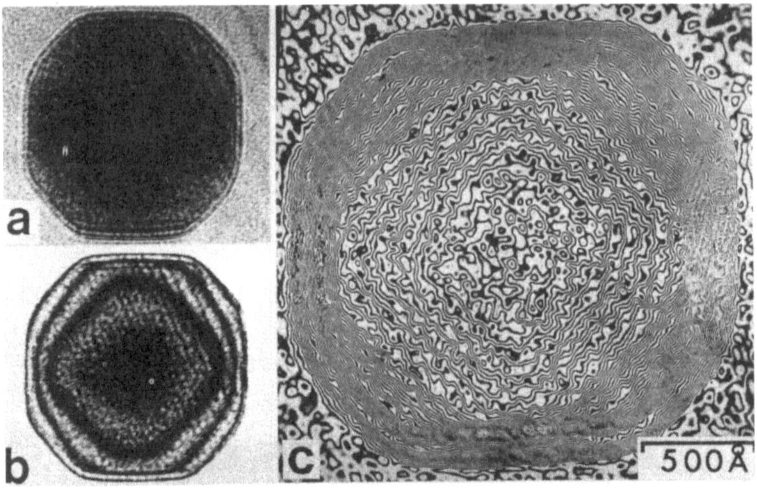

Fig.6. Interference micrographs of fine particle Beryllium.

(a) Reconstructed image, i.e. electron micrograph
(b) Conventional contour map
(c) Contour map phase-amplified 32times

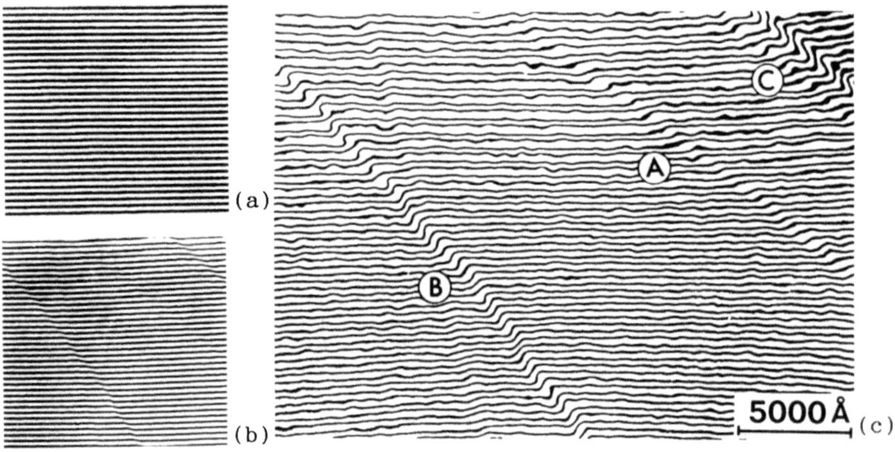

Fig.7. Interference micrographs of thin film molybdenite.

(a) Interferogram (phase amplification: x1)
(b) Interferogram (phase amplification: x4)
(c) Interferogram (phase amplification: x24)

this micrograph, it cannot be determined on which side of the film such surface steps exist. However, it can easily be known by reflection electron holography.

A non-amplified interferogram taken by reflection electron holography is shown in Fig.8. The sample is a single-crystalline GaAs. An electron beam obliquely illumates the GaAs-(110) surface, and excites Bragg reflections. The (880) beam is used to form a hologram between a reference wave reflected from an atomi-

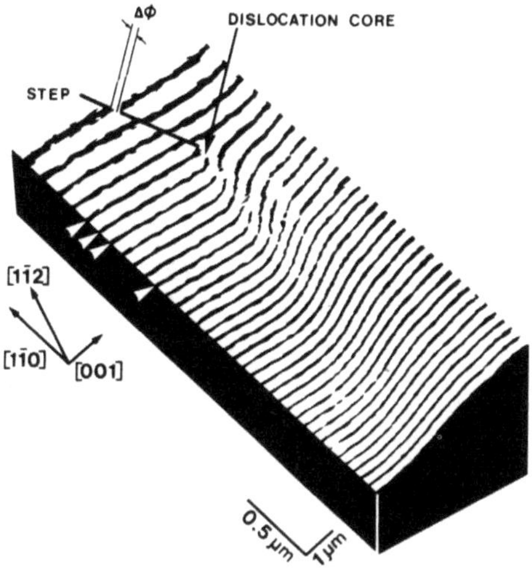

Fig.8. Interference micrograph of GaAs (110) surface.

cally flat region and an object wave from a region where a screw dislocation emerges on the surface. The optically obtained interferogram clearly reveals the surface topography surrounding the dislocation core. A one-fringe shift corresponds to only 0.5 Å in height. A monatomic step of 2 Å height is terminated by the core, and the relaxed surface topography can be observed in detail. For example, this interferogram reveals that the slope of the surface staircase around the core is not uniform, but concentrated along the [112] direction. The reason why such a gradual change in surface height (2 Å height change in a few μm region) can be measured comes from the short wavelength of an electron beam.

This feature of holographic electron interferometry can be further employed for observing surface topography.

References

[1] K. Takayanagi and K. Yagi, Trans. Jpn. Inst. Metals 24, 37 (1983).

[2] N. Osakabe et al., Surf. Sci. 97, 393 (1980).

[3] A. Endo, H. Daimon and S. Ino, to be published in Proceedings of the 12th International Congress for Electron Microscopy., Seattle 1990.

[4] A. V. Crewe et al., Rev. Sci. Instrum. 39, 576 (1968).

[5] D. Gabor, Proc. R. Soc. London, Ser. A 197, 454 (1949).

[6] A. Tonomura et al., Phys. Rev. Lett. 54, 60 (1985).

[7] A. Tonomura, Rev. Mod. Phys. 59, 639 (1987); see also A. Tonomura, Physics Today, April (1990) p.22.

[8] E. N. Leith and J. Upatnieks, J. Opt. Soc. Am. 52, 1123 (1962).

[9] N. Osakabe et al., Phys. Rev. Lett. 62, 2969 (1989).

[10] G. Möllenstedt and H. Düker, Naturwissenschaftan 42, 41 (1955).

[11] J. Endo et al., Proceedings of the 13th International Comission for Optics, edited by H. Ohzu (Organizing Committee, Sapporo, 1984) p.480.

Effect of Tip Apex on STM and STS

O. Nishikawa, M. Tomotori, and F. Iwawaki

Department of Materials Science and Engineering,
The Graduate School at Nagatsuta, Tokyo Institute of Technology,
4259 Nagatsuta, Midori-ku, Yokohama 227, Japan

Abstract. In scanning tunneling microscopy (STM) and scanning tunneling spectroscopy (STS) the number and arrangement of apex atoms of a scanning tip vary frequently. This paper demonstrates the variation of STM images with the number of apex atoms and the apex profile of a scanning tip. The demonstration extends to the effects of the adsorbed atoms at the tip apex and the apex profile on tunneling characteristics.

1. Introduction

Since the introduction of the scanning tunneling microscope (STM)[1, 2], a tacit understanding has been that only a single apex atom of the scanning tip participates in the electron tunneling and the tunneling characteristics are not affected significantly by the adsorbates at the apex. Then, Kuk and Silverman have demonstrated that the surface corrugation traced by a scanning tip of an STM decreases with the increase in the number of apex atoms by directly observing the arrangement of apex atoms with the field ion microscope (FIM)[3]. The study has stimulated the development of a combined instrument of an STM and an FIM[4].

The apex of a scanning tip often touches the specimen surface and picks up atoms and molecules on the surface. The mechanical contact also displaces the apex atoms and deforms the apex profile. The tip with the adsorbed and displaced atoms exhibits unfamiliar STM images. The FIM observation of the tip apex may be able to correlate the unfamiliar images with the arrangement of the adsorbates and displaced atoms. However, the FIM cannot identify the adsorbate nor clarify the effect of the adsorbates on the tunneling spectroscopy. Accordingly, the purpose of this paper is to demonstrate the variation of the STM images with the number and arrangement of apex atoms and to examine the effect of adsorbates on the tunneling characteristics by analyzing the apex atoms and field emitted electrons from a tip apex with the atom-probe (A-P)[5, 6] and the field emission electron spectrometer (FEES)[7, 8].

Eds.: A. Yoshimori · T. Shinjo · H. Watanabe

2. Experimental

An ultrahigh vacuum STM was employed. The STM was operated in a vacuum of $1-2\times10^{-10}$ Torr. A scanning tip was fabricated by electrochemically etching a W[111]-oriented single crystal filament [9]. After the introduction of the tip into the vacuum chamber of the STM, the sharpness, cleanliness and profile of the tip apex were inspected by observing field emission images projected on the screen installed in the chamber. The tips were cleaned by electron bombardment which heated and sharpened the apexes by the build-up effect of the W (111) plane [9].

The specimens were P-doped *n*-Si(111) wafers with resistivity of about 0.01Ω cm. The wafers were cut into rectangular pieces 5×15 mm, etched by the Shiraki method[10], and then resistively heated at 500-600℃ for several hours, keeping the vacuum of the STM chamber better than 2×10^{-9} Torr, and finally flashed at around 1200℃.

Topographic STM images were depicted passing a constant tunneling current in the range 0.2-1.0 nA and applying +2V to the specimen. One frame of the image was formed by 256× 128 pixels and a grey-scale image was obtained by expressing the vertical position of each pixel with 256 grey steps. The scanning time of one frame was 30-60s.

In order to examine the effect of the adsorbates and the tip profile on the tunneling characteristics, the STM was operated in two different scanning tunneling spectroscopy (STS) modes [11]. One of the modes is called dual-polarity tunneling imaging (DPTI). This mode exhibits two constant current images of the same surface area simultaneously by scanning each raster line twice successively, switching the specimen bias voltage from a positive voltage to a negative voltage, for example +2 to -2V, and vice versa.

The other mode is current imaging tunneling spectroscopy (CITS). In this mode, the tip scanned over the specimen surface, keeping the tunneling current constant, and *I-V* curves were obtained at each pixel, varying the sampling voltages 128 steps in the range of +3V to -3V while fixing the tip-specimen separation. One frame of the CITS image was 64× 64 pixels and it took about 90s to acquire 1MB of data. The electronic state density of a specimen surface was obtained by calculating $(dI/dV)/(I/V)$ from the *I-V* curve.

The effect of the deposited Si atoms on the tunneling characteristics of a metal tip apex was examined by the A-P with a FEES[7,8], Fig. 1. A Si layer was deposited on a Mo tip and the variation of the tunneling characteristics with the Si coverage was investigated by repeating the field evaporation of a few Si atoms and the acquisition of the energy spectra of field emitted electrons until the Mo substrate was exposed.

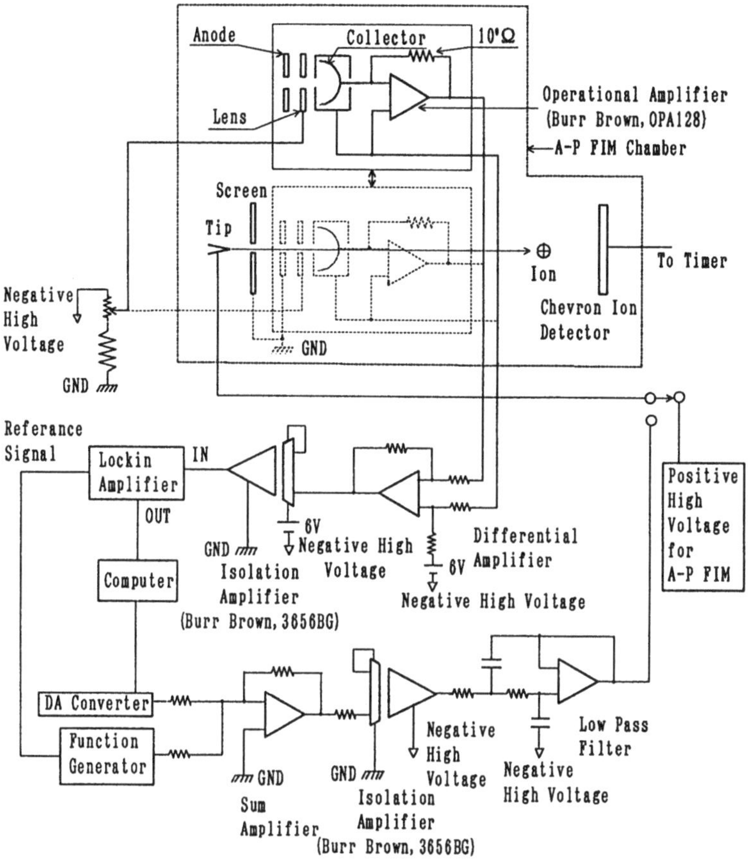

Fig. 1. Schematic diagram of an A-P with FEES. The electron collector can be removed during the A-P mass analysis of the deposited Si atoms.

3. Results and Discussions

The trace of a scanning tip with a single apex atom depicts the topograph of surface atoms. However, the number of apex atoms is not necessarily constant. Nishikawa et al. showed that two apex atoms on the W(111) plane depict anomalous ordered images[12], Figs. 2(a) and (b). These images may mislead observers to believe the finding of new surface structures unless they know the real surface and apex structures.

Not only the apex atoms but also the apex profile distorts the surface topograph. For example, the probed depth of the corner hole of the Si(111)-7X7 structure increases with the sharpness of the tip apex but the maximum depth probed by the [111]-oriented W tip is only 2.4Å ; 55% of the real depth of

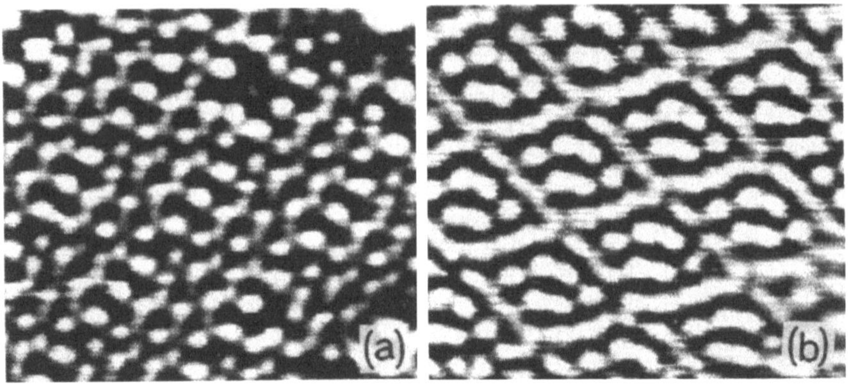

Fig. 2. Anomalous STM images of Si(111)-7×7 structure depicted by two apex atoms on the W(111) plane. (a)The two W atoms are nearest neighbours 4.5Å apart. (b)The two W atoms are second nearest neighbours 7.76Å apart. It appears as if the corner holes, dimer rows and rest atoms of the 7×7 structure are imaged bright.

the hole 4.38Å [12]. The result seems to be contradictory to the high vertical resolution of the STM because the tunneling current varies about 2% by the displacement of the tip apex from the specimen surface 0.01Å and makes it possible to depict the 3-dimensional surface topograph with atomically high resolution. However, it should be noted that this unique capability of the STM is valid only for relatively isolated, protruding surface atoms such as the adatoms of the 7×7 structure and the depth profile of a pit like the corner hole is hardly depicted precisely. Figures 3 (a) and (b) are schematic drawings of the scanning traces of the tips with different apex profiles. The figures show that even a sharp tip with tip radius 100Å fails to follow the shallow pit.

The tunneling characteristics also vary with the apex composition and profile. Nishikawa et al. reported that the DPTI images at positive and negative bias voltages were occasionally completely different[11]: while the normal Si(111)-7×7 structure was imaged at the bias voltage $V_s = -2$ V, an anomalous image as observed for a tip with double apex atoms as shown in Fig. 2(b) was exhibited at $V_s = +2$V. Park et al. also observed the polarity dependent DPTI images and interpreted the observed images as the results of two protruding chemically different apex atoms[13]. However, in their case, the anomalous image was observed for $V_s = -2$V and explained as the result of an adsorbed Si atom which had a high density of states above the Fermi level E_F and participated in the tunneling as one of the two apex atoms. In contrast, the anomalous image was observed at $V_s = +2$ V for the present study.

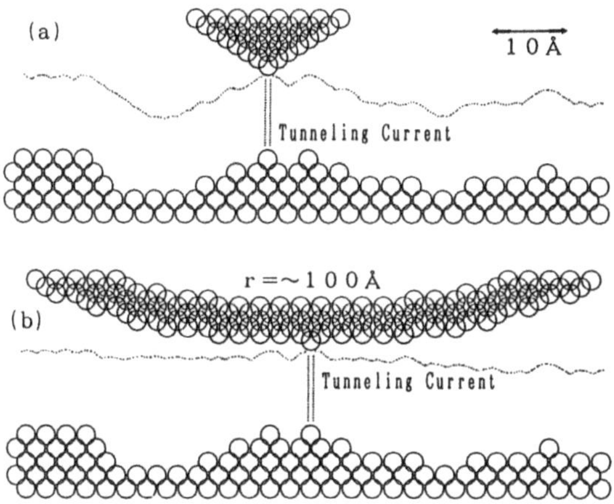

Fig. 3. Variation of the scanning traces of the tips with a single apex atom. (a)Cone angle of the tip apex is 115°. This apex profile is thermodynamically unstable. (b) The trace of the tip apex with the radius 100Å. At the pits the W atoms at the second layer are closer to the surface atoms than the apex atom and the tunneling currents start to flow between them.

This implies that one of the apex atoms has enhanced states below E_F. Such a situation could be realized by a hydrogen atom adsorbed beside a W atom which provides the energy state of about 1 eV below E_F [14]. Then, only the H-free W apex atom tends to receive a tunneling current from the Si surface atom for $V_S = -2V$ showing the regular reconstructed structure.

The CITS mode exhibits two kinds of images: one is topographic and the other is current images. The topographic images are the ordinary STM images depicted by keeping the bias voltage and tunneling current constant and the current images are obtained by converting the currents measured at a designated voltage of the *I-V* curve at each pixel to grey-scale images. CITS also showed misleading images, that is the completely reversed grey-scale current images of the topographic images[11]. The reversed images were explained as the result of tunneling from the dangling bonds in the corner hole and the regions surrounding the adatoms where Si-Si backbonds and additional Si-Si bonds were exposed to the vacuum[15]. This explanation is very contradictory because the normal and reversed images were observed alternately with the measurement being made under identical conditions. The probable explanation is the difference in the apex traces by the tip sharpness. Since the corrugation traced by a tip apex increases with the tip sharpness as shown in Fig. 3, the apex ascends high right above Si atoms and sinks deep into a valley between surface atoms, approaching the Si surface, while depicting a topographic image. Thus, the tunneling current from the valley of the Si surface is larger than that above the Si atom and the reversed CITS current image is depicted. This indicates that the tunneling characteristics vary with the tip profile and a sharp tip gives a reversed current image while a dull tip depicts a normal current image[11].

Up to this point we have demonstrated that STM images and STS characteristics sensitively vary with the atomic arrangements and compositions and the profiles of a tip apex. Some of the most promising techniques for the characterization of the tip apex are FIM observation, A-P mass analysis and FEES measurement. Accordingly, a combined instrument for A-P and FEES as shown in Fig. 1 has been developed[7, 8].

It is well known that a scanning tip frequently bumps against a specimen surface and picks up some of the surface atoms. For example, it is highly probable that a W tip scanning over a Si surface is partially covered by Si atoms and exhibits a different electronic state from that of a pure W tip. In order to examine the effect of the deposited Si atoms, Si was deposited on a Mo tip and the composition and electronic state of the Si covered surface were analyzed by the combined A-P and FEES instrument. The deposited Si atoms form clusters as in the case of Si on Au[16] and the cluster diameters are in the range of 10-50Å. In the A-P/FEES analysis the probe hole of the A-P was aimed at a cluster and the measurement of the FEES spectrum and the A-P mass analysis of the cluster was repeated until the Mo substrate appeared. Figures 4 and 5 show the detection sequence of field-evaporated ions and the variation of the FEES spectra with depth. An interesting finding is that a Si cluster as small as ~10Å in diameter exhibits the same semiconductive electronic state as Si on Au [16].

Heating the Si-covered Mo tip results in the formation of a silicide layer with a metallic electronic state. While the work function of the Si-covered Mo surface is about 10% higher than the Mo substrate, those of the silicides are found to be nearly 10% lower.

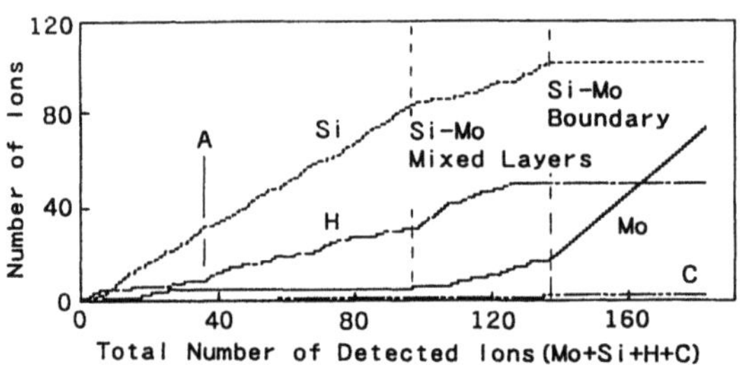

Fig. 4. Detection sequence of ions field-evaporated from a non-heated Si/Mo tip. The number of ions detected from one atomic layer is 10-20. Thus, the Si-Mo boundary is atomically abrupt.

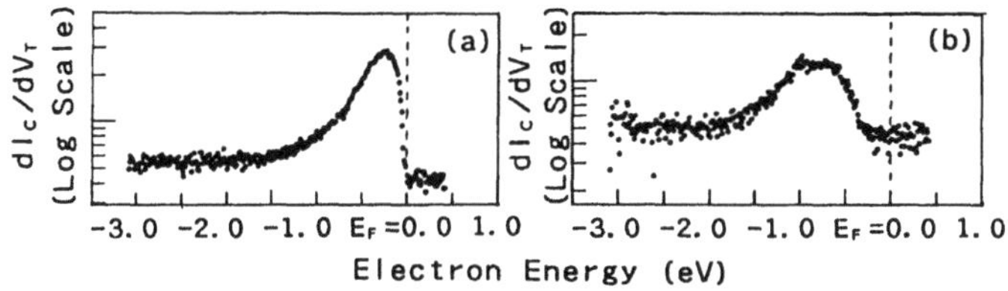

Fig. 5. FEES spectra of Si-covered Mo. I_C is the field emission current and V_T is the tip voltage. (a) Si substrate. The threshold voltage for the I_C detection is at the Fermi level F_E. (b) After Si deposition at A of Fig. 4. The spectrum is semiconductive with the threshold voltage at about -0.5V.

4. Conclusion

Variation of the STM images and the STS current images and spectra with the apex profiles and compositions of a scanning tip was demonstrated. The study indicates that the precise knowledge of the tip apex is an essential requirement for the promotion of surface studies fully utilizing the unique capability of the STM and STS and that the combined A-P and FEES instrument is highly useful for the characterization of the tip apex. Accordingly, the construction of a new combined STM, A-P and FEES instrument is in progress.

[1] G. Binnig, H. Rohrer, Ch. Gerber and E. Weibel, Phys. Rev. Lett. 50, 120 (1983).
[2] C. F. Quate, Physics Today 39, No. 8, 26 (1986).
[3] Y. Kuk and Silverman, Appl. Phys. Lett. 48, 1597 (1986).
[4] T. Sakurai, T. Hashizume, I. Kamiya, Y. Hasegawa, J. Matsui, S. Takahashi, E. Kono, and M. Ogawa, J. Vac. Sci. Technol. A 6, 803 (1988).
[5] E. W. Müller and S. V. Krishnaswamy, Rev. Sci. Instrum. 39, 83 (1968).
[6] O. Nishikawa, K. Kurihara, M. Nachi, M. Konishi, and M. Wada, Rev. Sci. Instrum. 52, 810 (1981).
[7] O. Nishikawa, H. Koyama, N. Kodama, and M. Tomitori, Colloque de Physique 50, C8-507 (1989).
[8] O. Nishikawa, H. Koyama, M. Tomitori, and F. Iwawaki, J. Vac. Sci. Technol. B 9, 789 (1991).
[9] O. Nishikawa, M. Tomitori, and F. Katsuki, J. Microscopy 152, Pt. 3, 637 (1988).
[10] A. Ishizaka and Y. Shiraki, J. Electrochem. Soc. 133, 666 (1986).
[11] O. Nishikawa, M. Tomitori, F. Iwawaki, and N. Hirano, J. Vac. Sci. Technol. A 8, 421 (1990).

[12] M. Tomitori, F. Iwawaki, N. Hirano, F. Katsuki, and O. Nishikawa, J. Vac. Sci. Technol A 8, 222 (1990).
[13] S. Park, J. Nogami, H. A. Mizes, and C. F. Quate, Phys. Rev. B 38, 4269 (1988).
[14] E. W. Plummer and A. E. Bell, J. Vac. Sci. Technol. 9, 583 (1972).
[15] R. J. Hamers, R. M. Tromp, and J. E. Demuth, Surf. Sci. 181, 346 (1987).
[16] Y. Kuk, M. F. Jarrold, P. J. Silverman, J. E. Bower and W. L. Brown, Phys. Rev. B 39, 11168 (1989).

Scanning Tunneling Microscopy of Stepped Si(111) Surfaces

H. Tokumoto and K. Miki
Electrotechnical Laboratory,
1-1-4 Umezono, Tsukuba, Ibaraki 305, Japan

Abstract. STM has been made on stepped Si(111) surfaces. On Si(111)-7x7 surfaces prepared by thermal flashing, only $[\bar{1}\bar{1}2]$ directed steps were found to be straight. This can be explained by applying a simple bond model to double Si atoms, which are removed simultaneously in a cluster form from a step edge during sublimation. An edge of the straight step coincided with an unfaulted half edge of the 7x7 unit on the adjacent upper terrace. This indicates the 7x7 formation starts from making dimers accompanied with the unfaulted half just at the step edge. The reversible heating-current-induced step motion was also observed on Si(111) surface, originating from the electromigration effect. On cleaved Si(111) surfaces, a flight of $[11\bar{2}]$ directed steps was found, whose appearance is also explained by the simple bond model for double Si atoms.

1. Introduction

A crystal surface can be regarded as *a new material* since it exhibits various structures different from the bulk structure [1]. From this viewpoint, many experimental and theoretical investigations have been performed so far and revealed many interesting surface structures on various materials. The presence of atomic steps could modify the structure on the surface resulting in new structures caused by strain fields. When we concentrate our attention on the behavior of the steps on the surface, scanning tunneling microscopy (STM) [2] is the most suitable way because of various advantages such as atomic resolutions in both vertical and horizontal directions and a real-space image of the structure. These advantages cannot be seen in other techniques such as low energy electron diffraction and reflection electron microscopy (REM). In this paper, we shall review our STM observation of the steps on the Si(111)-7x7 surface [3, 4] prepared by thermal flashing and on the Si(111)-2x1 surface [5, 6] prepared by cleaving, and we shall add some explanations to them.

2. Experimental

We have successfully constructed two types of ultra-high vacuum (UHV) STM systems having different tunneling units and cleaning mechanisms. [7, 8] For the one, the tunneling unit consists of a mechanical pusher for coarse positioning and a cubic-type tripod for *xyz* scanning of a tip. The sample is cleaned by thermal flashing on passing an electric current directly through the sample. [8] Another UHV-STM system [9] consists of three chambers which are connected to each other through gate valves to transfer a sample or an STM tip without breaking UHV. In a load-lock chamber, three

Eds.: A. Yoshimori · T. Shinjo · H. Watanabe

tips and three samples can be introduced into vacuum from air at the same time. In an STM chamber, an STM unit is fixed together with a facility for cleaving wafer samples. The unit consists of a differential screw for coarse positioning and a small tube-type piezoelectric scanner. The third chamber is for the sample preparation by thermal flashing. The base pressure is as low as $5x10^{-11}$ Torr for both systems. As for the tip, a mechanically ground Pt-tip is used without any special cleaning.

For the study of thermally cleaned surfaces, we prepared rectangular-shaped samples from an Sb-doped n-type Si(111) wafer. They were pre-cleaned chemically and then set in the UHV chamber. When the pressure of the chamber reached 10^{-10} Torr range, we outgassed the samples at 500-600 °C while maintaining the pressure below 10^{-10} Torr range. Then the samples were heated several times by passing a current directly through the sample at 1200 °C for 20-30 s. Each time the pressure was maintained in the 10^{-9} Torr range or less. After cooling the sample to room temperature, the STM measurement was carried out. For cleaving, we used a B-doped p-type Si($\bar{1}$01) wafer so as to obtain the (111) plane by cleaving perpendicularly to the wafer. The sample was moderately etched by a HF etchant and then set on the sample holder in the UHV chamber. When the pressure reached $5x10^{-11}$ Torr, the sample was cleaved by bending with a lever. The STM measurement was carried out with both constant and variable current modes around 0.1-0.5 nA. STM data are recorded either on a computer memory or on a video tape with 256x256 pixel resolution without any signal processing. No corrections were made on the distortion due to the interference between *xy* piezos for tip scanning which had been calibrated within an error of 10%.

3. Results and Discussion

3.1 Si(111)-7x7 Surfaces

Figures 1 (a) and (b) show typical STM images of the different parts of the same Si(111)-7x7 surface. These images clearly show twelve adatoms of a DAS

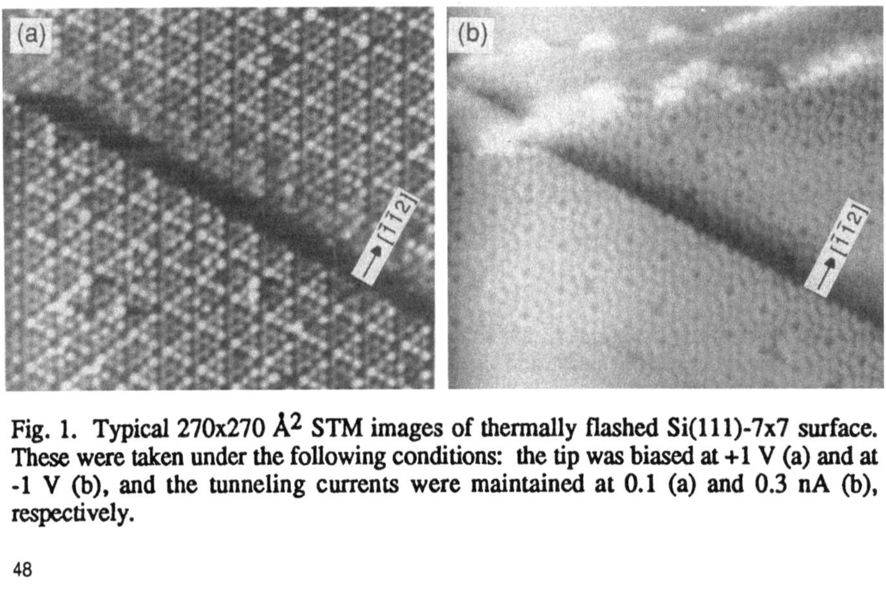

Fig. 1. Typical 270x270 $Å^2$ STM images of thermally flashed Si(111)-7x7 surface. These were taken under the following conditions: the tip was biased at +1 V (a) and at -1 V (b), and the tunneling currents were maintained at 0.1 (a) and 0.3 nA (b), respectively.

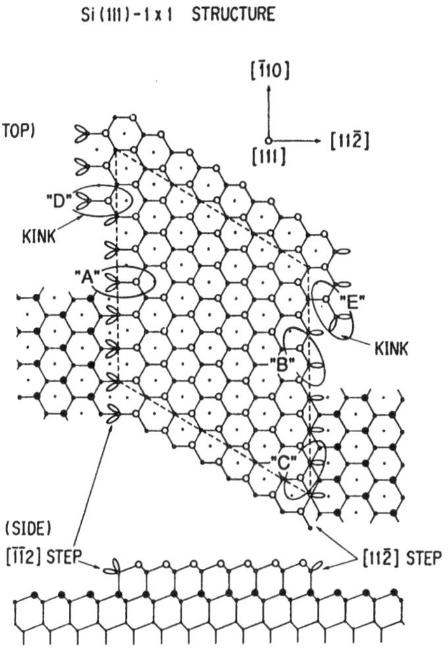

Fig. 2. Si(111)-1x1 structure model with two types of steps, [1̄1̄2] and [112̄] directed steps, in the most stable configuration. Rabbit ear shapes represent dangling bonds and the dotted line indicates the size of 7x7 structure. Five types of clusters of two Si atoms are denoted by capital letters, "A", "B", "C", "D", and "E".

(dimer/adatom/stacking-fault model) structure.[10] The lower parts of these images are higher than the upper parts. We can see two types of atomic steps [3]: one is straight [1̄1̄2] directed steps of ~3 Å height; the other is non-straight steps having no definite directions. From these figures, it is clear that the straight step edge coincides with the side of the unfaulted half on the adjacent upper terrace, whose image is darker than that of the faulted half, but does not coincide with the edge of the 7x7 unit on the lower terrace.

We have already discussed the appearance of the straight [1̄1̄2] directed steps based on the conventional simple bond model and have come to contradicting conclusions[3]. There we considered only a single Si atom involved during the sublimation process. Figure 2 shows the most stable 1x1 structure of Si(111) surface with two types of atomic steps, the [1̄1̄2] and [112̄] directed steps, which are realized at high temperatures such as 1200 °C. Each atom at the step edge has two dangling bonds (db's) at the [1̄1̄2] directed step and one db at the [112̄] directed step. If the single Si atom process is dominant during sublimation at either step site, the resulting structure is different from the beginning and is unlikely to be a stable configuration. In order to obtain the same kind of stable configuration even after the sublimation process, two Si atoms should be removed in a cluster form at the same time. We call these two Si atoms double Si atoms, hereafter. This type of phenomena has already been found in the Si growth with a molecular beam epitaxy technique: the analysis of the RHEED oscillation during crystal growth has revealed that the double Si atoms incorporate at the same time to form a new atomic layer. [11] First we shall discuss the removal of the double

Si atoms from the straight steps with no kink sites. At the [$\bar{1}\bar{1}$2] directed steps, three bonds should be broken to remove the double Si atoms and the method of breaking is unique ("A" in Fig. 2; no domain structure), which would make the step edge smooth, whereas at the [11$\bar{2}$] directed steps, four bonds should be broken and there are two methods, "B" and "C" in Fig. 2, which could form domains and make the step edge rough. Secondly we shall discuss the atoms at the kink sites. For both steps, three bonds should be broken to remove the double Si atoms as shown by "D" and "E" in Fig. 2 and their means of removing the double Si atoms are exactly the same. When we compare the method of removal from the kink site with that of the straight step site, the number of breaking bonds is the same but the surroundings of the double Si atoms are different: at a kink site, there exist three db's ("D") while at the straight edge, there are two db's ("A"). If the one extra db at the kink site could result in the reduction of the breaking energy of the adjacent bond, the removal of the double Si atoms at the kink site would occur more frequently than that at the straight steps and make the steps straight. In the case of the [11$\bar{2}$] directed steps, it is clear that the kink-site double Si atoms ("E") would be easily removed due to the smallness of the breaking bonds compared with those at the straight parts. At the same time, the double Si atoms at the steps are also removed frequently due to high temperature treatment. If we take both processes discussed above into account, the appearance of the straight [$\bar{1}\bar{1}$2] directed steps can be explained as a result of the removal of the double Si atoms.

Next we shall discuss the coincidence of the step edge with the side of the unfaulted half of DAS on the upper adjacent terrace. This is consistent with the previous REM result that the 7x7 domain grows from the step edge [12]. This would be closely related to the phase transition from 1x1 to 7x7 during cooling. In order to form the 7x7 structure characterized by the DAS model, it is obvious that the forming of corner holes and dimers requires a lot of elastic energy compared with the other parts: among them, more energy is required for the faulted half than the unfaulted half due to the different atomic layer stacking. This indicates that the unfaulted half is energetically more stable than the faulted one and among the unfaulted half, the corner holes and dimers could be formed with less energy loss just at the step edges due to the smaller number of bonds. It is also likely that the temperature just at the edge would be lower than the other parts of the terraces and hence the phase transition could start just from the step edge.

In the present experiment, we passed a dc electric current through a sample in order to raise the sample temperature to around 1200 °C. It is pointed out by an REM study [13] that the dc electric current causes the shifting of steps depending on the current direction against step direction and the sample temperature. In fact, this phenomenon has also been observed in the STM measurement. [4] Here we shall briefly summarize the STM observations. Figure 3 shows STM images of the Si(111) surface which exhibit the effect of the dc current. The height gets lower from the left to the right and arrows indicate the directions of the currents in these particular images. The dc step-down current increases the number of steps and makes their space even, as shown in Figs. 3 (a) and (b), whereas the step-up current decreases the number of steps and causes the step bunching, as seen in Figs. 3 (c) and (d). There are several possibilities for causing the step motion: the effect of contamination such as SiC which clumps the step motion, the strain effect introduced by clamping the sample, and the electromigration effect. The first two possibilities can be ruled out for the present case since the sample temperature of 1200 °C is high enough to sublimate all contaminants from the surface and the strains are not dependent on the direction of the current. [4] The whole STM data shown above are consistent with the results by REM

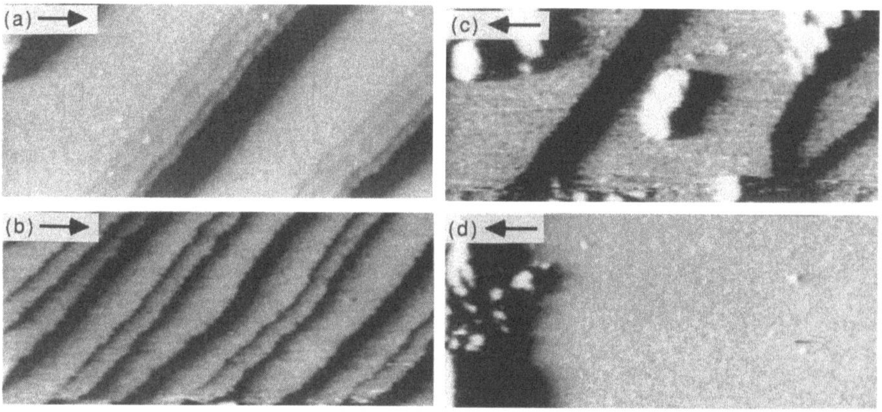

Fig. 3. Typical 2160 x 1080 Å^2 STM images of Si(111) surfaces heated at 1200 °C by flowing dc electric current. The current directions are shown by arrows. The tip was biased at +1 V and the tunneling current was maintained at 0.1 nA. (a) A standard image of a cleaned surface; (b) after flowing step-down current for 5 min.; (c) after flowing step-up current for 3 min.; (d) after flowing step-up current for 3 min. more. These images show the heating-current-induced shifting of steps.

[13] and therefore we would ascribe the heating current-induced step motion to the electromigration effect. On the other hand, STM images for the current perpendicular to the step direction did not show the change in step arrangement. From the above results, the electromigration effect can occur only for the step-up or step-down current. However, the present experiment was carried out at 1200 °C where both sublimation and electromigration effects were involved. In order to understand these phenomena microscopically, more extensive STM studies should be conducted on effects of ac current and heat by radiation, dynamic behavior at various temperature *etc.*

3.2 Si(221)-3x1 Surfaces

We can also obtain clean surfaces of solids by cleaving samples. The surface structures are, however, different from those prepared by thermal flashing. In the case of cleaving, the stress may play an important role in forming surface structures. This means that the cleavage gives us a variety of structures depending on its method. Therefore we shall describe the present way of cleaving here. [6] The sample was held to a cleavage stand by a coil-spring suspended metal plate and the stresses were applied slowly onto one end of the sample. With increasing stresses the sample bends more, and gets cleaved above the critical stress. During this procedure, the head side will experience the tensile stresses while the tail side will experience the compressive stresses. Both the optical microscope and STM images showed that the tensile region was fairly smooth while the compressed region was rough and exhibited many stripes, the *river pattern*. In the present study, the tensile region was investigated extensively.

High-resolution STM images of the cleaved Si(111) surface usually consist of two types of terraces [5, 6]: triangular-shaped terraces and long and narrow terraces. On the triangular terraces, there exist the atomic rows running along one of the <110> directions which correspond to the dimer rows originating from the Si(111)-2x1

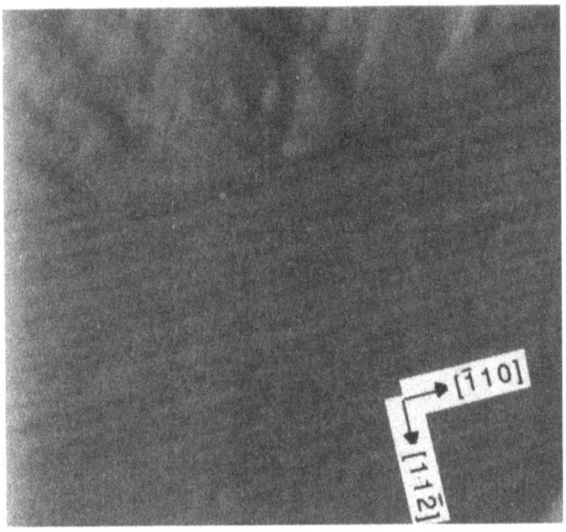

Fig. 4. Typical 1800x1800 Å^2 STM image of Si(221) surface which was occasionally obtained from stepped Si(111) surface. The tip was biased at +1 V and the tunneling current was maintained at 0.1 nA.

reconstruction. The step edges associated with the long and narrow terraces are parallel to the [$\bar{1}$10] direction and directed toward the [11$\bar{2}$] direction. This feature is clearly seen in an STM image (Fig. 4) for a particular part of the cleaved Si(111) surface. In this image, there exists a flight of [11$\bar{2}$] directed steps, whose surface is assigned to (221) from the width of 10 Å and the step height of 3.1 Å. Here it should be noted that the straight [11$\bar{2}$] directed steps appeared in the present case in contrast to the [$\bar{1}\bar{1}$2] directed steps in the case of thermal flashing and also found by Henzler [14], Pandey [15], and Feenstra and Stroscio [16].

Here we shall discuss the origin of [11$\bar{2}$] directed steps and discrepancies among many studies. To do this, we shall use again the structure model given by Fig. 2 since the structure just before cleaving should be the 1x1 structure. In our case of Si($\bar{1}$01) wafer and the [$\bar{1}$01] stress, the resulting stresses induced in the head side of the sample should be the tensile stresses whose direction is in the [111] direction and whose strength is getting weaker toward the [$\bar{1}$01] direction. When the inhomogeneous stresses are applied along the [111] direction slowly (*quasi-statically*) at room temperature, the atomic bond breaking will start at the head end of the sample and proceed toward the inside of the sample, keeping the local equilibrium. Once a kink is formed occasionally at a certain [11$\bar{2}$] step site, the double Si atoms at the kink site ("E") can be more easily removed than the other double Si atoms ("B" or "C") due to the appreciable difference in the number of breaking bonds. Then the kink can move very easily leaving the straight [11$\bar{2}$] directed steps. In the case of the [$\bar{1}\bar{1}$2] step sites, the number of bonds to be broken is the same for the double Si atoms at kink sites ("D") and the other ("A"). So in the quasi-equilibrium states, there would be no appreciable difference between them. Neither is a difference expected for two clusters of the double Si atoms at the same [$\bar{1}\bar{1}$2] directed steps, which makes step edges rough. Therefore we expect no long straight [$\bar{1}\bar{1}$2] directed steps. One of the possible reasons for the discrepancies among many studies is the different stresses in the sample since both their ways of cleaving and the crystallographic directions of the their samples are different from ours.

When a part of the (221) surface was magnified up to atomic scale resolution, regular rows running parallel to the [11$\bar{2}$] direction were observed on the terraces of 10

Å wide. [5, 6] The spacing between the rows measures about 0.52 nm which is appreciably smaller than the distance between the dimer rows in the Si(111)-2x1 structure. We have already discussed the origin of this new structure and propose a model as the Si(221)-3x1 structure. [6]

4. Summary

In this paper, we have presented STM observation of stepped Si(111) surfaces prepared by thermal flashing and cleaving. Some new attempts were made to explain the origin of new findings of steps on this Si(111) surface. However, since none of them can explain all the experimental results, more extensive work is necessary to solve these problems.

Acknowledgements

We would like to thank S. Wakiyama, H. Murakami, M. Okano, Y. Morita, S. Okayama, K. Kajimura, and M. Ono for their support in the STM measurements, K. Sakamoto and T. Sakamoto for helpful discussion on the Si surface structure, and Seiko Instruments Inc. for help in constructing a UHV-STM.

References

1. An international conference from the viewpoint of "*Surface as a New Material*" was held on July, 1990 at Osaka, Japan, sponsored by The Yamada Science Foundation.
2. G. Binnig, H. Rohrer, Ch. Gerber, and E. Weibel, Phys. Rev. Lett. **49**, 57 (1982).
3. H. Tokumoto, K. Miki, and K. Kajimura, J. Cryst. Growth **99**, 1329 (1990).
4. H. Tokumoto, K. Miki, H. Murakami, and K. Kajimura, to appear in March/April issue of J. Vac. Sci. Technol. (1991).
5. H. Tokumoto, S. Wakiyama, K. Miki, and S. Okayama, Appl. Phys. Lett. **56**, 743 (1990).
6. H. Tokumoto, S. Wakiyama, K. Miki, H. Murakami, S. Okayama, and K. Kajimura, to appear in March/April issue of J. Vac. Sci. Technol. (1991).
7. K. Kajimura, H. Bando, K. Endo, W. Mizutani, H. Murakami, M. Okano, S. Okayama, M. Ono, Y. Ono, H. Tokumoto, F. Sakai, K. Watanabe, and S. Wakiyama, Surf. Sci. **181**, 165 (1987).
8. H. Tokumoto, K. Miki, H. Murakami, N. Morita, H. Bando, A. Sakai, S. Wakiyama, M. Ono, and K. Kajimura, J. Microscopy **152**, 743 (1988).
9. H. Tokumoto, S. Okayama, K. Miki, H. Murakami, M. Okano, K. Sakamoto, T. Sakamoto, H. Kawanami, E. Suzuki, S. Wakiyama, K. Ando, K. Ohmura, K. Watanabe, A. Homma, K. Ishihara, A. Inoue, F. Sakai, Y. Kobayashi, S. Watanabe, and F. Matsumoto, *Bulletin of the Electrotechnical Laboratory* **54**, 136 (1990).
10. K. Takayanagi, Y. Tanishiro, M. Takahashi and S. Takahashi, Surf. Sci. **164**, 367 (1985).
11. T. Sakamoto, N.J. Kawai, T. Nakagawa, K. Ohta, and T. Kojima, Appl. Phys. Lett. **47**, 617 (1985).
12. N. Osakabe, K. Yagi and G. Honjo, Jpn. J. Appl. Phys. **19**, L309 (1980).
13. A.V. Latyshev, A.L. Aseev and A.B. Krasilnikov and S.I. Stenin, Surf. Sci. **213**, 157 (1987).
14. M. Henzler, Surf. Sci. **19**, 159 (1970); *ibid.* **36**, 109 (1973).
15. K.C. Pandey, Phys. Rev. Lett. **47**, 1913 (1981); *ibid.* **49**, 223 (1982); Physica **117B** and **118B**, 761 (1983).
16. R.M. Feenstra and J.A. Stroscio, Phys. Rev. Lett. **59**, 2173 (1987).

Energetics of Surface Atomic Processes

T.T. Tsong
Physics Department, The Pennsylvania State University,
University Park, PA 16802, USA

Abstract. The energies needed in various surface atomic processes such as surface diffusion, binding of small atomic clusters on the surface, interaction between two adsorbed atoms, dissociation of an atom from a small cluster or from a surface layer, binding of kink site atoms or atoms at different adsorption sites to the surface etc. can be derived from a field ion microscope image analysis and a kinetic energy measurement of field desorbed ions using the time-of-flight atom-probe field ion microscope. These energies can be used to understand the transport of atoms on the surface in atomic reconstructions, crystal growth, adsorption layer superstructure formation, and why an atomic ordering at the surface is energetically favored. Mechanisms of various atomic processes can also be clarified by these studies.

1. Introduction

To understand a surface phenomenon and an atomic ordering at the surface and interface, it is best to consider the atomic interactions and the free energy of the system. Using the field ion microscope (FIM) and the atom-probe FIM [1], it is possible to observe directly different atomic processes and to derive atomic interactions and energies needed in these processes. I will describe briefly here results of these FIM and atom-probe FIM measurements, and how they can be used to understand why some of the surface reconstructions and atomic ordering are energetically favored.

2. Surface Diffusion

In the field ion microscope, single metal and semiconductor atoms on a metal surface can be directly imaged. Thus by heating the surface to a desirable temperature for a desirable duration without applying the image field, surface diffusion of single adsorbed atoms in the absence of the field can be

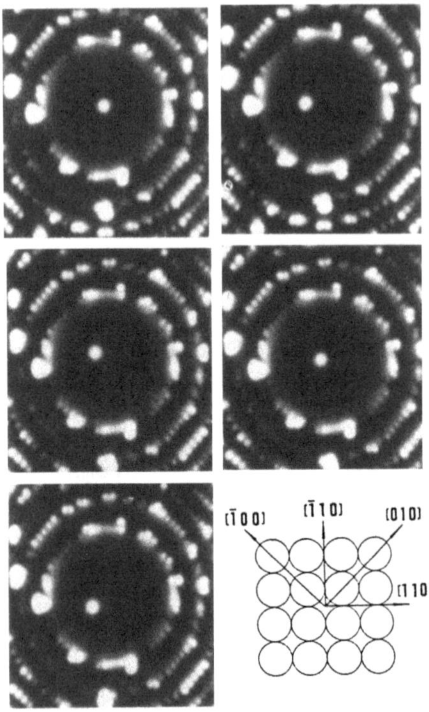

Fig.1 Random walk of an Ir atom on Ir(001)

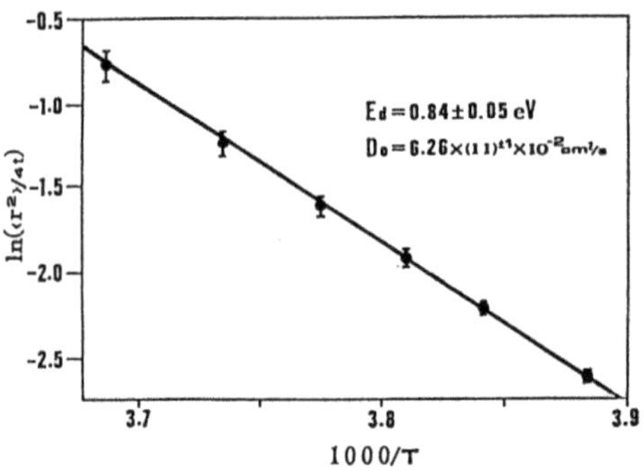

Fig.2 Arrhenius plot for Ir on Ir(001)

observed and studied directly [2]. Fig. 1 shows the random walk diffusion of one Ir adatom on an Ir (001) surface. By plotting the logarithm of the mean square displacement against the inverse temperature, a linear plot is expected as shown in Fig. 2. From the slope and the intercept, the activation energy E_d and the pre-exponential factor, or the frequency factor, are derived. Data available from FIM studies are still limited. For bcc metals, only diffusion of metal and silicon adatoms on W surfaces has been studied. Fcc metals studied include Pt, Ir, Rh, and Ni surfaces etc. with a variety of metal adatoms.

The activation energies of surface diffusion for metal and Si adatoms on bcc tungsten {110}, {112} and {123} surfaces range from about 0.5 to 1.0 eV. Contrary to our intutition, the smoothest {110} surface exhibits the highest E_d among these surfaces. On the rough surfaces such as {111} and {001}, the activation energies are very high. Adatoms do not diffuse until a temperature is reached that is high enough for some of the surface layers to start to dissolve; thus it is difficult to study surface diffusion on these surfaces. On fcc metal surfaces, E_d varies over a wide range, from less than 0.2 eV for Rh on Rh {111} to near 1.0 eV for Ir on Ir {001}. As expected, E_d is higher for a rougher surface. For the frequency factor, ν_0 always come out to be $10^{13\pm1}$ s^{-1}, which corresponds to a diffusivity D_0 of about $10^{-3\pm1}$ cm^2/s. In other words, the entropy factor is very small.

The question is, what is the mechanism of surface diffusion and what is the physical significance of E_d? In the past, it has been assumed that surface diffusion is achieved by hopping of the adatom over a saddle point of the surface potential profile of the adatom-surface interaction, or by an atomic-hopping mechanism. E_d represents the potential barrier height of the saddle point. Recent theoretical and experimental studies [3-5] point to another mechanism of surface diffusion, namely by an atomic-replacement. On the {001} surface of some fcc metals such as Pt and Ir, surface diffusion of an adatom of some species occurs by replacing a substrate atom. This conclusion is reached from the jumping direction of the adatom which can be found by digitizing the position of an adatom after each heating of the surface from which the jump-direction of the adatom is revealed. A jump here means either an atomic hop or a displacement resulted from an atomic replacement. Fig. 3 shows a visited-site-lattice (the thin grid) of an Ir adatom diffusing on an Ir (001) surface (the thick grid). This visited site lattice is found to correspond to the c(2×2) net of the substrate lattice, thus the adatom

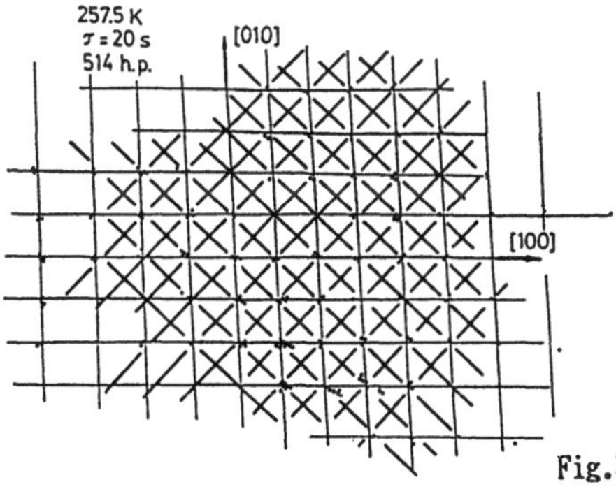

Fig.3 A visited-site-lattice for Ir on Ir(001)

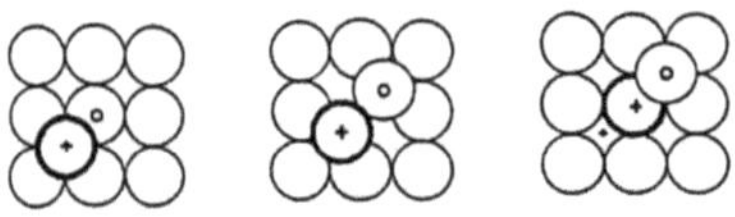

Fig.4 Atomic-replacement mechanism for diffusion of Ir atoms on Ir(001)

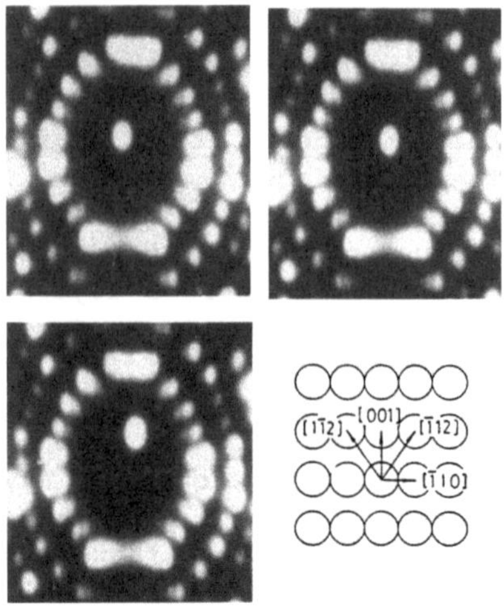

Fig.5 2-D random walk : Ir on Ir(110)

can jump only along the <001> directions over the atop site of substrate atoms. Since the saddle point of the adatom-substrate interaction cannot be at the atop-site, it is concluded that the diffusion is achieved by an atomic-replacement mechanism as illustrated in Fig. 4. It is also interesting to note that, except for some systems at high temperatures, most experimental displacement distributions of single atom diffusion are consistent with theoretical distributions based on discrete nearest-neighbor random walk of the adatoms. In the case of Ir on Ir {001}, the jump length is 3.8 Å along the <001> directions [3].

An atomic-replacement mechanism is also known to occur for some metal adatoms on the {110} surface of some fcc metals [6]. This surface have a deep atomic trough structure. If diffusion occurs by atomic hopping, one can expect it to be only along the [110] surface channel direction. Diffusion of adatoms, however, is found to be both along and across these surface channels (Fig. 5). In fact, for Ir on the Ir {110} surface, the cross-channel diffusion has a lower E_d of 0.71 eV compared to 0.80 eV for the along-channel diffusion [7]. Thus the energy needed to replace a substrate atom by an adatom is slightly lower than the energy needed for the adatom to hop over the potential barrier along the surface channel.

Diffusion of small atomic clusters has also been studied. W, Re and Ir adatoms in the neighbor channels of the W {112} surface tend to form a stable chain-like structure with the atoms having only two bond lengths, presumably 4.47 Å and 5.32 Å. Diffusion of the chain is accomplished by coupled but nearly random hopping of individual atoms. The atomic hopping process has also been studied in great detail for diffusion of W-dimers on the W {110} surface [8]. It is also concluded that atoms hop nearly independently although at high temperatures, these atomic hops are correlated. E_d for dimers differs from that of single atoms by only ~0.1 to 0.3 eV, the difference in the bond energies of the stretched and equilibrium bonds. In some special cases, it may be slightly smaller for dimers. D_0 for small clusters may be one to two orders of magnitudes smaller than that for single atoms, but because of the limited accuracy of FIM measurements for D_0, an accurate value of this difference has not been established.

For the systems where diffusion occurs by the atomic-exchange mechanism, cluster diffusion also behaves very differently. An example is the diffusion of Ir-dimers on the Ir {110} surface [9]. A dimer with two Ir atoms in the neighbor surface channels is unstable since one of the adatoms invariably crosses the channel by an atomic-exchange, and combines with the other adatom to form a closely-packed dimer with its bond parallel to the surface channel. Contrary to single Ir adatoms on this surface, E_d for the dimer diffusing along the channel is lower than that diffusing across the channel, 1.05 eV vs 1.18 eV. Ir-dimers are also unusually stable, having a binding energy of 1.10 eV as compared to less than 0.3 eV for most metal dimers on the W {110} surface. Because of the high stability of Ir-dimers on this surface, it is believed that Ir-dimers diffuse by concerted atomic motion of the two atoms, not by hopping of each atom. Concerted motion of atoms in clusters has also been invoked to explain atomic steps in the (1×1) to (1×2) surface reconstruction of Pt and Ir {110} surfaces [10].

It appears that both the (1×1) to (1×2) reconstruction of the fcc {110} surface and the (1×1) to (1×5) reconstruction of the Ir {001} surface are favored because of the extreme stability of the nearest-neighbor bond and the closely packed [110] atomic rows. The (1×5) structure of the Ir {001} can be considered to be formed by squeezing six [110] atomic rows into five row spacings. As the space is not quite wide enough, these rows are slightly buckled and shifted to have a quasi-hexagonal atomic arrangement.

3. Atomic Interactions and Cluster Formation

When there are many adatoms on a surface, these atoms may encounter each other during their diffusion. If their interaction is attractive, a cluster will be formed. Fig. 6

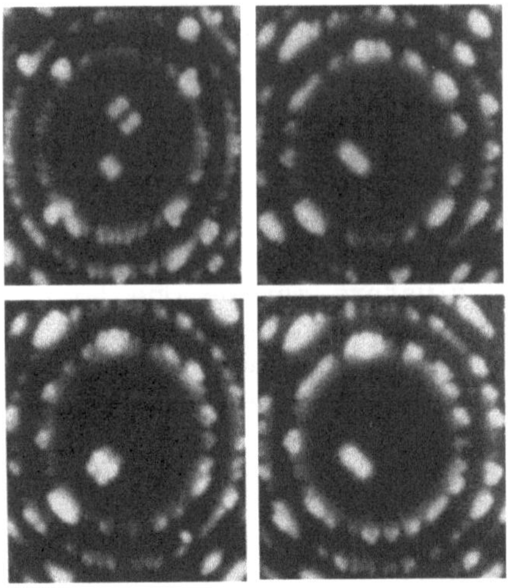

Fig.6 Structures of a 6-atom cluster

shows the formation of a 6-atom Ir cluster on an Ir (001) surface formed by combining three dimers, and the one-dimensional to two-dimensional structure transformation of this 6-atom cluster. If the temperature is high enough, the cluster will dissociate. The adatom-adatom interaction can be derived by depositing two adatoms on a surface and measuring the pair-distribution directly from field ion images taken as the surface is repeatedly heated to a high temperature without the applied field and then quenched. An interaction potential can be obtained if the adatom-adatom interaction is weak and the plane boundary is reflective to adatom diffusion. It is found that the adatom-adatom interaction of metal adatoms on the W {110} surface is very weak, of only about 0.1 eV, or only about one tenth a chemical bond strength, and there is a repulsive region and a small oscillatory tail which can extend to near 10 Å in distance [11]. Despite a possible many-body effect, the accuracy of the pair energy additivity approximation is good enough to predict a superstructure of a Si adsorption layer on the W {110} surface [12]. Silicon atoms form a c(2×2) structure of the conventional unit cell of the substrate, in agreement with the measured Si-Si pair interaction which has two relative minima at the two bond lengths and directions corresponding to the two sides of the c(2×2) unit cell.

The internal binding energy of a cluster can be derived from the temperature dependence of the dissociation time of the cluster as well as the relative probabilities of observing the cluster in different atomic structures. The binding energy of a diatomic cluster of metal atoms on a metal surface

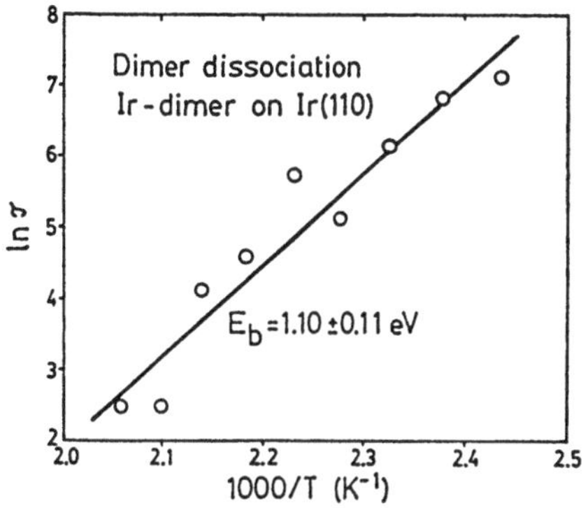

Fig.7 Dissociation of Ir-dimers on Ir(110)

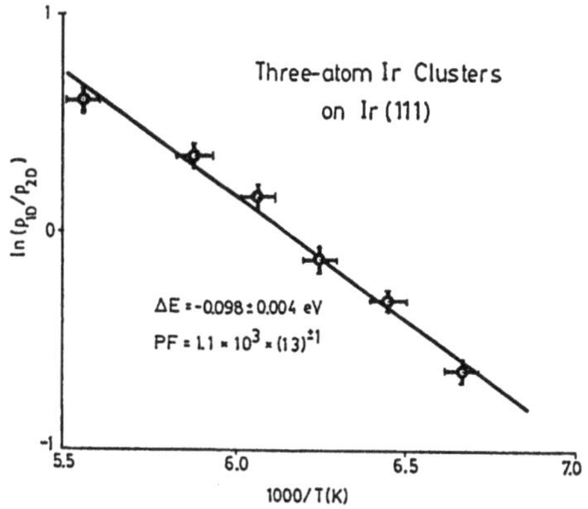

Fig.8 Structure changes of clusters

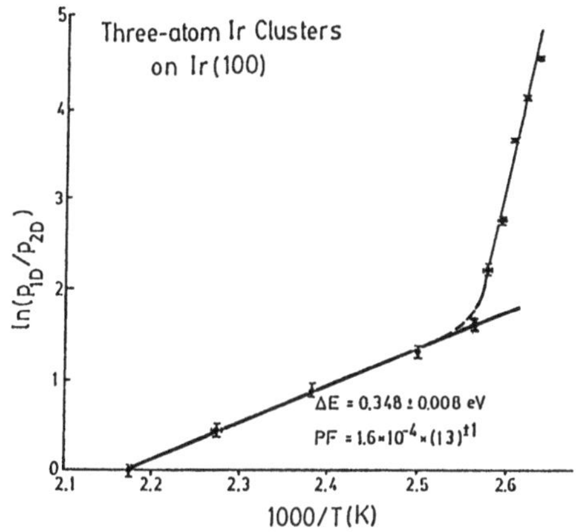

Fig.9 Structure changes of clusters

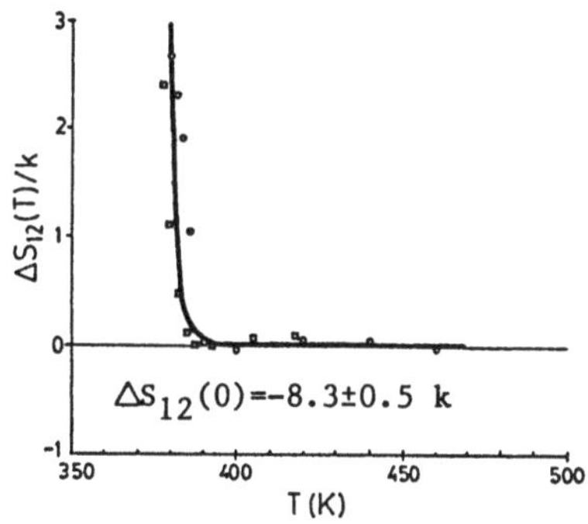

Fig.10 Entropy factor in structure transformation

is usually on the order of only about 0.2 eV for dimers although in the case of the Ir-dimer in the same atomic-channel of the Ir {110} surface, the binding energy is as high as 1.10 eV, or comparable to the strength of a chemical bond (Fig. 7). Whereas one normally would think that a cluster of an fcc metal on its surface should form a structure with the closest possible bond configuration, for Ir and Pt atoms on their {001} surfaces, linear-chain structures are often more stable than 2-D island structures [13]. For Pt-clusters on the Pt {001} surface, the stability switches oscillatory with the number of atoms in the cluster [14].

A quantitative measurement of the difference in the binding energies of 3-atom Ir clusters on the Ir {111} and {001} surfaces has recently been made [13]. On the {111} surface, 3-atom Ir clusters with the closely-packed 2-D structure are more stable than the 1-D chain structure. The energy difference is about 0.10 eV as shown in Fig. 8. On the {001} surface, the situation is reversed, or the 1-D linear-chain structure is more stable with a lower binding energy of 0.335 eV as shown in Fig. 9. In addition, there is a significant deviation from the linear behavior in the Arrhenius plot which can be interpreted to have a temperature dependent entropy term (Fig. 10).

4. Binding Energy of Atoms on the Surface

Another important question is what is the binding energy of a surface atom with the substrate, and how is this energy related to the atomic configuration of the adsorption site? In the past, the binding energy of surface atoms has been derived by a temperature programmed thermal desorption method [15]. When a surface is heated, before surface atoms are desorbed, they will diffuse on the surface. The site of desorption is unknown. Thus the energy measured is not atomic site specific. I will describe two new methods with which the site-specific binding energy of surface atoms can be determined. A low temperature method uses a high resolution pulsed-laser time-of-flight atom-probe (Fig. 11) to measure the kinetic energy of low temperature field evaporated ions. From the onset flight-time, i.e. the flight-time of the most energetic ions, the binding energy of the surface atoms can be derived [16]. This method is most convenient for measuring the binding energy of kink site atoms since, in low temperature field evaporation, atoms are gradually removed from kink sites at the lattice steps.

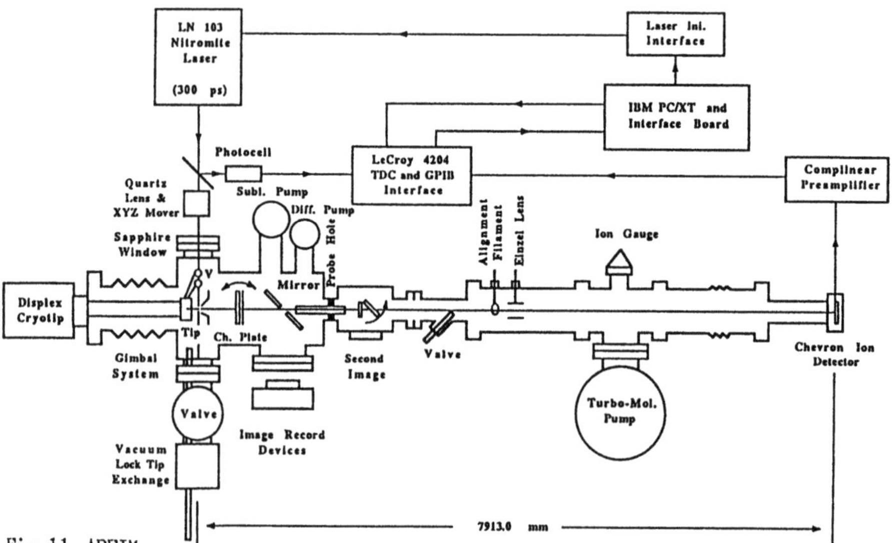

Fig.11 APFIM

A question often asked is whether or not the applied field will change the binding energy of the surface atoms drastically to make the measurement unreliable. Our measurement of the binding energies of kink site atoms of metals, E_k, all agree with their cohesive energies to within ±0.2 eV as shown in Table 1. This result indicates that the polarization binding is offset by a reduction in the crystal binding by the same applied field because of the denuding of the binding electrons. At the present time, there is still the question of why the reduction in the crystal binding cancels almost exactly the polarization binding.

Table 1 Binding Energies

Metal	Measured E_k (eV)	E_c (eV)	Difference (eV)
Fe	4.50±0.16	4.32	+0.18±0.16
Co	4.34±0.08	4.41	-0.07±0.08
Ni	4.35±0.04	4.46	-0.11±0.04
Rh	5.41±0.17	5.78	-0.37±0.17
W	9.00±0.49	8.90	+0.10±0.49
Pt	5.62±0.12	5.84	-0.22±0.12
Co in Pt_3Co	4.98±0.16	unknown	
Pt in Pt_3Co	5.53±0.17	unknown	

The binding energy of self-adsorbed atoms on a surface plane can also be derived by measuring the dissociation energy of kink site atoms from the plane edges. It can be easily shown that the binding energy of self-adsorbed atom E_b is equal to $(E_c + E_d - E_{dis})$ where E_c is the cohesive energy, E_d is the activation energy of self-diffusion and E_{dis} is the dissociation energy of kink site atoms to terrace sites. The latter energy can be determined by measuring the temperature dependence of the dissolution time of a surface layer as shown in Fig. 12(a) and 12(b) for the Ir (001) surface in a preliminary experiment. From this data, the binding energy of the Ir adatom on the Ir {001} surface is determined to be $E_b = (6.94 + 0.88 - 1.40)$ eV = 6.42 eV. The accuracy of this measurement is ~±0.2 eV, but can be improved by further studies.

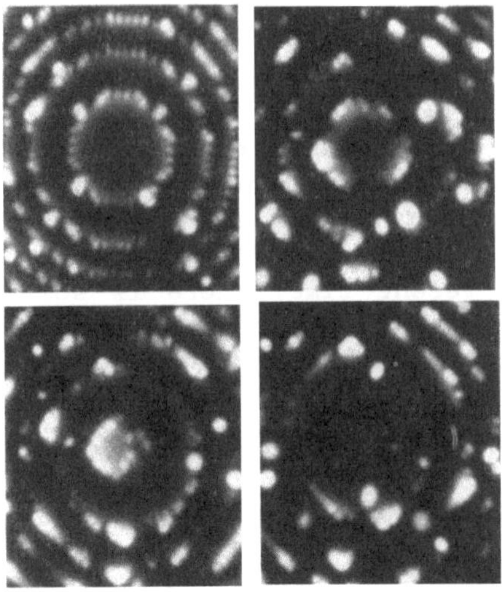

Fig.12(a) Dissociation of an Ir(001) layer

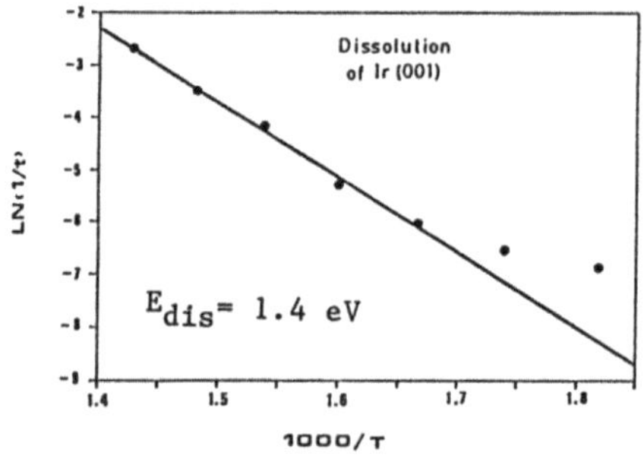

Fig.12(b) T-dependence of the dissociation time

5. Summary

Surface diffusion of single atoms and small atomic clusters on flat metal surfaces can occur by either atomic hopping or atomic replacement with substrate atoms. Atoms in a cluster can displace by jumping of individual atoms or by a concerted motion of the atoms. For surfaces where atomic reconstructions have been observed, atomic replacement diffusion and concerted atomic motion seem to occur also.

Beside diffusion parameters, it is now possible also to measure the binding energies and statistical weights or entropies of many other surface atomic processes, and the binding energies of atoms in different adsorption sites with the substrate. These energies are derived under the best controlled conditions possible, and are therefore most suitable for comparing with theoretical results.

Acknowledgements

This work was supported by the National Science Foundation. The help of Mr. Chong-lin Chen in the preparation of this manuscript is very much appreciated.

References

1. E. W. Müller and T. T. Tsong, "Field Ion Microscopy, Principles and Applications", Elsevier, New York (1969); T. T. Tsong, "Atom-Probe Field Ion Microscopy", Cambridge Univ. Press, Cambridge (1990).
2. a. G. Ehrlich and K. Stolt, Ann. Rev. Phys. Chem. 31, 603 (1980); b. T. T. Tsong, Rpt. Progress Phys. 51, 759 (1988).
3. C. L. Chen and T. T. Tsong, Phys. Rev. Lett. 64, 3147 (1990).
4. G. L. Kellogg and P. J. Feibelman, Phys. Rev. Lett. 64, 3143 (1990).
5. P. J. Feibelman, Phys. Rev. Lett. 65, 729 (1990).
6. a. D. W. Bassett and P. R. Weber, Surface Sci. 70, 520 (1978). b. J. D. Wrigley and G. Ehrlich, Phys. Rev. Lett. 44, 661 (1980).
7. T. T. Tsong and C. L. Chen, submitted to Phys. Rev. B.
8. T. T. Tsong and R. Casanova, Phys. Rev. B22, 4632 (1980).
9. T. T. Tsong and C. L. Chen, to be published in Surface Sci.
10. G. J. Gao and T. T. Tsong, Phys. Rev. B36, 2547 (1987); T. T. Tsong and Q. J. Gao, Surface Sci. 182, L257 (1987).
11. T. T. Tsong and R. Casanova, Phys. Rev. B24, 3063 (1981).
12. T. T. Tsong and R. Casanova, Phys. Rev. Lett. 47, 113 (1981).
13. C. L. Chen and T. T. Tsong, Phys. Rev. B42, 1464 (1990).
14. G. L. Kellogg, 37th Intern. Field Emission Symp. (1990).
15. J. B. Taylor and I. Langmuir, Phys. Rev. 44, 4236 (1933).
16. J. Liu, C. W. Wu and T. T. Tsong, to be published in Surface Sci.

Surface and Interface Structural Analysis by Coaxial Impact Collision Ion Scattering Spectroscopy (CAICISS)

M. Katayama[1], *B.V. King*[1(a)], *R.S. Daley*[1(b)], *R.S. Williams*[1(b)], *E. Nomura*[2], *and M. Aono*[1;2]

[1]RIKEN, Hirosawa 2-1, Wako-shi, Saitama 351, Japan
[2]Aono Atomcraft Project, Kaga 1-7-13, Itabashi-ku, Tokyo 173, Japan
(a)On leave from Newcastle University, New South Wales, Australia
(b)On leave from the University of California, Los Angeles, USA

The epitaxial growth of CaF_2 on a Si(111) substrate has been monitored in situ with a novel form of low-energy ion scattering spectroscopy, or coaxial impact-collision ion scattering spectroscopy (CAICISS), which can "see" not only the near-surface one or two atomic layers but subsurface several atomic layers by virtue of its 180° experimental scattering angle. In the course of this study, it has been found that an extra compound is formed at the CaF_2/Si(111) interface in a certain condition. Through the composition and structure analysis of the extra compound with CAICISS, the extra compound has been identified to be $CaSi_2$ epitaxially grown on the Si(111) substrate. To our knowledge, this is the first "bulk" crystral structure analysis with a low-energy ion beam. In this way, a new field which should be called "low-energy ion-beam crystallography" has been opened.

1. Introduction

CaF_2 is an attractive insulator in connection with the fabrication of 3-dimensional integrated circuits and semiconductor-on-insulator (SOI) structures because of a small lattice mismatching between CaF_2 and Si (only 0.6% at room temperature). For the growth of high-quality epitaxial CaF_2 films on Si(111) substrates, it is of importance to understand the structure and electronic properties of the CaF_2/Si(111) interface. A number of studies have been carried out to study the crystalline quality and electronic properties of thin CaF_2 films epitaxially grown in the initial stage of the formation of the CaF_2/Si(111) interface with Rutherford backscattering spectroscopy [1,2], electron microscopy [3] and X-ray photoemission [4].

Recently, some of the present authors and co-workers have developed a novel form of low-energy ion scattering spectroscopy, or coaxial impact-collision ion scattering spectroscopy (CAICISS) [5,6]. In CAICISS, a pulsed-beam low-energy ion source and a time-of-flight energy analyzer are put coaxially so as to take the experimental scattering angle at 180° . Contrary to the common sense that low-energy ion scattering spectroscopy is extremely sensitive to the surface one or two atomic layers, CAICISS can "see" not only the surface atomic layers but subsurface several atomic layers in spite of the use of a low-energy ion beam. This is because those ions and neutrals which are backscattered from a sample at the 180° scattering angle, effectively retracing the trajectories of incident ions, are selectively detected. This makes it possible to analyze not only surface structures but subsurface "bulk" structures.

In our recent study, we have applied CAICISS to monitor the epitaxial growth of CaF_2 on a Si(111) a substrate in situ. In the course of the study, we have found that a certain compound [7,8] with little fluorine is formed at the CaF_2/Si(111) interface in a certain condition and that the compound is $CaSi_2$ epitaxially grown on the Si(111) substrate. To our knowledge, this is the first "bulk" crystal structure analysis with a low-energy ion beam. In other words, a new field which should be called "low-energy ion beam crystallography" has been opened.

2. Experimental

The apparatus of CAICISS has been described elsewhere [5,6]. The apparatus is equipped with low-energy electron diffraction (LEED) optics and three shuttered Knudsen cells for molecular-beam epitaxy (MBE) which are directed towards the sample position during CAICISS measurements. The Si(111) substrate was cleaned by electron-beam heating from behind at about 1200 ℃, resulting in a clean surface with a sharp 7 x 7 LEED pattern. The temperature of the substrate during the epitaxial growth of CaF_2 was kept at 700 ℃ by electron beam heating from behind, the energy and current of the electron beam being controlled during the epitaxial growth; a positive voltage of 200-50 V was applied to the substrate and the current of the electron beam was 20-5 mA. All CAICISS measurements were done using a pulsed-beam of He^+ with an energy of 2 keV and a pulse width of 100ns.

3. Results and discussion

Two kinds of CaF_2 films with different thicknesses were epitaxially grown on the Si(111) substrate. One is a thick film with a thickness of several 100 ML (1 ML corresponds to 7.84×10^{14} molecules/cm^2) and the other is a medium thick film with a thickness of several 10 ML. For these two films, the intensity of He particles (ions and neutrals) scattered form Ca atoms was measured with changing the angle between the axis of CAICISS and the surface plane, α, in the (110) plane (see the inset of Fig.1). The results for the two films are shown in Figs.1a and b. In the case of the thick film (Fig.1a), the observed intensity variation shows peaks at α = 36°, 74°, 106°, 134° and 154°. This intensity vatiation is essentially identical to that observed for the (111) face of single-crystal CaF_2 (not shown), indicating simple (111) epitaxial growth of CaF_2. On the other hand, the intensity variation for the medium thick film (Fig.1b) shows extra peaks at α = 60°, 110°, 120° and 140° in addition to the peaks mentioned above, indicating the coexistence of an extra compound with a different structure. After the measurements of Fig.1b, the medium thick film was heated at 750 ℃ for 20 min by electron beam heating from behind and similar measurements were made after the heating. The observed intensity variation is shown in Fig.2a. As we see, the intensity variation has only the extra peaks and no peaks attributed to CaF_2 are observed. This means that the medium thick film completely changed into the extra compound by the heating at 750 ℃. CAICISS spectra of the extra compound showed no peak due to F atoms, indicating the absense F atoms. It has been found that the extra compound is $CaSi_2$ epitaxially grown on the Si(111) substrate as follows. For this purpose, computer simulations corresponding to the result shown in Fig.2a were performed for the structure of $CaSi_2$. $CaSi_2$ has a trigonal-rhombohedral structure (a =3.855 A, c = 30.6 A) [9] in which hexagonal Si double layers and a hexagonal Ca layer are alternatively stacked (see Fig.3). It is known that a small lattice mismatching between $CaSi_2$ and Si (only 0.4% at room temperature) allows the epitaxial growth of high-quality $CaSi_2$ on Si(111) [10]. The result of the computer simulation for the structure is shown in

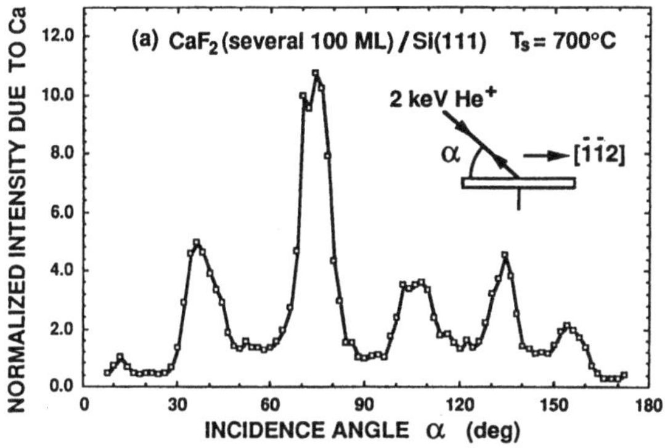

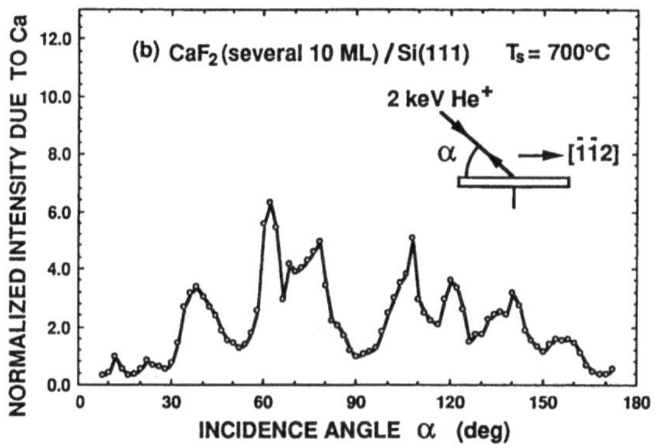

Fig.1 Intensities of He particles scattered from Ca atoms in CaF_2 films epitaxially grown on a Si(111) substrate with thicknesses of (a) several 100 ML and (b) several 10 ML as a function of the angle between the axis of CAICISS and the surface plane, α. As for the scattering geometry, see the inset.

Fig.2b. As we see, the result of the computer simulation agrees well with the experimental result shown in Fig.2a. It is thus found that the extra compound is $CaSi_2$ epitaxially grown on the Si(111) substrate.

In the computer simulation mentioned above, we actually made simulations increasing the number of atomic layers that are taken into account from 1 to 10. Selected results of the simulations are shown in Figs.4a-d. As we see, the result for 10 layers (Fig.4d) is essentially the same as that for 7 layers (Fig.4c). Namely, the computer simulations give a convergent result when we take more than 7 layers into account. From this result and the fact that the convergent result agrees well with the experimental result (to be discussed in more detail), we can conclude that CAICISS can "see" more than 7 atomic layers from the surface in spite of the use of the low-energy ion beam of 2 keV. This stems from the 180° experimental scattering angle used in CAICISS.

The reason why the convergent result shown in Fig.4d is slightly different from the result shown in Fig.2b which agrees more with the experimental result (Fig.2a) is as follows. The results of the computer simulations depend on whether the surface is terminated with layer A, B, C, D, E or F in Fig.3. The result shown in Fig.4d was obtained for a surface terminated with layer D, while the result shown in Fig.2b was obtained for a surface terminated with

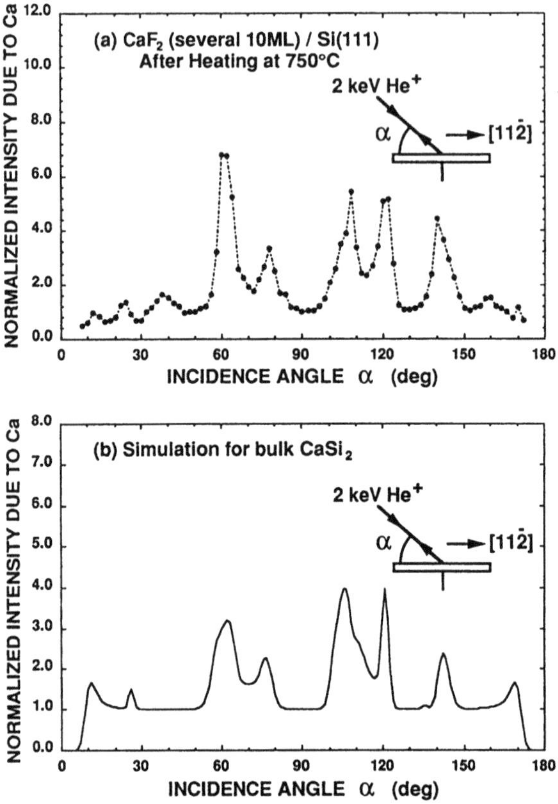

Fig.2 (a) Intensities of He particles scattered from Ca atoms in the extra compound (see text) as a function of the angle between the axis of CAICISS and the surface plane, α .(b) Result of a computer simulation corresponding to the experiment shown in Fig. 2a for $CaSi_2$ put epitaxially on Si(111).

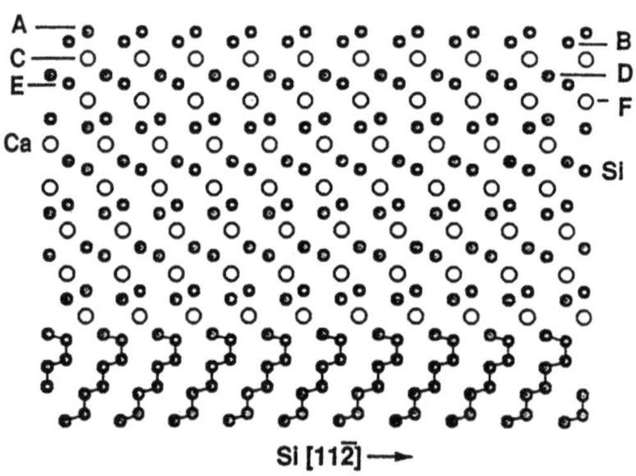

Fig.3 Structure of $CaSi_2$ epitaxially grown on a Si(111) substrate.

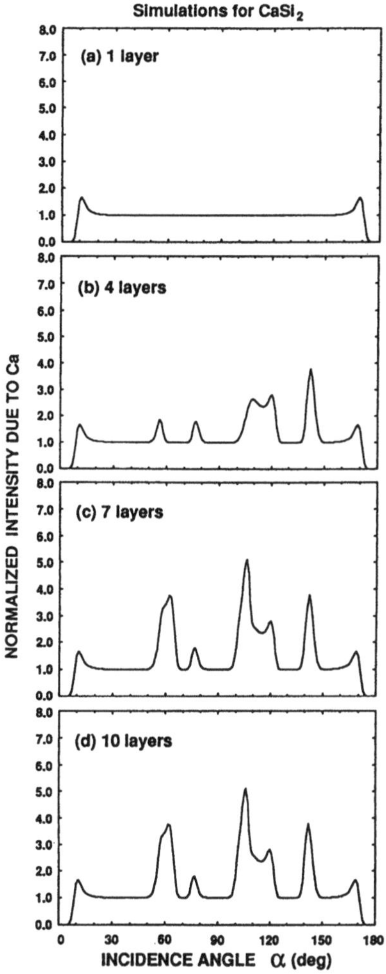

Fig.4 Results of computer simulations for $CaSi_2$ put epitaxially on Si(111), which were made increasing the number of atomic layers that were taken into account from 1 to 10.

layers C and F in the proportion of 1:1. Since the latter agrees well with the experimental result (Fig.2a), the surface of the extra compound $CaSi_2$ seems to be terminated with Ca layers, or layers C and F in Fig.3.

As for the reason why the extra compound $CaSi_2$ is formed on the Si(111) substrate in spite of the deposition of CaF_2, we are now doing experiments to clarify the reason. In this paper, we report only the fact that the extra compound $CaSi_2$ was formed on a Si(111) substrate when CaF_2 was deposited in a certain condition (described already) and rather emphasize that CAICISS easily identified the extra compound to be $CaSi_2$. In this way, a new field which should be called "low-energy ion-beam crystallography" has been opened.

4. Summary

In the course of a study to monitor the epitaxial growth of CaF_2 on a Si(111) substrate with CAICISS, we found that an extra compound was formed at the CaF_2/Si(111) interface in a certain condition. The extra compound contained no F atoms as CAICISS spectra showed. For the extra compound, CAICISS spectra were measured changing the angle between the axis of CAICISS and the surface plane. The result agreed well with a corresponding computer simulation for $CaSi_2$ put epitaxially on Si(111). In this way, the extra compound was identified to be $CaSi_2$ epitaxially grown on the Si(111) substrate. To our knowledge, this is the first "bulk" crystal structure analysis with a low-energy ion beam. This means that a new field which should be called "low-energy ion-beam crystallography" has been opened.

References

[1] T. Asano and H. Ishikawa, J. Appl. Phys. 55, 3566 (1984).
[2] S. Hashimoto, L. J. Schowalter, R. W. Fathauter, and W. M. Gibson, Mat. Res. Soc. Symp. Proc. 56, 247 (1986).
[3] R. W. Fathauer and L. J. Schowalter, Appl. Phys. Lett. 45, 519 (1984).
[4] M. A. Olmstead, R. I. G. Uhrberg, R. D. Bringans, and R. Z. Bacharach, Phys. Rev. B 35, 7526 (1987).
[5] M. Katayama, E. Nomura, N. Kanekama, H. Soejima, and M. Aono, Nucl. Instr. Meth. Phys. Res. B33, 857 (1988).
[6] M. Aono, M. Katayama, E. Nomura, T. Chasse, D. Choi, and M. Kato, Nucl. Instr. Meth. Phys. Res. B37/38, 264 (1989).
[7] B. V. King, M. Katayama, R. S. Daley, R. S. Williams, E. Nomura, and M. Aono, in proceedings of the 11th International Vacuum Congress and the 7th International Conference on Solid Surfaces, to be published.
[8] M. Katayama, B. V. King, R. S. Daley, E. Nomura, and M. Aono, to be published.
[9] J. Evers, J. Solid State Chem. 28, 369 (1979),
[10] J. F. Morar and M. Wittmer, Phys. Rev. B37, 2618 (1988).

Part II

Surface Ordering

Lattice Gas Simulation of the Surface Reconstruction of Si(111) and Ge(111)

J. Kanamori and Y. Sakamoto

Department of Physics, Faculty of Science, Osaka University, Toyonaka, Osaka 560, Japan

The surface reconstruction of Si(111) and Ge(111) is discussed by use of the lattice gas model which incorporates essential features of the Takayanagi reconstruction in the former case. Each particle which represents an adatom and its bonding mates on the substratum is assumed to take either of two states, one corresponding to the normal substratum and the other to the stacking faulted one; the dimer formation in the second layer is taken into account in terms of interaction constants between particles. The model reduces to a simple lattice gas model applicable to Ge(111) for an appropriate choice of parameters. Theoretical analyses and Monte Carlo simulation calculations of the model explain the interrelation between the (7×7) reconstruction of Si(111) and the c(2×8) one of Ge(111), the absence of the stacking fault in the (1×1) phase of Si(111) and yield a consistent interpretation of the LEED and RHeed diffuse scattering pattern in the (1×1) phase of both the surfaces.

1. Introduction

The reconstruction of the (111) surface of Si and Ge is characterized by the extra atoms called adatoms above the surface layer whose presence was established by the STM experiments first for Si(111)[1] and later for Ge(111)[2]. The adatoms decrease the number of dangling bonds of surface atoms by forming the chemical bonds with three atoms on the substratum layer. The (7×7) reconstruction of Si(111)[1] and the c(2×8) one of Ge(111) [3] correspond to ordered arrangements of them, while the high temperature (1×1) phase to their disordered arrangements. Since adatoms are situated on the top of the lattice sites of the second layer, the reconstructions may be modeled as a lattice gas on the triangular net with interactions giving rise to the ordering at low temperatures. In this paper we summarize our recent work which explains various features of the reconstruction on the basis of this assumption. [4-9]

We assume first a simple lattice gas without paying attention to the reconstruction of the substratum layer. The ordered arrangements of (7×7) and c(2×8) which may look quite different are interrelated by the fact that they can be generated from the (2×2) state by introducing discommensuration walls running either in three directions or parallel in one direction (see Fig.1.) By a simple general argument which will be described in the next section (section 2) we can show that the (7×7) state cannot be lower in energy than the (2×8) if the energy is expressed in terms of the energy required for the generation of the walls and the energy of the vertexes, i.e. the wall crossings. This argument gives a theoretical support to the Takayanagi model [10] of the (7×7) state in which the substratum of one of the (2×2) domains separated by a discommensuration wall is a stacking fault and atoms on the second layer at the wall make dimers by use of the dangling bonds produced by the stacking fault. If the energy required to gen-

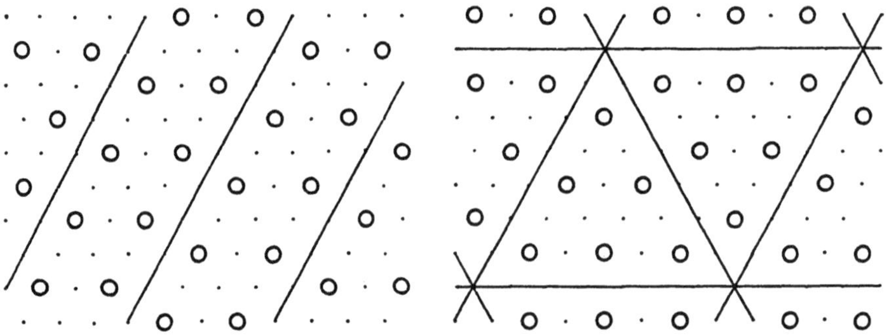

Fig. 1

The c(2x8) and (7x7) states of the simple lattice gas. Particles are shown by open circles and vacant sites by dots; discommensuration walls are shown by lines. The particle arrangement in each region within the walls is the (2x2) states in which particles occupy the third neighboring sites, while the second neighboring pairs are formed across the walls.

erate this Takayanagi wall is lower than that of the discommensuration walls in a simple lattice gas model, the (7×7) state can be of lowest energy. [5]

In the case of the c(2×8) arrangement where the discommensuration walls run parallel unidirectionally, we can not construct the arrangement of adatoms with the Takayanagi walls only; we need a wall of another kind in-between two Takayanagi walls to realize it. [5] Since the latter wall seems to cost a high energy, we conclude that the Takayanagi mechanism is not operative in the c(2×8) of Ge(111). Recently the absence of the stacking faults in Ge(111) seems to be established experimentally [11], which supports this conclusion.

Assuming the simple lattice gas model without stacking faults for Ge(111), we have carried out a Monte Carlo simulation calculation to investigate the phase transition to the (1×1) phase and the temperature dependence of the short range order after the transition. The RHEED [12] and LEED [13] experiments on the (1×1) phase of Ge(111) immediately above the transition temperature observed the appearance of rather sharp diffuse peaks close to the superlattice Bragg spots of the (2×2) state which moved away in the direction towards the superlattice spots of the ($\sqrt{3} \times \sqrt{3}$) state with temperature. [12, 13] Our Monte Carlo simulation calculation shows that the particle arrangement above the transition may be regarded as a domain structure of the (2×2) arrangements separated by an irregular honeycomb-like network of discommensuration walls. [6-9] We can present a plausible explanation for this characteristic short range order.

The mechanism of lowering the energy of the Takayanagi wall requires an ordered arrangement of adatoms. Thus we are led to the prediction that the stacking faults will be absent in the high temperature (1×1) phase of Si(111) where adatoms are disordered [5]. This prediction seems to be supported experimentally, though no detailed investigation is reported. In order to establish the underlying mechanism, we have extended the lattice gas model to incorporate essentials of the Takayanagi mechanism of the stabilization of the (7×7) state. [8] We assume that each particle has a degree of freedom corresponding to the choice of either the normal substratum or the stacking faulted one. We may represent the degree of freedom by a pseudospin having two states. The interactions between particles are assumed to be dependent on the states of interacting particles thus defined. In a range of interaction parameters, we can obtain the (7×7) state as the ground state. By carrying out a Monte Carlo simulation calculation on the model, we can demonstrate that the stacking faults decrease in the (1×1) phase. The model can be reduced to the simple lattice gas model if we make appropriate choices of parameters. Thus we can construct a unified model of the c(2×8) and (7×7) reconstructions.

Section 2 summarize our analysis of the interrelation between the c(2×8) and (7×7) states of the simple lattice gas model. Section 3 deals with the extended lattice gas model with which the reconstructions of both Ge(111) and Si(111) can be discussed from a unified point of view. In section 4 we discuss the results of Monte Carlo simulation calculations of the models. Section 5 summarizes the discussions in preceding sections.

2. Analysis of the Ground States of the Simple Lattice Gas Model

We note first of all that adatoms will not make the nearest neighboring pairs on the triangular lattice, since they will not share a dangling bond of a substratum atom. Thus we assume in the following discussion that the nearest neighboring pairs are forbidden with a very large repulsive interaction energy. We can determine rigorously the ground state of the lattice gas for a given set of interaction parameters without making any assumption for the choice of particle configuration, provided that the range of interactions is finite. [14,4,5] In the following, however, we confine ourselves to those states which can be generated from the (2×2) state by introducing the discommensuration wall.

We choose the (2×2) state in which particles occupy mutually third neighboring sites on the triangular lattice as the reference state. Since the unit cell of the (2×2) state contains 4 lattice sites, there are four kinds of the (2×2) arrangements. As can be seen from Fig.1 particles across a discommensuration wall separating the (2×2) regions of different kinds make the second neighboring pairs. Denoting the interaction energy of the k-th neighboring pairs of particles by V_k, we obtain the energy per unit lattice spacing required for generating the wall as

$$\varepsilon_{wall} \equiv -w = (1/2)V_2 - V_3 + V_4 - 2V_6 + \cdots. \tag{1}$$

The energy of the state having parallel walls running at intervals of m lattice spacings which is actually the c(2×2m) can be expressed as

$$E_{2\times 2m} = E_{2\times 2} - Nw/m, \tag{2}$$

while that of the states with walls running in three directions at intervals of n lattice spacings is given by

$$E_{n\times n} = E_{2\times 2} - 3Nw/n + Nv/n^2, \tag{3}$$

where N is the total number of lattice sites and v is the energy of a wall vertex.

For the states under consideration m should be an even integer satisfying $m \geq 2$ and n an odd integer satisfying $n \geq 3$. With $w>0$ Eq.(2) yields the minimum energy for the smallest m,i.e. m=2. If we include the sixth neighbor and more distant interactions, we need to add an interaction energy between neighboring walls to the r.h.s. of Eq.(2). For example, if V_6 only is considered, the nearest neighboring walls in the c(2×4) with m=2 has the interaction energy V_6 per unit spacing, while the state c(2×8) with m=4 has no such an energy. If we delete every other wall in the c(2×4) to obtain c(2×8), the energy change is given by

$$\Delta E \equiv -2V_6 + w = -V_2/2 + V_3 - V_4 \ . \tag{4}$$

Thus if $\Delta E<0$ and $w>0$, i.e.

$$V_3 - V_4 < V_2/2 < V_3 - V_4 + 2V_6 \tag{5}$$

hold, the c(2×8) will be of lowest energy among the (2×2m) states. Here we neglected more distant neighbor interactions than the sixth one. The inequality (5) indicates that the sixth neighbor interaction should be repulsive ($V_6>0$).

On the other hand we obtain the minimum energy of the (n×n) state from Eq.(3) by a simple calculus.[5] By minimizing the r.h.s. of Eq.(3) with respect to n, we can see easily

$$\mathrm{Min}[E_{n\times n}] = E_{2\times 2} - 3Nw/2n_{min} \tag{6}$$

with $n_{min} \equiv 2v/3w$. If $n_{min}=7$, $E_{7\times 7}$ is given by

$E_{7\times 7} = E_{2\times 2} - 3Nw/14 > E_{2\times 8} = E_{2\times 2} - Nw/4$ if $w > 0$. A more detailed analysis concludes that $E_{7\times 7}$ is always higher than $E_{2\times 8}$ except for v/w=35/4 where the energies of the (7×7), (5×5) and c(2×8) are degenerate. [5]

The above analysis shows that we need an additional energy to stabilize the (7×7) state in the simple lattice gas model.

3. A Lattice Gas Model for the Takayanagi Reconstruction

We denote the wall energy of the Takayanagi model [10] by -w' in order to distinguish it from -w of the simple lattice gas. w'>w can be expected in a favorable case because of the dimer formation in the second layer and a relaxation of the elastic strains arising from the bonds with adatoms. The atomic configuration of the Takayanagi wall allows these two effects to help each other. In order to incorporate the effects, we represent an adatom by an equilateral triangle whose vertexes correspond to its bonding mates on the substratum layer. If the substratum layer is a stacking fault, the orientation of the triangle is different from that of the triangle corresponding to the normal layer, as is illustrated in Fig.2. Thus the triangle is a convenient representation of the two states of a pseudospin mentioned in the introduction. The triangles cannot occupy the nearest neighboring sites of the triangular lattice as before irrespective of their orientations; if they are of the same orientation, they have to share a substratum atom as bonding mate and otherwise substratum atoms are too close to be realized.

If the triangles of different orientations occupy the second neighboring sites, we have two situations. The line connecting the two sites passes either the vertexes pointing to each other or the centers of the facing edges. The first case is forbidden, since again the substratum atoms have

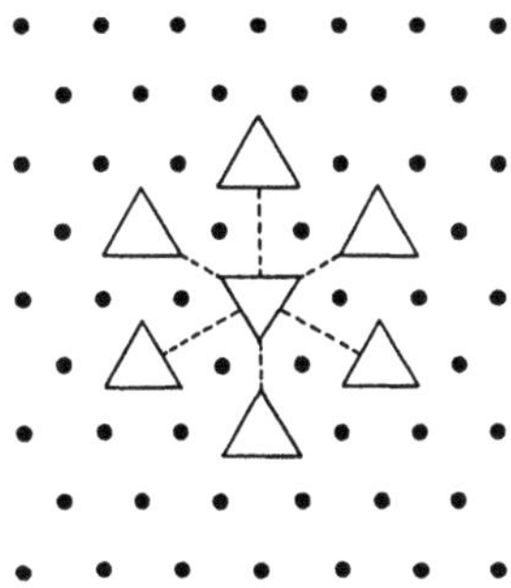

Fig. 2a Two orientations of triangles and the second neighbor interaction of triangles of different orientations.

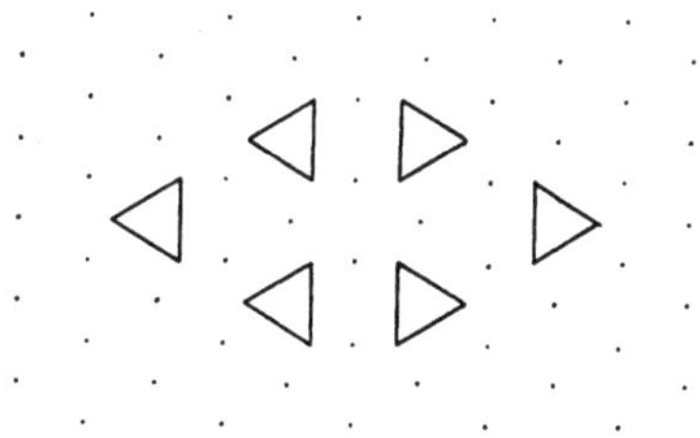

Fig. 2b The (5×5) state of the Takayanagi structure. The triangle configuration of a unit cell is shown. The wall runs in-between two groups of triangles.

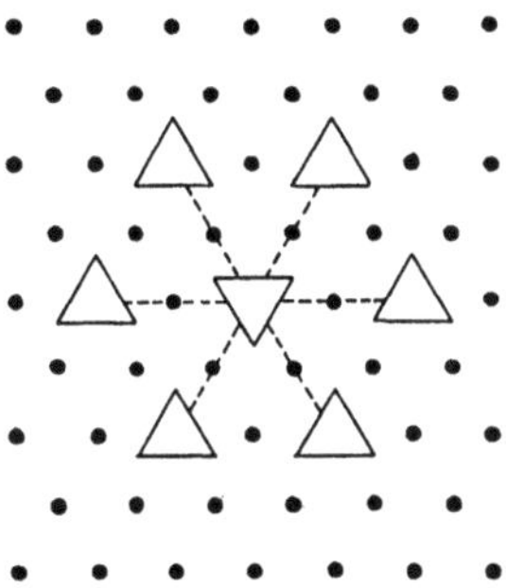

Fig. 3 The third neighbor interaction of triangles of different orientations.

to come too close to each other (see Fig.2.) The second case which is found at the wall separating the normal region from the stacking faulted one can be assumed to be energetically more favorable than the second neighboring pairs of the triangles of the same orientation, corresponding to a different interaction constant.

Figure 3 shows the third neighboring pairs of the triangles of different orientations. In this case we may assign the same interaction energy to all the directions of the interaction line. The interaction between the triangles of different orientation will be higher in energy than the third neighboring pair of the triangles of the same kind, since intervening second layer atoms will lose a bond.

We call hereafter the triangles on the normal substratum normal triangles and those on the stacking fault inverted ones. Confining ourselves to up to third neighbor interactions, we write the energy of our extended lattice gas model as

$$E= N_1\mu_1 + N_2\mu_2 + V_2p_2 + \bar{V}_2q_2 + \bar{V}_3p_3 + \bar{V}_3q_3 , \tag{7}$$

where the first two terms represent the sum of chemical potentials of particles with N_1 denoting the number of normal triangles and N_2 that of inverted triangles; P_2 and P_3 denote the number of the second and third neighboring pairs of the triangles of the same kind, respectively with V_2 and V_3 denoting the interaction constants; q_2 is the number of the allowed second neighboring pairs of the triangles of different kinds; q_3 is the number of the third neighbors of the triangles of different kinds; the corresponding interaction constants are represented by $\overline{V}_2$ and $\overline{V}_3$.

The energy of the Takayanagi wall, -w' is given by

$$-w' = (1/2)\overline{V}_2 - V_3 \tag{8}$$

apart from the chemical potential contribution. Thus if $\overline{V}_2$ and the chemical potential difference $\mu_2 - \mu_1$ are small enough, the (7×7) can be lower in energy than the c(2×8). An obvious shortcoming of the model is its neglect of other features of the reconstruction; for example the structure of the wall vertex where a first layer atom is missing in the Takayanagi reconstruction [10] will require an additional energy for a faithful representation of the energetics of the arrangement. Thus if we make the energy comparison among the (n×n) and other ordered states, the incorrect representation of the vertex energy is inexcusable. Our primary purpose, however, is to discuss the nature of the (1×1) phase in which, as will be discussed in the next section, we expect the disappearance of the stacking faults, i.e. the inverted triangles. We believe that the present model is meaningful to that purpose.

4. Results of Monte Carlo Simulation Calculations

We have carried out Monte Carlo simulation calculations with the simple lattice gas for the surface reconstruction of Ge(111) and with the extended one for Si(111) both below and above the transition. [6-9] The Ge(111) case can be discussed in principle with the latter model by choosing a suitable set of parameters which do not favor the appearance of the Takayanagi walls. For simplicity, however, we assume the simple lattice gas in the Ge case.

4.1. Ge(111)

The energy of the simple lattice gas adopted in the calculation is expressed as

$$E = V_2 p_2 + V_3 p_3 + V_6 p_6 , \tag{9}$$

where we assume $V_2 > 0$ in accordance with the discussion given in section 2. We assume $V_4 = 0$ and $V_5 = 0$ for simplicity and also because their effect is mainly a mere renormalization of V_2 and V_3. The third and sixth neighbor interactions are chosen to satisfy the condition (5). The concentration of particles is assumed to be slightly less than 1/4. The nearest neighboring pair is forbidden. We have carried out the Monte Carlo calculation for the lattice sizes of 24×24 and 48×48 sites with periodic boundary conditions. Ensemble averages are taken over 30,000 Monte Carlo steps.

Referring details of the calculation to references [6,7], we describe here a part of the result which, we believe, is the most interesting. Figure 4 shows an example of the particle arrangement obtained by the Monte Carlo simulation in the (1×1) phase at a temperature immediately above the first order transition from the c(2×8) state, where $V_3 = 0$ and $V_6/V_2 = 0.3$ are assumed. The arrangement consists of irregular shape domains of the (2×2) arrangements. The discommensuration walls which are of the same nature as those in the c(2×8) state run in three directions at irregular intervals. The vertexes where the walls meet are either the triplets of mutually second neighboring particles or voids where some articles are missing.

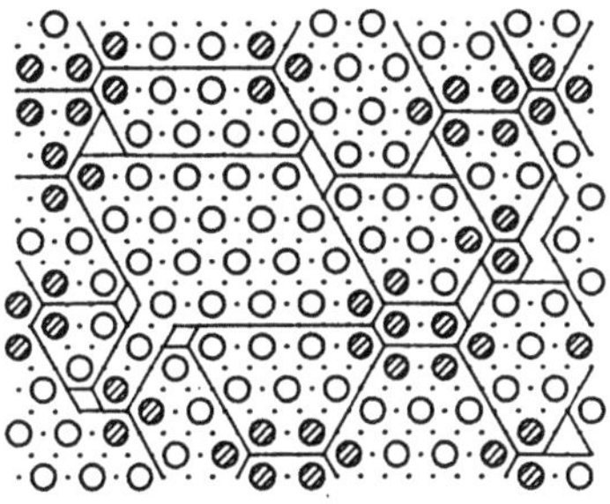

Fig. 4a An example of the particle arrangement of the (1×1) phase of the lattice gas simulating Ge(111) immediately above the transition temperature. Those particles which participate in the mutually second neighboring triplets are hatched.

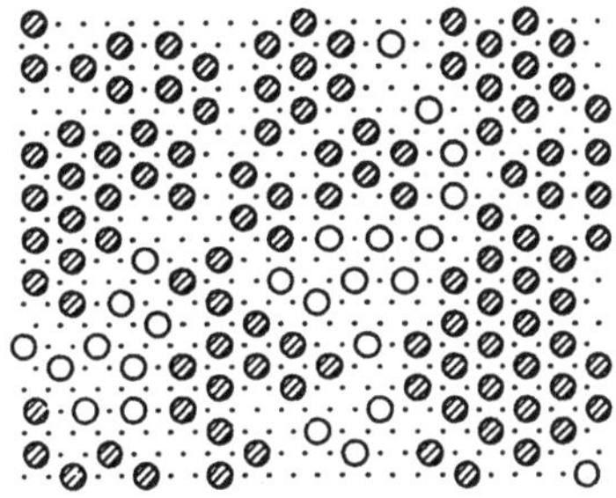

Fig. 4b The particle arrangement at high temperature.

The analysis of possible particle arrangements under the condition of no nearest neighboring pair (the hard hexagon condition) carried out previously [14,4] shows that a finite entropy can be obtained at the concentration x=1/4 only when the triplets of mutually second neighboring particles are generated. Thus we need the wall vertexes, in order to gain the entropy. The triplet corresponds to the local concentration x=1/3; note that the ($\sqrt{3}\times\sqrt{3}$) in which every particle participates in triplets has the concentration x=1/3. The triplets are, therefore, always accompanied by voids to make the average concentration equal to 1/4. This reasoning explains the honeycomb-like irregular domain structure just above the transition. We know from the discussion made by Bak et al. [15] on the ordered state of rare gas adsorbates on graphite that if both the wall energy and the vertex energy are negative, the stable wall structure is the honeycomb arrangement of the walls. In our case, the free energy for the appearance of the vertexes becomes negative in the (1×1) phase and the random distribution of the vertexes associated with the finite entropy makes the domain structure irregular to some extent; the repulsion between the walls produced by the sixth neighbor interaction, however, seems to make the size of the domain uniform compared with the situation in the absence of it [6,7].

As was discussed in detail in [7] the domain structure explains quite well the diffuse scattering pattern of RHEED [12] and LEED [13] which is characterized by the appearance of rather sharp diffuse peaks situated closer to the superlattice Bragg points of the (2×2) than the superlattice spots of the c(2×8). With increasing temperature the diffuse peaks move away from the superlattice Bragg spots of the (2×2) towards the superlattice spots of the ($\sqrt{3}\times\sqrt{3}$) state and at the same time decrease the peak height to fade away within 200°. The calculation yields a qualitatively similar temperature variation. In the calculation diffuse spots reappear around the superlattice Bragg spots of the ($\sqrt{3}\times\sqrt{3}$) state at higher temperatures, which correspond, as will be discussed below, to the case of Si(111). The calculated rate of the change with temperature, however, seems to be slower than the observed one under the condition of the fixed particle number. We suspect that a decrease of the particle number, i.e. the number of adatoms with temperature accelerates the change.

4.2 Si(111)

We carried out a Monte Carlo simulation calculation on the extended lattice gas model with the energy expression given by Eq.(7). The cases where $\mu_1=\mu_2$, V_3/V_2=0.2, 0.4, and 0.5, $\bar{V}_2/V_2$=0.1, and $\bar{V}_3/V_2$=2.0 have been investigated so far with the lattice of 21×21 sites with the periodic boundary condition. [9] In each case the calculation was continued to 200,000 Monte Carlo steps because of a large fluctuation of particle configurations. The calculation with larger sizes of the lattice is in progress at present.

The calculation is admittedly preliminary. The choice of the 21×21 lattice and a fixed number of triangles corresponding to a concentration slightly less than 12/49 but larger than 6/25 select the (7×7) state as the ordered state among the (n×n). The wall energy parameter $w'=V_3-(1/2)\bar{V}_2$ is much larger than $w=V_3-(1/2)V_2$ to exclude the possibility of the c(2×8). Thus the calculation yields the (7×7) as the ground state. We start with the (7×7) state at low temperatures and by increasing temperature find the transition to the (1×1) state in which our main interest lies.

In the (1×1) state an imbalance between the numbers of the triangles of different orientations is concluded. Figure 5 shows an example of the temperature dependence of the numbers of the triangles. If we take into account $\mu_1<\mu_2$, this imbalance will increase and the stacking faults will practically disappear. The decrease or disappearance of the stacking faults concluded from the Monte Carlo simulation calculation confirms the prediction made in [5]. It is also consistent with observation, though no detailed report has been made. The triplets of mutually second neighboring particles which are essential for the gain of the entropy cannot be formed by the triangles of different orientations, since the forbidden pair of vertex-vertex pointing triangles is involved inevitably in such a triplet. [5] Thus the decrease of the stacking faults results in a gain of entropy; the mechanism is independent of the choice of parameters.

We conclude also that the characteristic domain structure found in the case of Ge(111) immediately above the transition does not exist in the case of Si(111). The (1×1) phase of Si(111) immediately above the transition corresponds to that of Ge(111) at a temperature far above the transition where the main structure almost collapses. This is because the (7×7) state can persist to a higher temperature due to the stabilization by the Takayanagi mechanism than ordered states of the simple lattice gas such as the c(2×8). The fact that the diffuse spots of RHEED [16] and LEED [17] appear around the superlattice Bragg points of the $(\sqrt{3}\times\sqrt{3})$ state in

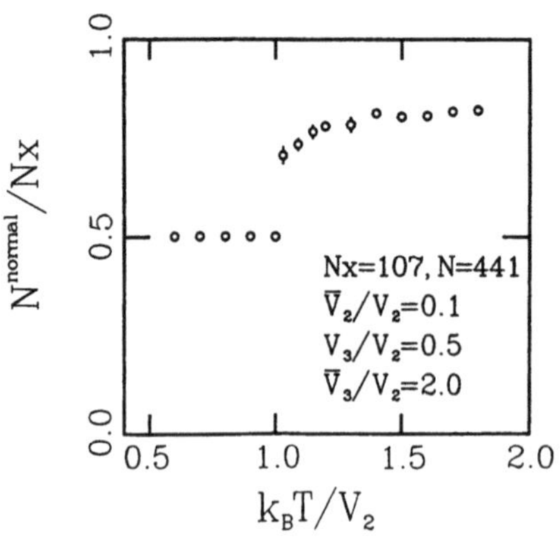

Fig. 5 Temperature dependence of the numbers of the particles (triangles). Nx is the total number of the particles; N the total number of lattice sites; N^{normal} is the number of the particles on the normal substratum layer.

the case of Si(111) is consistent with this conclusion (see the discussion for Ge(111)). For more details we refer to [9] and a future publication.

5. Summary

We have discussed that we can construct the lattice gas models for the reconstruction of Ge(111) and Si(111). Our discussion starts with an analysis of the ground states of the lattice gas model which gives a theoretical support to the Takayanagi structure of the (7×7) state of Si(111), since the simple lattice gas applicable to Ge(111) cannot yield the (7×7) state as the ground state. We propose an extended lattice gas model for Si(111) which is capable of taking into account an essential part of the Takayanagi structure.

Monte Carlo simulation calculations on the models reveal the nature of short range correlation in the high temperature (1×1) phase of both Ge(111) and Si(111). A characteristic domain structure of the particle arrangement is found for Ge(111) immediately above the transition temperature. It explains well the result of the diffuse scattering experiments. We can give also a plausible argument for the underlying mechanism for the generation of the domain structure.

In the case of Si(111) the absence of the stacking faults in the (1×1) phase is concluded by the simulation calculation in accordance with the previous theoretical prediction. A consistent explanation for the diffuse pattern of LEED and RHEED for this case is also obtained.

Acknowledgment

The present research was supported by the Grant-in-Aid for Scientific Research on Priority Areas, "Surface as New Material".

References

1. G. Binnig, H. Rohrer, Ch. Gerber and E. Weibel: Phys. Rev. Lett. 50 (1983) 120.
2. R. S. Becker, J. A. Golovchenko and B. S. Schwartzentruber: Phys. Rev. Lett. 54 (1985) 2678.
3. P. W. Palmberg and W. T. Peria: Surf. Sci. 6 (1967) 57.
4. J. Kanamori and M. Okamoto: J. Phys. Soc. Jpn. 54 (1985) 4636.
5. J. Kanamori: J. Phys. Soc. Jpn. 55 (1986) 2723.
6. Y. Sakamoto and J. Kanamori: J. Phys. Soc. Jpn. 58 (1989) 2083.
7. Y. Sakamoto and J. Kanamori: J. Phys. Soc. Jpn. 58 (1989) 1512.
8. J. Kanamori and Y. Sakamoto: Proc. of The 26th Yamada Conf. on "Surface as a New Material" July 2-6, 1990 Osaka, Japan, to be published.
9. Y. Sakamoto and J. Kanamori: Proc. of The 26th Yamada Conf. on "Surface as a New Material" July 2-6, 1990 Osaka, Japan, to be published.
10. K. Takayanagi: J. Microsc. 136 (1984) 287.
11. P. M. J. Maree, K. Nakagawa, J. F. van der Veen and R. M. Tromp; Phys. Rev. B38 (1988) 1585; R. Feidenhans'l, J. S. Pedersen, J. Bohr, M. Nielsen, F. Grey and R. L. Johnson: Phys. Rev. B38 (1988) 9715.
12. T. Ichikawa and S. Ino: Solid State Commun. 34 (1980) 349.
13. R. J. Phaneuf and M. B. Webb: Surface Sci. 164 (1985) 167.
14. J. Kanamori: Solid State Commun. 50 (1984) 363.
15. P. Bak, D. Mukamel, J. Villain and K. Wentowska: Phys. Rev. B19 (1979) 1610.
16. S. Ino: Jpn. J. Appl. Phys. 16 (1977) 891.
17. H. Iwasaki, S. Hasegawa, M. Akizuki, Sung-Te Li, S. Nakamura and J. Kanamori: J. Phys. Soc. Jpn. 56 (1987) 3425.

Surface Dynamics on Si(001) Studied with a High Temperature Scanning Tunneling Microscope

D. Dijkkamp, E.J. van Loenen, and H.B. Elswijk

Philips Research Laboratories,
P.O. Box 80000, 5600 JA Eindhoven, The Netherlands

Abstract. We have recently developed a new, thermally stable STM, which allows the observation of surfaces at elevated temperatures, up to at least 800 K. With this instrument, we have observed the evolution of growth induced mosaic step patterns into the equilibrium step distribution at 750 K in real time. We deduce a value of 2.0 eV for the kink detachment energy. In addition, we have for the first time directly observed the migration of atom-like species on the Si(001) surface. At temperatures around 350 K, diffusion takes place exclusively along the dimer rows.

1. Introduction

Surface diffusion of atoms is an important phenomenon, especially in Molecular Beam Epitaxy and other growth processes. Even simple growth models involve several parameters such as the activation energies for surface diffusion and for atom detachment from a kink, for which experimental data are scarce and conflicting. For the technologically important Si(001) surface, neither the activation energy for surface diffusion nor the amount of diffusion anisotropy or even the nature of the diffusing species are known.

Very recently, several researchers have used a Scanning Tunneling Microscope (STM) to obtain information about surface dynamics. When using a STM operating at ambient temperature, one either has to restrict oneself to surfaces which show dynamic features at room temperature (RT) (e.g. self diffusion of metals [1]), or one has to take refuge in the technique of flash annealing, where one observes changes in static features at RT, after heating the sample to a given temperature for a given amount of time [2]. The latter technique is also used in

Field Ion Microscopy (FIM) to study atomic diffusion on metal surfaces, except that in this case the measurement takes place at liquid nitrogen temperature. Obviously, to obtain reliable results it is necessary that dynamics during sample heating and cooling transients can be neglected. This makes the technique well suited for FIM, where the small heat capacity of the tip ensures rapid heating and cooling, but at the same time limits its usefulness in STM somewhat, because of the generally much larger sample size. More importantly, with interrupted annealing it is not possible to study the exact same surface area in the STM, contrary to FIM. To overcome these limitations, it is desirable to use a STM operating at high temperatures.

2. Experimental

We have recently modified one of our STMs to be able to observe Si surfaces at elevated temperatures. The general design of the instrument has been described in detail elsewhere [3,4]. The modifications consist of:

i) A second electrical lead towards the sample (which is a Si strip of 5 x 10 x 0.5 mm). This allows resistive heating by direct current, using a floating power supply. A resistive divider in parallel with the sample is used to provide tip-sample bias voltage in an approximately symmetric way.

ii) A Ta heat shield mounted between sample and z-piezo tube, with the tip wire protruding through a small hole in it.

The original design of the STM contains a high symmetry (symmetric sample holder, xy-piezo cross rather than half cross) which should help ensure thermal stability.

In addition, we have mounted a small Si evaporator in our chamber, which can evaporate Si (from a hot Si strip) on the sample while it remains in the tunneling position. Typical deposition rates are 1 - 10 monolayers (ML) per minute. It is necessary to lift the sample slightly away from the tip to deposit material in the scan area.

3. Results and Discussion

Fig. 1 shows four successive images of approximately the same area on a Si(001) surface held at 575 K. The time lapse between

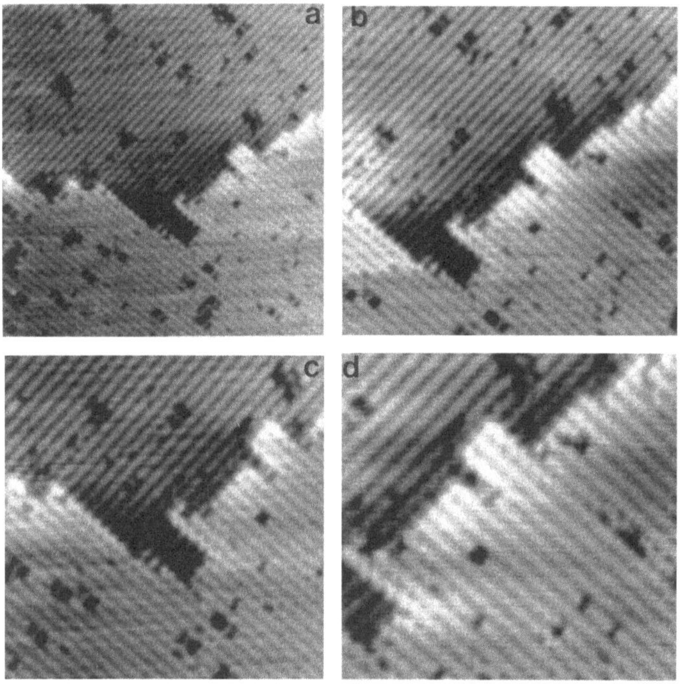

Figure 1: Four subsequent images of a Si(001) surface kept at a temperature of 575 K containing a B-type step. Time lapse between images 200 s; dimer row spacing 0.77 nm.

images is about 200 seconds. This figure demonstrates the feasibility of both atomic resolution and thermal stability at 575 K. The area shown contains a B-type (dimer rows perpendicular to the step edge) atomic step with a number of kinks. From a comparison of the images it is clear that the kinks are quite stable. Using a typical frequency factor of 10^{13} s^{-1} we can estimate a lower limit of 1.9 eV for the kink detachment energy (which is roughly the sum of kink binding energy and activation energy for surface diffusion).

In another experiment, we have first prepared a single domain surface by depositing approximately 0.5 ML of Si at 750 K on vicinal Si(001) (0.5^{o} misoriented towards [110]), as described in [5]. Up to a temperature of about 700 K we observed little tendency of returning to the equilibrium double domain structure, on a time scale of about one hour. This observation is in qualitative agreement with the data published

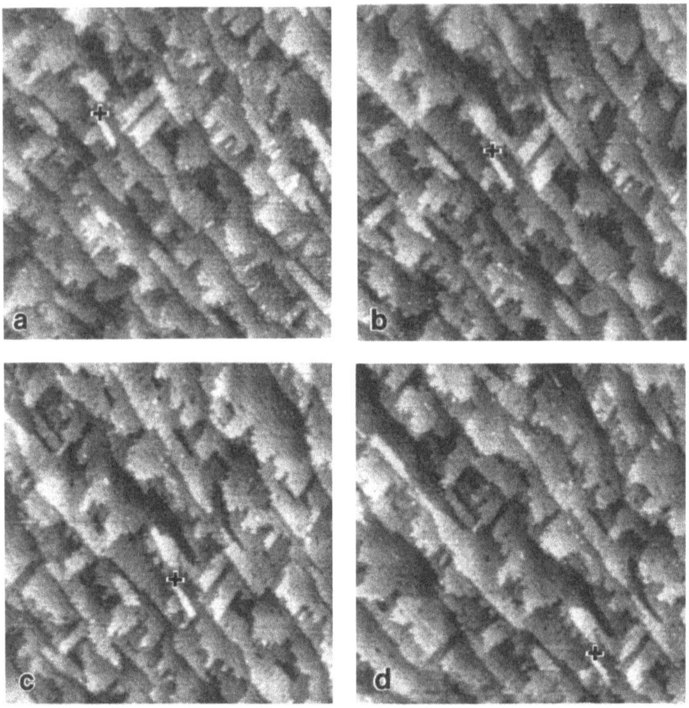

Figure 2: Four subsequent images of a 180 x 180 nm^2 area on a Si(001) surface at 750 K, after growth of about 2 ML at 675 K. The crossmark designates a fixed position on the sample. Time lapse between images 150 s.

by Sakamoto et al.[6], who observed recovery of the double domain surface (using RHEED) only at temperatures above 750 K. From the temperature dependence of the recovery time they deduced a kink detachment energy of 2.4 eV.

In a subsequent experiment, we deposited a few MLs of Si on a clean double domain sample at a temperature of 675 K, slightly lower than required for formation of a single domain surface. The result is a mosaic two domain surface, with little resemblance to the original vicinal stepped surface. Fig. 2 shows a sequence of four images of this surface, obtained at 750 K, with 150 seconds intervals. Again the thermal drift is low enough to be able to observe about the same area in all images. The crossmark in each frame designates a fixed position on the sample. Although the overall morphology appears quite stable, careful observation reveals that some patches are

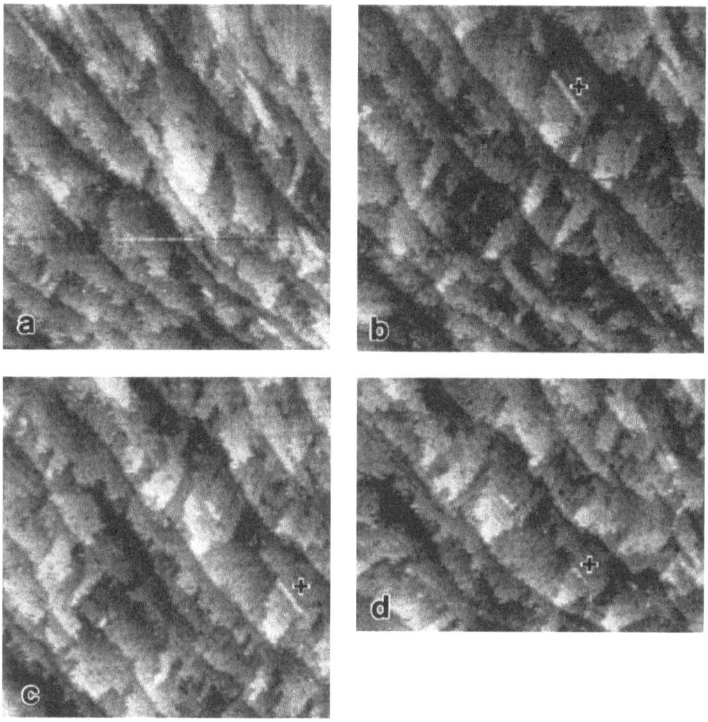

Figure 3: Four subsequent images of the same surface as shown in Fig. 4, taken immediately after increasing the substrate temperature to 800 K. Time lapse between images 150 s; size ~ 130 x 130 nm^2 (a-c), ~ 130 x 100 nm^2 (d). Crossmark in (b-d) indicates a fixed spot on the surface.

disappearing, whereas other areas are growing. Note, for example, the area just above the crossmark. Other examples can be found in the images. Faint black lines are visible in mutually perpendicular directions on terraces separated by a single atomic step. These are missing dimer rows, leading to a 2 x 8 superstructure, which is caused by slight Ni contamination of the sample [7].

We subsequently increased the temperature to 800 K, and measured the sequence of images shown in Fig. 3, with 150 seconds intervals. These images were taken between 10 and 15 minutes after increasing the temperature, and are therefore still somewhat distorted by thermal drift. One stable feature is marked by a crossmark to guide the eye. Quite clearly, the surface is now rapidly returning to the equilibrium stepped structure, with alternating straight (A-type) and rough (B-type) monoatomic steps. From the time lapse between the images of Fig. 2d and 3a, and using a typical length scale of 15 nm (the average step spacing in equilibrium) we deduce an activation energy of (2.0 ± 0.2) eV for kink detachment. This value is in good agreement with the value of (1.8 ± 0.2) eV we deduced [8] from a comparison of Monte-Carlo simulations with island coarsening experiments [2], but somewhat lower than the above mentioned result of Sakamoto et al. [6] (2.4 eV).

As a next step, we have attempted to observe the atomic diffusion directly, at a more moderate temperature of 340 K. Fig. 4 shows a sequence of four images taken with 300 seconds in between, of a Si(001) surface on which we had deposited approximately 0.005 ML of Si atoms at RT. The deposited material is visible as small white protrusions. Qualitatively,

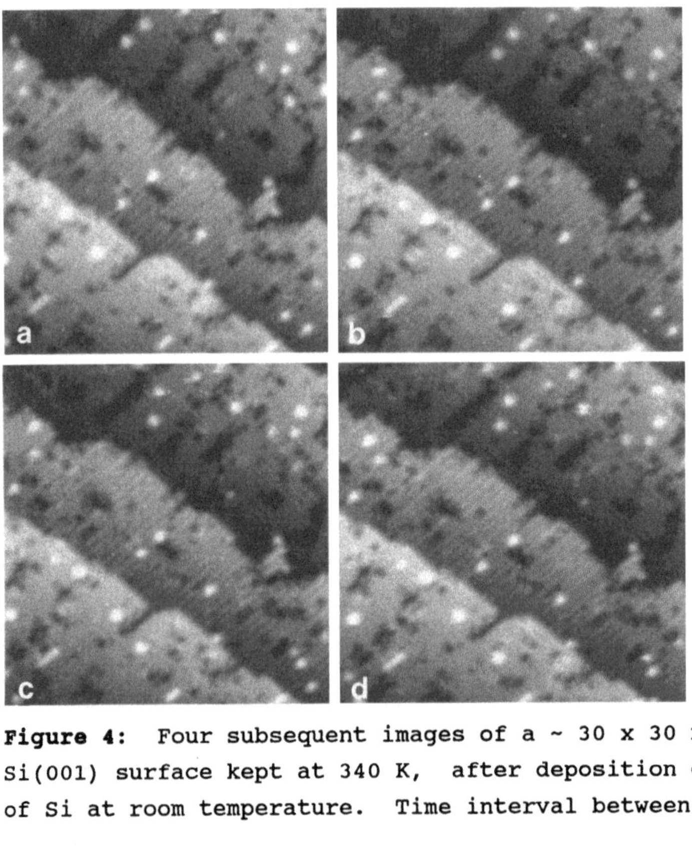

Figure 4: Four subsequent images of a ~ 30 x 30 nm^2 area on a Si(001) surface kept at 340 K, after deposition of ~ 0.005 ML of Si at room temperature. Time interval between images 300 s.

one can observe changes in the positions of some of these protrusions between the images. It is obvious that movement occurs predominantly along the dimer rows of the underlying substrate, in agreement with recent theoretical predictions [9]. In fact, after careful analysis of these and similar images we could not observe any movement in the perpendicular direction. We estimate a lower limit of 10^3 for the diffusion anisotropy. We have analyzed the data quantitatively by counting the distribution of atomic displacements, as will be described in detail elsewhere. From this, we determine a diffusion constant of 1×10^{-17} $cm^2 s^{-1}$ at 340 K and (assuming a frequency factor of 10^{13} s^{-1}) an activation energy for diffusion along the dimer rows of (1.0 ± 0.1) eV.

At present, we cannot be sure that the diffusing entities are Si atoms. In fact, preliminary evidence indicates that they actually consist of 2 or more atoms. In any case, their diffusion may well be the rate limiting step for surface diffusion on the Si(001) surface. The number of 1.0 eV is in excellent agreement with the value deduced from the Monte Carlo simulations [8] mentioned above.

4. Concluding remarks

In this paper we have demonstrated the feasibility of STM measurements at temperatures up to at least 800 K. From our observations of the Si(001) surface we have deduced a kink detachment energy of (2.0 ± 0.2) eV and an activation energy for surface diffusion of (1.0 ± 0.1) eV, assuming a typical frequency factor of 10^{13} s^{-1} in both cases. The diffusion anisotropy at 340 K is at least a factor 10^3, with diffusion occurring preferentially along the dimer rows. The nature of the diffusing species is as yet unknown.

References

1. J. Schneir, R. Sonnenfeld, O. Marti, P.K. Hansma, J.E. Demuth and R.J. Hamers, J.Appl.Phys. **63**, 717 (1988).
2. M.G. Lagally, R. Kariotis, B.S. Swartzentruber and Y.-W. Mo, Ultramicroscopy **31**, 87 (1989).

3. E.J. van Loenen, D. Dijkkamp and A.J. Hoeven, J. Microscopy **152**, 487 (1988).
4. A.J. Hoeven, E.J. van Loenen, P.J.G.M. van Hooft and K. Oostveen, Rev.Sci.Instrum. **61**, 1668 (1990).
5. A.J. Hoeven, J.M. Lenssinck, D. Dijkkamp, E.J. van Loenen and J. Dieleman, Phys. Rev. Lett. **63**, 1830 (1989).
6. K. Sakamoto, K. Miki and T. Sakamoto, Thin Solid Films **183**, 229 (1989).
7. H. Niehus, U.K. Koehler, M. Copel and J.E. Demuth, J. Microscopy **152**, 735 (1988).
8. H.B. Elswijk, A.J. Hoeven, E.J. van Loenen and D. Dijkkamp, Proceedings STM'90, to be published in J.Vac.Sci.Technol. A.
9. G.H.L. Brocks, P.J. Kelly and R. Car, to be published.

Phase Transition of Si(001) Studied by LEED

Y. Murata, M. Kubota, and T. Tabata

Institute for Solid State Physics, The University of Tokyo,
Roppongi, Minato-ku, Tokyo 106, Japan

Abstract. The temperature dependence of LEED patterns on the clean Si(001) surface has been studied. A phase transition between c(4x2) and 2x1 structures is observed at about 200 K and is the order-disorder transition of the asymmetric dimer configuration. The streak pattern appears clearly until 400 K and the change in width and length of the streak is specified by a quasi-one-dimensional Ising system with 1/2 spin. These results are described by the ordered buckling arrangement of the asymmetric dimer configuration in the dimer row well above the transition temperature and by a strong interaction between the adjacent dimers in the dimer row.

1. Introduction

The phase transition of the clean surface is one of the most specific surface phenomena. Various phase transitions have been found on semiconductors and metals. The Si(001) surface shows a phase transition between 2x1 and c(4x2) structures at about 200 K [1]. This phase transition is similar to that of the Ge(001) surface [2] but slightly different. The transition of Si(001) is the most typical system in surface science, since this transition can be clearly interpreted by a simple model [3] which is closely correlated with the asymmetric configuration for the dimers on the Si(001) surface.

On the other hand, the asymmetric configuration for the dimers on the Si(001) surface is questioned with the aid of scanning tunneling microscopy (STM) observations, where the symmetric dimer arrangement is dominantly seen in the vacancy-free region of the clean surface and the asymmetric configuration can be observed only in the vicinity of dimer vacancies [4]. Then, a recent calculation including an antiferromagnetic arrangement of the spins within the dimer shows that the symmetric configuration for the dimers is more stable than the asymmetric one [5]. Then, we discuss this problem through our recent observations of the striped low energy electron diffraction (LEED) patterns [6].

2. Transition between 2x1 and c(4x2)

Previously, we observed that the Si(001) surface changes from the 2x1 structure to the other ordered phase of the c(4x2) structure at low temperatures. Precise measurements of the temperature dependence of the LEED intensity show that this transition is the order-disorder transition [1]. Figure 1 shows the temperature dependence of the integrated intensity of the LEED spots. The intensities of the integral and half-order spots are well lined up on the Debye-Waller slope. On the other hand, the intensity of the quarter-order spot deviates from the slope above 150 K, decreases sigmoidally and disappears at about 300 K. No hysteresis was observed in this sigmoidal change with increasing and decreasing temperature. These results demonstrate that the transition is second order.

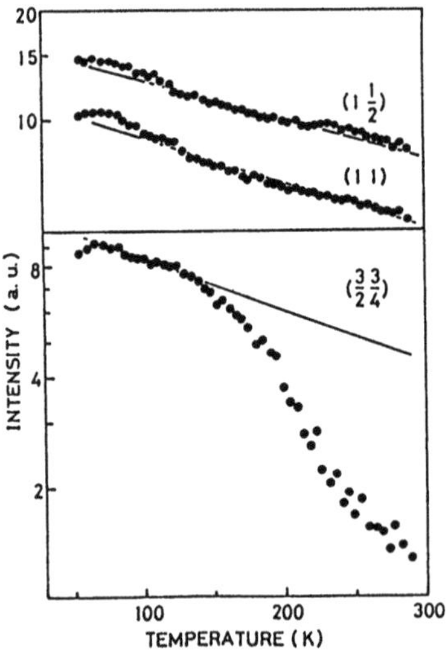

FIG. 1. Temperature dependence of the integrated intensity of LEED spots at normal incidence with E_p=132 eV. Straight lines represent the Debye-Waller slope.

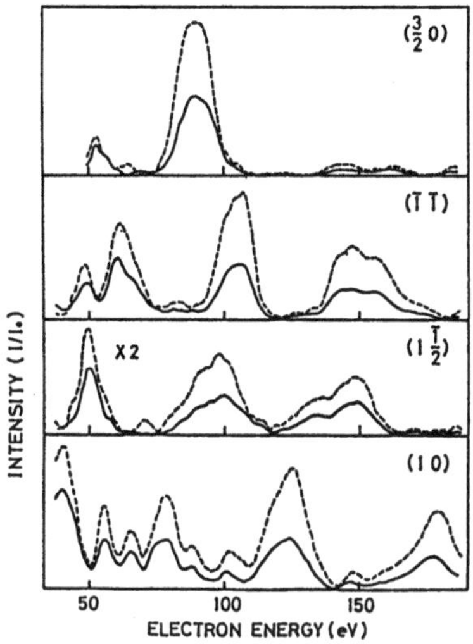

FIG. 2. Typical I-V curves of normal incidence at T=300 K (solid lines) and at T=46 K (broken lines).

Figure 2 shows integrated intensity versus incident electron energy (I-V) curves at room temperature and at 46 K. Both curves for each spot are very similar except for the effect of the Debye-Waller factor. This shows that the local atomic arrangement in a 2x1 unit mesh remains unchanged before and after the transition. This also confirms that the transition is an order-disorder one, and that the 2x1 phase at room temperature is not due to ferromagnetic ordering but due to paramagnetic disordering, when the asymmetric dimer configuration is specified by an Ising system with 1/2 spin.

3. Precise measurements of the striped LEED pattern

Streaks in the LEED pattern, which are correlated with the c(4x2) structure, appear at temperatures well above the transition. We precisely observed the streak LEED patterns. Since the intensity of the streak patterns is very weak as compared with that of the diffraction spots, a clear pattern with good signal-to-noise ratio was obtained by adding 256 digitized patterns observed independently. Each digitized frame was divided into 512x512 sections and each section had a dynamic range of 8 bits. Figure 3(a) shows a picture of a

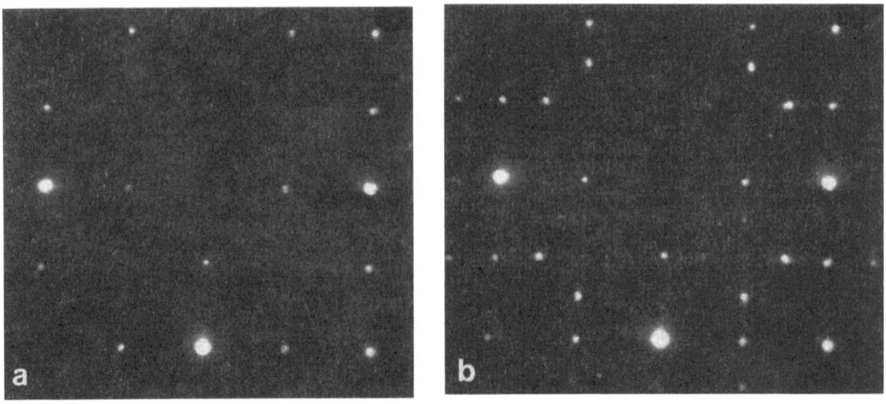

FIG. 3. LEED patterns on the clean Si(001) surface. E_p=36 eV. Normal incidence. (a) Striped pattern at 350 K and (b) c(4x2) structure at 180 K.

FIG. 4. Contour of a striped LEED pattern on Si(001) shown in Fig. 3(a). T=350 K. E_p=36 eV.

typical striped LEED pattern observed at 350 K and Fig. 3(b) shows a c(4x2) pattern observed at 180 K. Incident energy E_p is 36 eV. Figure 4 shows a part of the contour of the pattern shown in Fig. 3(a). Clear streaks appear along the ‹110› directions through the half-order spots at 350 K. When a single-domain surface is obtained in which the (1/2 1) spot vanishes, only the streak connecting the (1/2 1) and (1/2 1/2) positions remains in Fig. 4.

Figure 5 shows the temperature dependence of a bird's-eye view for LEED patterns of Si(001) at E_p=36 eV. Weak streaks are already observed at 450 K. The intensity of the streaks is enhanced at 400 K. Clear quarter-order spots representing long-range order appear at 320 K and are steeply enhanced near

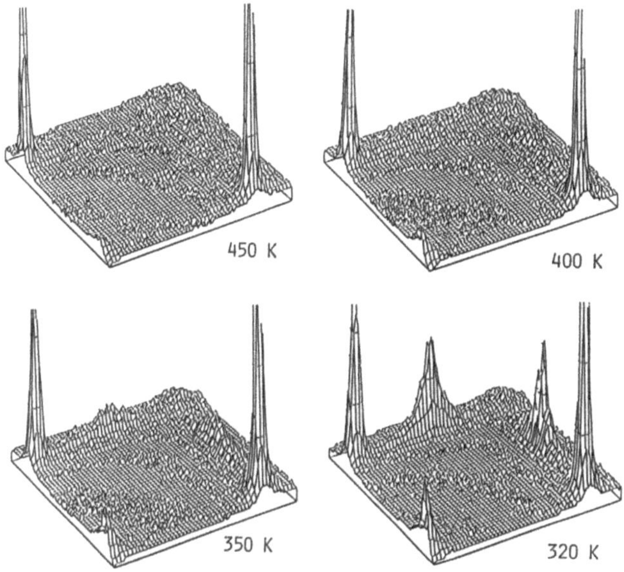

FIG. 5. Bird's-eye view of LEED patterns on Si(001) as a function of specimen temperature [6]. E_p=36 eV. Normal incidence.

the transition temperature. The streak pattern related to short-range order remains at temperatures well above the transition. Width and length of the streaks are unchanged over the temperature range at which the streaks are clearly observed [6].

4. Discussions

The streak pattern appears clearly at 400 K, which is well above the transition temperature. This result shows that ordering in the dimer row remains after the disappearance of the long-range order between the dimer rows. Another significant experimental result is that length and width of the streak remain constant in a wide temperature range. This result is clearly different from the order-disorder transition on the two-dimensional Ising system with 1/2 spin (in which the width of the streak changes with temperature and the temperature dependence of the width gives us the critical exponents [7]) and also different from the pure one-dimensional one [8]. In the quasi-one-dimensional case, the

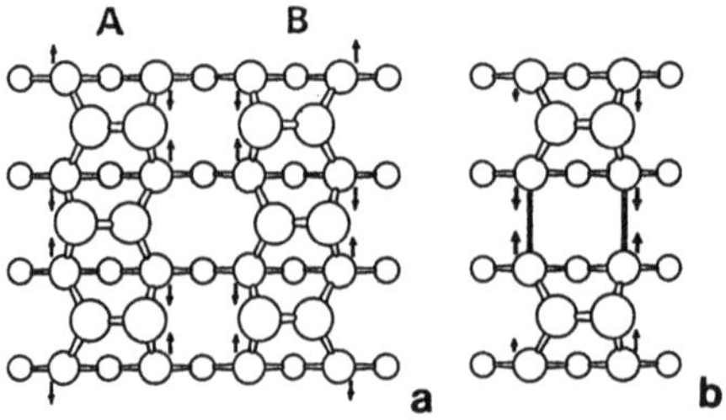

FIG. 6. Top view of the Si(001) surface of (a) c(4x2) structure and with (b) dimer vacancy. The arrows show the displacement vector of the second-layer atoms.

width of the Lorentzian which corresponds to the streak is expected to be constant around the transition. Thus, the present case shows that a quasi-one-dimensional transition occurs in a wide temperature range. These results can be interpreted by a model in which the structure change occurs cooperatively, keeping antiferromagnetic ordering in the dimer row. That is to say, A B transition of a flip-flop type occurs in the higher temperature range, where A and B are both arrangements for antiferromagnetic ordering in the dimer row, as shown in Fig. 6(a).

The cooperative structural transformation between A and B arrangements shows a strong coupling between dimers in the dimer row, as compared with the interaction between dimer rows. This result is clearly different from that obtained by a Monte Carlo simulation [9], in which the same coupling constants obtained by the tight-binding approximation are used as used in the renormalization group calculation [3]. The simulated result shows that the spin system is isotropically randomized. Therefore, the coupling constants for the spin Hamiltonian, V, H, and D (Fig. 7) describing the experimental results are considered to be very different from those estimated by the tight-binding caluculation and, especially, the experimental value of V/H is much larger than the calculated value.

The values of V/H and V/D calculated by the tight-binding approximation are close to those given by a classical scheme of the dipole-dipole interaction. The interaction potential of the dipole-dipole interaction is given by

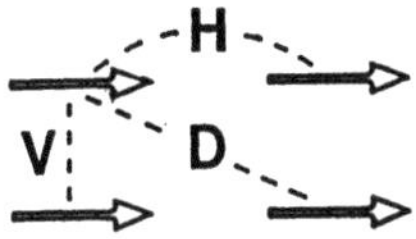

FIG. 7. Coupling constants used in the spin Hamiltonian, V, H, and V. Asymmetric dimer configuration is schematically illustrated by arrows.

$$U=[(p_1\cdot p_2)-3(p_1\cdot r)(p_2\cdot r)/r^2]/r^3,$$

where p_1 and p_2 are dipole moments corresponding to Ising spins and r is the position vector from p_1 to p_2. From this equation, the following relation is obtained in the present system: $H=-V/4=-0.25V$ and $D=-(7/25\sqrt{5})V=-0.125V$. These values of H/V and D/V are very close to those calculated by the tight-binding approximation: H/V=-0.38 and D/V=-0.15 [3].

A much larger value of V/H describing the experimental results than that calculated by dipole-dipole interaction is caused by the stabilization of antiferromagnetic ordering in the dimer row as compared with paramagnetic disordering. This ordering is stabilized by the feasible displacement of the second-layer atoms which is caused by the buckling arrangement while keeping Si-Si bond length constant. The direction of the displacement is parallel to the dimer row and opposite to that of the adjacent second-layer atoms, as shown in Fig. 6(a). In contrast, the displacement of the second layer atoms is hard in the ferromagnetic ordering. A similar argument is discussed in the Ge(001) surface [10].

When the dimer vacancy exists, the displacement vectors of two adjacent second-layer atoms at the dimer vacancy are directed in the same direction, as shown in Fig. 6(b). Then, the asymmetric dimer configuration near the vacancy is frozen against the flip-flop transition between A and B arrangements. The STM observation is slowly carried out as compared with the flip-flop motion and then the time-averaged image is obtained, in which the asymmetric dimer configuration is averaged into the symmetric-dimer image in the vacancy-free region. On the other hand, the asymmetric dimer configuration is kept in the vicinity of the dimer vacancy, since the flip-flop motion is prohibited.

Acknowledgements. This work was supported by a Grant-in-Aid for Scientific Research from the Ministry of Education, Science and Culture.

References

[1]T.Tabata, T.Aruga, and Y.Murata, Surf. Sci. **179**, L63(1987).
[2]S.D.Kevan and N.G.Stoffel, Phys. Rev. Lett. **51**, 1872(1984).
[3]J.Ihm. D.H.Lee, J.D.Joannopoulos, and J.J.Xiong, Phys. Rev. Lett. **51**, 1872(1982).
[4]R.M.Tromp, R.J.Hamers, and J.E.Demuth, Phys. Rev. Lett. **55**, 1303(1985).
[5]E.Artacho and F.Yandurain, Phys. Rev. Lett. **62**, 2491(1989).
[6]M.Kubota and Y.Murata, Phys. Rev. B to be submitted.
[7]R.F.Willis, in *Dynamical Phenomena at Surfaces, Interfaces and Superlattices*, ed. by F.Nizzoli, K.-H.Rieder, and R.F. Willis(Springer, Berlin, 1985) p.126.
[8]M.E.Fisher, Am. J. Phys. **32**, 343(1964).
[9]A.Saxena, E.T.Gawlinski, and J.D.Gunton, Surf. Sci. **160**, 618(1985).
[10]M.C.Payne, M.Needels, and J.D.Joannopoulos, J. Phys.: Condens. Matter 1, SB63(1989).

Angle-Resolved Electron Spectroscopy Study of the Surfaces and Interfaces of Silicon

S. Kono

Department of Physics, Faculty of Science, Tohoku University, Sendai 980, Japan

Abstract. Three types of angle-resolved electron spectroscopy for the study of solid surfaces have been demonstrated. They are angle-resolved ultraviolet photoelectron spectroscopy (ARUPS), X-ray photoelectron diffraction (XPD) and micro-probe Auger electron diffraction (μ-AED). Electronic structures of a single-domain Si(001)c(4x2) surface are revealed by ARUPS. Arrangements of Cs atoms on the Cs/Si(001) system are studied by XPD and LEED and an arrangement of In atoms in the Si(111)4x1-In surface is studied by μ-AED.

1. Introduction

Electron spectroscopy is suitable for the study of solid surfaces because the mean free path of electrons in solids is so short that most of the "non-loss" peaks in electron energy distribution curves convey information about solid surfaces. Electron spectroscopy becomes more versatile when angle-resolved detection of electrons is utilized. There are many forms of application of angle-resolved electron spectroscopy to the study of solid surfaces. Angle-resolved ultraviolet photoelectron spectroscopy (ARUPS) makes it possible to determine the surface-state dispersions. X-ray photoelectron diffraction (XPD) and Auger electron diffraction (AED) make it possible to determine the atomic position at surfaces. A newly developed technique of μ-probe Auger electron diffraction (μ-AED) is suitable for the determination of atomic positions of a small region (~μm in dimension) of surfaces. In the following sections, examples of ARUPS, XPD and μ-AED are presented in which the properties of Si(001), Si(111) and/or their metal-submonolayer interfaces are studied.

2.1 ARUPS study of Si(001)2x1 and c(4x2) surfaces[1]

It is generally accepted that the Si(001) surface is reconstructed into 2x1 periodicity due to pairing of the topmost surface Si atoms. However, the actual configuration of the paired Si atoms (dimers) remains unclear. The presence of buckled dimers (asymmetric dimers) was first suggested by Chadi[2]. Total-energy calculations showed later that a 2x1 arrangement of buckled dimers is more stable than that of symmetric dimers[3-5](cf. fig.1(a)). It was later pointed out that the 2x1 periodicity is not a ground state of the Si(001) surface and a higher-order periodicity should occur at about 250 K[6,7]. It was recently known that the Si(001) surface shows a phase transition from 2x1 to c(4x2) at about 200 K[8].

Eds.: A. Yoshimori · T. Shinjo · H. Watanabe

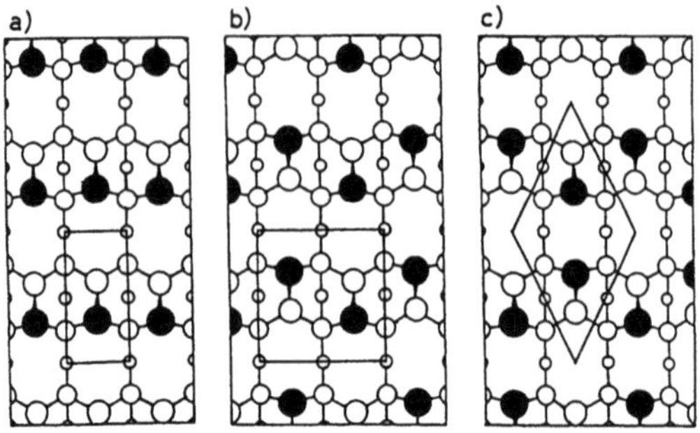

Fig.1--- Schematics of buckled-dimer models for Si(001)2x1, p(2x2) and c(4x2) surfaces. Large solid and open circles are outward- and inward-buckled Si atoms, respectively.

However, recent STM studies showed a 2x1 arrangement of symmetric dimers in "ideal" parts of the 2x1 surface[9,10]. It was also concluded from photoemission spectra of Si 2p core-levels for the 2x1 surface that there is no large and asymmetrical charge distribution within each of the dimers[11]. Although ARUPS spectra for the valence bands of the 2x1 surfaces are in principle consistent with those expected for the "ferromagnetic" arrangement of asymmetric dimers (i.e. Fig.1(a)), there remain additional surface bands that are not expected for an ideal "ferromagnetic" arrangement of asymmetric dimers[12]. In view of these circumstances, it is of great interest to measure ARUPS spectra of a single-domain Si(001) c(4x2) surface at a low temperature[1].

ARUPS spectra for the single-domain Si(001)2x1 and c(4x2) surfaces are shown in Figs.2(a) and (b). The polar angle (θ) of photoelectron emission is changed so as to scan the symmetric axes of the 2x1 and c(4x2) surface Brillouin zones (SBZ) (cf. inset of Fig.2(e)). Unpolarized He I light (21.2 eV) was incident at 45° polar angle opposite to the photoelectron emission. The ARUPS spectra for the 2x1 surface are shifted by 0.16 eV toward the lower binding energy (E_b) side in order to align bulk peaks at $E_b \geqq \sim 4$ eV. The shift is to compensate a difference in band bending for the two surfaces. A dramatic change in the spectra between the 2x1 and c(4x2) surfaces is found for the $\bar{J} - \bar{2}$ direction which is the direction perpendicular to the dimer row. For the $\bar{J} - \bar{I}$ direction, most of the structures in the c(4x2) spectra are already visible in the 2x1 spectra although the structures are much sharper for the c(4x2) surface.

Recently, Zhu et al.[13] made an *ab initio* calculation of the surface electronic structure of the c(4x2) surface based on the buckled-dimer model for the Si(001) surface. The configuration of the buckled dimers is an "antiferromagnetic" order for the c(4x2) surface (cf. Figs.1(c)). The surface-state (SS) bands due to dangling bonds of dimers of the c(4x2) surface are depicted as solid lines in Figs.2 (c) and (d) for two kinds of optimized geometries[13]. Experimental SS bands (circles) deduced from Figs.2 (a) and (b) and others not shown are compared with the theory; a solid shift of experimental curves toward the lower E_b side by 0.38 eV has been applied. Very good agreement of overall dispersion is achieved in Fig.2(d). Although overall agreement is very good, the absolute E_b for experiment does not agree with that for theory. In the experiment, E_b of the top of the occupied SS band is ~0.5 eV, whereas that of the theoretical SS for the two geometries is practically zero in both cases.

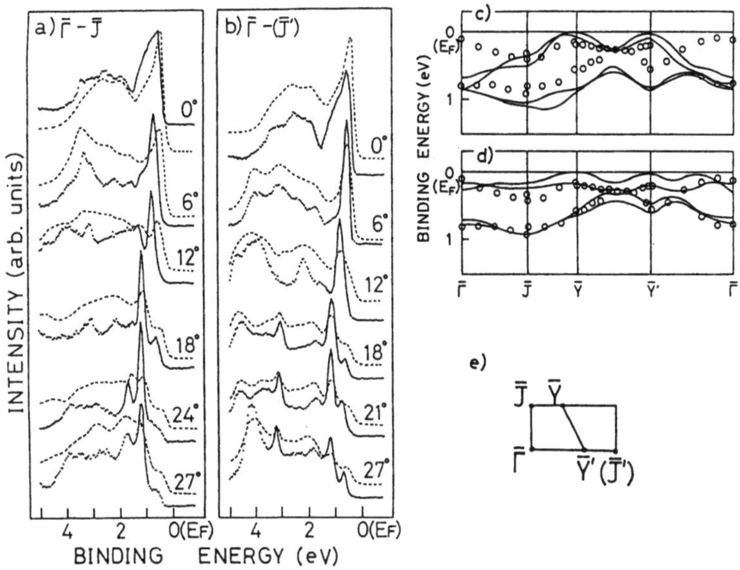

Fig.2--- (a),(b); ARUPS spectra of the single-domain Si(001)2x1 (dashed curves) and Si(001)c(4x2) (dots and solid lines) surfaces. (c), (d); Comparison of the experimental surface-state dispersion (circles) for the Si(001)c(4x2) surface with the results of recent *ab initio* calculations[13] for two kinds of atomic positions (see text). (e); Symmetry points of the surface Brillouin zones of the 2x1 and c(4x2) surfaces.

One reason for this disagreement might be related to the self-energy correction of the SS which has been reported to open up the gap between the occupied and empty SS bands for the Ge(111)1x1-As surface[14]. Another point to be mentioned is the fact that the experimental bands are in better agreement with the calculated bands for a c(4x2) structure in which atomic positions are optimized for the 2x1 periodicity with the "ferromagnetic" arrangement (cf. Fig.1 (a)) of buckled dimers. Detailed theoretical analyses including the self-energy correction are needed to facilitate further quantitative comparison.

On the basis of the present results, the following image of "ideal" parts of the Si(001) surface emerges although detailed descriptions are left to the original paper[1]. In the 2x1 surface at ambient temperature, the Si-dimers are buckled and buckling direction fluctuates with a tendency to the "antiferromagnetic" order within each of the dimer rows. This leads to a nominal "2x1" LEED pattern with weak half-order streak lines. ARUPS spectra of this surface consist of two kinds of dangling-bond SS bands; one is a band similar to the one expected for a seemingly "ferromagnetic" order of buckled dimers and the other is "additional"[15]. Below 200 K, the correlation of buckled direction among the dimer rows results in the c(4x2) arrangement. The dangling-bond SS bands for this surface appear to be in qualitative agreement with those of calculations by Tsukada and co-workers based on c(4x2) "antiferromagnetic" order of buckled dimers[13].

2.2 XPD study of the Cs/Si(001) system[16]

We have previously studied the structures of alkali atoms (K and Cs) at their saturation on Si(001) surfaces at around room temperature by XPD[17,18]. A main conclusion was that the alkali atoms form two kinds of array (double layer) with vertical separation of ~1 Å. This is in contrast to a long believed notion of the presence of one-dimensional metallic alkali chains on the Si(001) surface[19].

The properties of alkali adsorption on Si(001) surfaces at low and intermediate coverage are not so much studied as compared to those at saturation. We have recently made LEED and XPD studies of the structure of Cs-adsorbed Si(001) surfaces as a function of Cs coverage, θ, from 0 ML to more than 1 ML, where 1 ML is the surface atom density of a truncated Si(001)1x1 surface[16]. For this purpose, a wide-terrace single-domain Si(001)2x1 substrate was used with a temperature variation between ~90 K and ~1270 K just as in the ARUPS study of §2.1.

Table I shows the periodicities of the Cs-adsorbed Si(001) surfaces as θ is changed from 0 to more than 1 ML for the substrate temperatures, T_s, of ~110 K and ~320 K. For the clean substrate, the periodicities of c(4x2) and 2x1 plus streak(x2) are found for the two T_s's. At these intermediate coverages, there appear many surface orders that have not been reported so far which are streaky x3, 4x6, 2x6 and 8x4. The 2x6 structure is remarkable in the sense that it reversibly transforms to 2x3 for the T_s of ~320 K; the 2x3 order has been previously reported[18,20]. Other remarkable surface orders are streaky x3 and 8x4 at θ of ~1/6 and 1/2 ML, respectively. The existence of these new periodicities might be a consequence of the wide-terrace single-domain substrate.

Azimuthal XPD patterns of Cs 3d core-levels were measured for most of the surfaces listed in Table I. A main finding in the XPD experiments is the fact that XPD patterns for θ <1/2 ML are qualitatively different from those of the Cs saturated 2x1 surface whereas XPD patterns for $\theta \geqq 1/2$ ML show a feature that is characteristic of the Cs saturated 2x1 surface[16]. This indicates that the double-layer structure starts to occur at θ=1/2 ML and it develops up to 1 ML although two dimensional periodicities are different from the final 2x1. For θ <1/2 ML, the XPD patterns indicate that each of the surfaces is composed of a single-layer of Cs atoms. Curves with solid dots in Fig.3 show experimental XPD patterns for the surfaces at θ <1/2 ML, which are the x3 streak, 2x3 and 2x6 surfaces. θ_t and ϕ are the take-off angle (measured from the surface) and the azimuthal angle of photoelectron emission, respectively. The ordinate is the anisotropy of photoelectron intensity defined as $(I_{max}-I)/I_{max}$. These XPD patterns show no indication of a forward peaking of photoelectron diffraction characteristic of the double-layer model of Cs-adsorption[16].

Table I. Periodicities of the Cs-adsorbed Si(001) surfaces as observed by LEED as a function of coverage as determined from XPS for the two substrate temperatures given.

θ (ML)	0	1/6	1/4	1/3	1/2	~2/3	1	>1
Orders at ~110 K	c(4x2)	Streak x3	4x6	2x6	8x4	High Back 2x1	2x1	High Back 2x1
Orders at ~320 K	2x1+ x2 Streak	Streak x3	---	2x3	8x4	High Back 2x1	2x1	No more adsorption

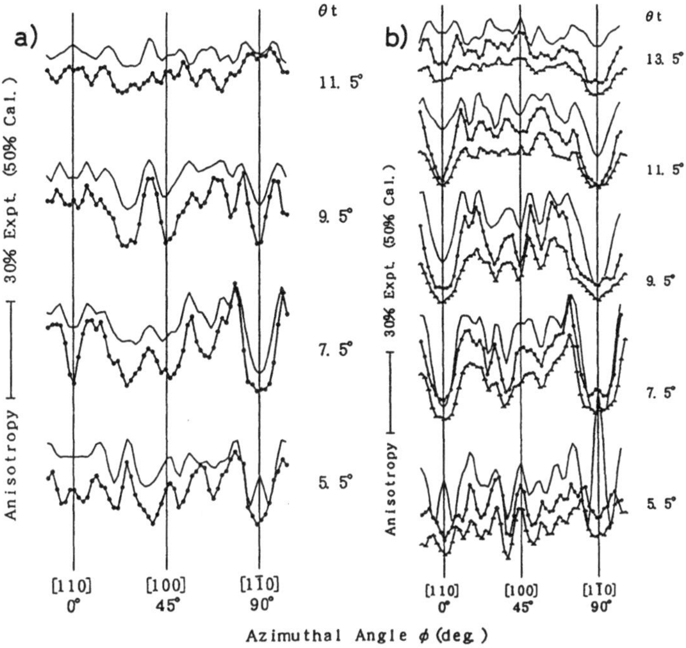

Fig.3--- Comparison of theoretical XPD patterns with those of experiments for the x3 streak, 2x3 and 2x6 surfaces of the Cs/Si(001) system. a)The x3 streak surface; dots are experimental points and solid curves are theory for the Cs overlayer model of Fig.4. b) Circular and rectangular dots are experimental points for the 2x6 and 2x3 surfaces, respectively. Solid lines are theoretical curves for a structural model for the 2x3 periodicity in which paired Cs atoms are arranged on every 2x3 site[16].

The XPD patterns in Fig.3 were kinematically analyzed taking into account the information about the periodicities and coverages from LEED and XPS. Figure 4 shows a model for the x3 streak surface in which a Cs atom is placed on every x3 site along the Si dimer rows (the registry of the Cs atoms to the Si dimer rows is not determined in this analysis) and these x3a_0 arrays are separated by 2a_0 (a_0 being the unit length of the substrate) with no correlations among the Cs arrays. In this way, the x3 streak LEED patterns are reproduced. It was also found from the XPD analysis that the nearest neighbor distance between Cs atoms in the neighboring x3a_0 arrays is not 2a_0 but $\sqrt{5}a_0$ as shown in Fig.4. XPD patterns as calculated for this model are shown as solid lines in Fig.3(a). Formalism and parameters in the kinematical calculations are similar to those described previously[18]. The theoretical XPD patterns are in very good agreement with the experimental patterns except for ϕ=~90° at θ_t=5.5°. For ϕ=~90° at θ_t=5.5°, the path of photoelectrons is nearly parallel to the Cs arrays so that multiple scattering of electrons may occur, which explains the discrepancy between the theory and experiment. The relationship of the x3a_0 Cs arrays to the substrate is not determined from the present XPD analysis.

It is surprising to see in Fig.3(b) that the XPD patterns for the 2x3 and 2x6 surfaces are very similar to each other, the only difference being the higher anisotropy for the 2x6 surface. This

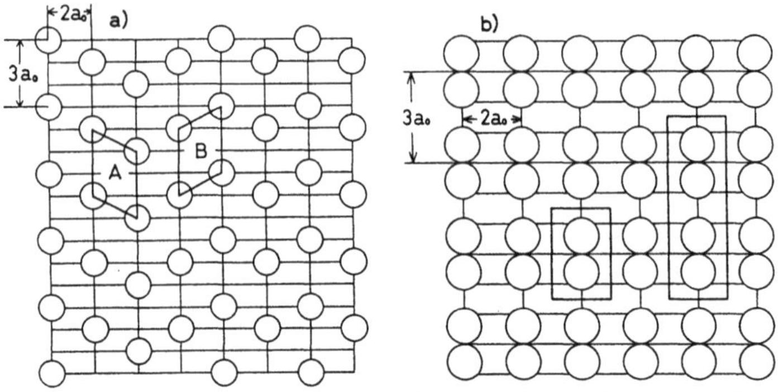

Fig.4--- a) Model for the x3 streak surface of Cs/Si(001). Circles are Cs atoms and rectangles are the 2x1 unit cells of the substrate. A and B are two possible relationships between the two neighboring x3a_0 Cs arrays, where a_0 is the unit length of a 1x1 surface. b) Model for the 2x3 and 2x6 surfaces of Cs/Si(001) (see text).

definitely shows a similarity of the two surfaces. Figure 4(b) shows a model for the 2x3 and 2x6 surfaces. For the 2x3 surface, pairs of Cs atoms are placed on every 2x3 site and, for the 2x6 surface, the pairs of Cs atoms may have a small amount of modulation in positions (for example clockwise and counterclockwise rotations about the centers of the Cs pairs) to become a 2x6 periodicity[16]. Theoretical XPD patterns for the 2x3 surface are shown as solid curves in Fig.3(b)[16]. The theoretical curves are in very good agreement with the experimental ones except for ϕ=~90° at θ_t=5.5°. The reason for this discrepancy may be the same as in the case of the x3 streak surface. The reason why the anisotropy for the 2x6 surface is higher than that for the 2x3 surface may be that the positions of Cs pairs in the 2x3 surface might be fluctuating with time (for example rotational vibration about the centers of the Cs pairs) while the positions of Cs pairs are fixed in the 2x6 surface. The modulation and fluctuation in positions in Cs pairs are expected to be very small because the XPD patterns for both the 2x3 and 2x6 surfaces are very alike[16].

2.3 μ-AED study of the In/Si(111) system[21]

Figure 5 shows main components of the newly developed μ-AED apparatus[22]. From a field emission electron gun, an electron probe beam of an order of tens of Å in diameter is obtained. Images of SEM (scanning electron microscopy) and dark field images of μ-probe RHEED (reflection high energy electron diffraction) can be obtained with a spatial resolution of ~100 Å. The working distance of the sample to the end of a pole-piece of a magnetic lens is ~4 cm so that many devices like an electron energy analyzer, an Ar ion gun and a metal deposition source can be used at the sample position for SEM and μ-probe RHEED. The electron energy analyzer is a special one (named constant take-off angle analyzer) by which electrons can be energy- and angle-analyzed for an angular range of ~70° in azimuth at a constant (but changeable) take-off angle. μ-probe Auger electron diffraction patterns can be obtained by exciting Auger-electrons with the μ-probe beam and by measuring Auger-electron intensities with the constant take-off angle analyzer.

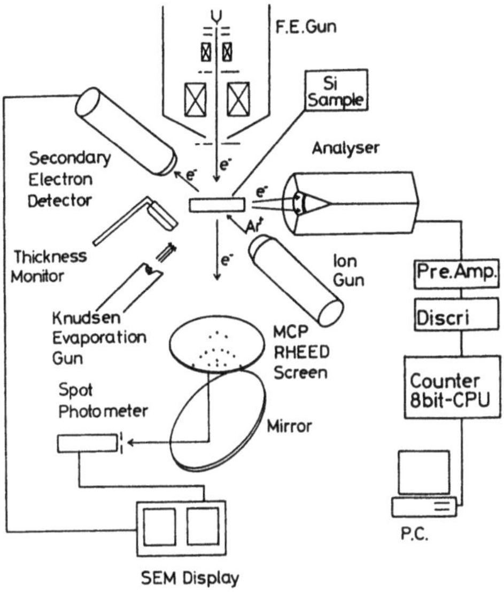

Fig.5--- Elemental components of the newly developed apparatus of micro-probe Auger electron diffraction (μ-AED)[22].

The In/Si(111) system was chosen as a test system for the μ-AED apparatus. Much work has been done on this system, especially, the Si(111)$\sqrt{3}$x$\sqrt{3}$-In surface is one of the best understood metal/Si(111) overlayer systems[23]. A Si(111)4x1-In surface is obtained when a monolayer of In atoms is deposited on a Si(111)7x7 surface and subsequently annealed. There are at least three domains of the 4x1 surface on a Si(111) surface because of the symmetries of the reconstruction and the substrate. It is difficult for conventional surface-techniques to investigate the properties of the 4x1 surface if all three domains are present on the sample surface. μ-AED is best suited for the study of this kind of surface, since one can probe a single domain of the reconstruction if the domain-size is wide enough for the μ-AED; the smallest domain-size which the present μ-AED apparatus can deal with is a few μm.

Figure 6 shows μ-AED patterns of In MNN Auger electrons measured for a single domain of a S(111)4x1-In surface[21]. The 4x1 surface was made by depositing 2 ML of In onto a room temperature Si(111)7x7 substrate and by a subsequent annealing at 450 ℃ for a few minutes. Dark field images of μ-RHEED showed the existence of μ-domains of the 4x1 reconstruction and most of the domains had a uniform orientation of the 4x1 periodicity in a certain area. μ-AED patterns in Fig.6 were measured from an area of ~300x30 μm^2 with an incident probe beam at 9 KV. The take-off angles are the polar angles of electron emission as measured from the surface. The abscissa corresponds to the azimuthal angles of electron emission, whose relationship to the substrate orientation is also marked. It is surprising to see in Fig.6 that peaks in the diffraction patterns do not change their azimuthal positions with changes in take-off angles. These peaks are expected to be forward peakings of electron scattering so that there must be scatterers of In atoms along these directions. A concrete structural model can only be obtained from a quantitative analysis of the diffraction patterns. Results of our kinematical analysis will be presented elsewhere[21]. Briefly, the

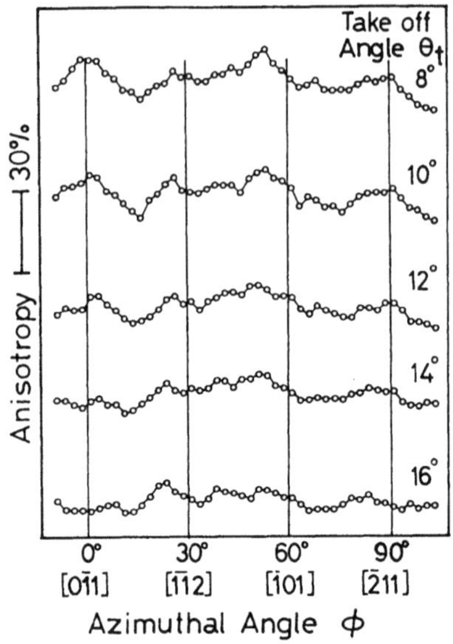

Fig.6--- μ-AED patterns of In MNN Auger lines from a single-domain Si(111)4x1-In surface[21]. These patterns were obtained using the μ-AED apparatus of Fig.5 with a specially developed electron energy analyzer (constant take-off angle analyzer).

analysis is based on several conditions provided by the present μ-probe study and other work, like scanning tunneling microscopy[24]. It was also assumed that In atoms form an overlayer on a nearly truncated Si(111) surface.

References

1. Full account of the work for the c(4x2) surface is published; Y. Enta, S. Suzuki and S. Kono, Phys. Rev. Lett. 65(1990)2704.
2. D.J. Chadi, Phys. Rev. Lett. 43(1979)43.
3. J. Ihm, M.L. Cohen and D.J. Chadi, Phys. Rev. B21(1980)4592.
4. W.S. Verwoerd, Surf. Sci. 99(1980)581.
5. M.T. Yin and M.L. Cohen, Phys. Rev. B24(1981)2303.
6. J. Ihm, D.H. Lee, J.D. Joannopoulos and J.J. Xiong, Phys. Rev. Lett. 51(1983)1872.
7. A. Saxena, E.T. Gawlinski and J.D. Gunton, Surf. Sci. 160(1985)618.
8. T. Tabata, T. Aruga and Y. Murata, Surf. Sci. 179(1987)L63.
9. R.J. Hamers, R.M. Tromp and J.E. Demuth, Phys. Rev. B34(1986)5343.
10. R.J. Hamers and U.K. Köhler, J. Vac. Sci. Technol. A7(1989)2854.
11. D.H. Rich, T. Miller and T.-C. Chiang, Phys. Rev. B37(1988)3124.
12. Y. Enta, S. Suzuki, S. Kono and T. Sakamoto, J. Phys. Soc. Jpn. 59(1990)657, and references therein.
13. Z. Zhu, N. Shima and M. Tsukada, Phys. Rev. B40(1989)11868.
14. M.S. Hybertsen and S.G. Louie, Phys. Rev. Lett. 58(1987)1557.
15. K. Inoue, M. Nakayama and H. Kawai, Surf. Sci. in press and these proceedings.

16. Details of the work will be published by T. Abukawa, T. Okane and S. Kono; Surf. Sci. in press.
17. T. Abukawa and S. Kono, Phys. Rev. B37(1988)9097.
18. T. Abukawa and S. Kono, Surf. Sci. 214(1989)141.
19. J.D. Levine, Surf. Sci. 34(1973)90.
20. Y. Enta, T. Kinoshita, S. Suzuki and S. Kono, Phys. Rev. B39(1989)1125.
21. Details of the work will be published by N. Nakamura, K. Anno and S. Kono; Surf. Sci. in press.
22. Details of the μ-AED apparatus will be published by S. Kono, N. Nakamura, K. Anno and T. Teruyama.
23. T. Kinoshita, H. Ohta, Y. Enta, Y. Yaegashi, S. Suzuki and S. Kono, J. Phys. Soc. Jpn. 56(1987)4015 and references therein.
24. Sang-il Park, J. Nogami and C.F. Quate, J. Microscopy 152(1988)72.

Ordering and Electronic Structure of the Si(001) Surface

M. Nakayama

College of General Education, Kyushu University,
Ropponmatsu, Chuo-ku, Fukuoka 810, Japan

Electronic states and ARUPS spectra of Si(001) surface are calculated for several arrangements of the surface buckled dimers to analyze a recent experiments by Enta et al. Electronic states are calculated by the LCAO model which simulates a recent pseudopotential calculation. At the low temperature phase of c(4×2) structure, the observed intensity of the spectra can be explained when both the height of and distance between the dimers are modulated alternately along the rows. At the room temperature phase of "(2×1)" disordered structure, the ARUPS spectra have distinct peaks at nearly the same position as the ordered phase, if the dimer rows are correlated with moderate range.

§1. Introduction

The reconstruction of the Si(001) surface is one typical example of the interplay of the atomic and electronic structures of solid surfaces[1,2]. It has been generally accepted that neighbouring Si atoms form a dimer, say along the [110] direction. The dimers are arranged to form a row along the [$\bar{1}$10] axis so that the surface is at least (2×1) reconstructed. The buckling of the dimer was proposed by Chadi[3]: one of the atoms moves up and the other down to reduce the total energy. This picture was confirmed by a pseudopotential calculation of the total energy by Yin and Cohen[4], which shows that the energy is further lowered by a displacement of the dimer along the [110] direction. Precise atomic configuration and the arrangement of the dimers are governed by the electronic structure of the π-like dangling bond orbitals and are still a matter of controversy. In this article we shall show that ARUPS spectra, particularly its intensities, strongly depend on the arrangements of the surface dimers so that we can get information on the latter through the analysis of experiments.

Observed dispersion of one of the branches in the top region of the ARUPS spectra had been thought to agree

Eds.: A. Yoshimori · T. Shinjo · H. Watanabe

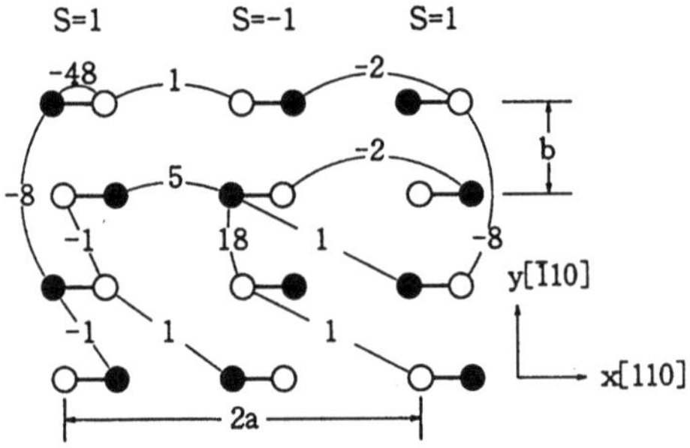

Fig.1

Top view of the surface atoms of the standard model (SM) structure of the Si(001) and LCAO parameters. Solid and open circles denote up and down atoms, respectively. Side view is shown in Fig.2b. Lines connecting atoms denote the LCAO off-diagonal parameters and numbers show the values in 0.01eV.

well with the calculated surface band of (2×1) structure[5]. We call the branch the normal(N) branch. Detailed experiments of ARUPS spectra, however, have revealed that there exists an additional branch[6], which will be called the extra or Umklapp(U) branch. The nature of the extra branch has been speculated as due to the double domain structure of the surface or the defects on the surface. Recently, Enta et al. prepared a wide-terrace single-domain surface of good quality and performed an ARUPS experiment on the surface[7]. They observed two branches at room temperature, which clearly shows that the extra branch reflects the intrinsic nature of the surface.

The LEED experiment by Tabata et al. revealed an order-disorder transition on the surface[8]. Below about 200K, the surface is c(4×2) reconstructed. As a model for the c(4×2) structure Zhu et al.[9] assumed that the sense of the buckling alternates along the directions both parallel and perpendicular to the dimer row ("antiferromagnetic" ordering) with the same local configurations of the dimer and subsurface atoms as that of the (2×1) structure of Yin and Cohen. The structure of the surface atoms is shown in Fig.1. We call the structure the standard model (SM) structure. They calculated the total energy and showed that the c(4×2) structure has indeed the lowest energy, though the difference from the p(2×2) structure, where all the antiferromagnetic rows have the same phase, is very small. As the temperature is raised to room temperature, a detailed LEED experiment shows that the quarter spot of the c(4×2) phase elongates along the direction perpen-

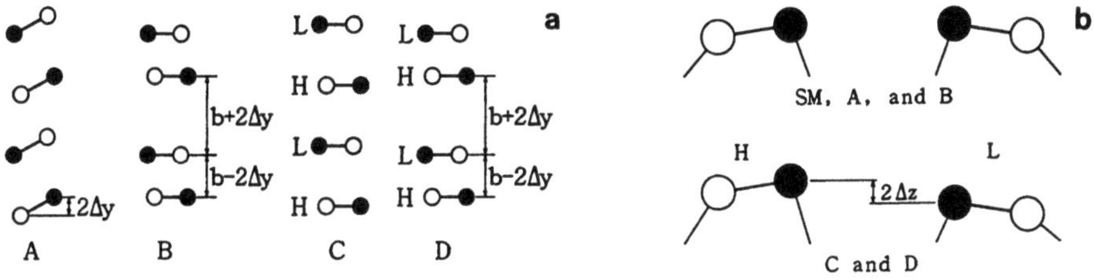

Fig.2

Configurations of the surface atoms for a few model structures.
2a: Top view of the surface atoms. 2b:side view of the dimers.

dicular to the row but still has a rather narrow width along the row so that a streak pattern is observed[8]. This fact suggests that the so-called (2×1) phase is not a perfectly ordered phase but a disordered phase where the correlation among the buckling is quite anisotropic: the buckled dimer is still ordered antiferromagnetically along the row with finite but rather long range, while the correlation of the phase between the rows has much shorter range. Such a picture is also shown in a numerical simulation of the surface structure by Kakitani and Yoshimori [10].

Very recently, Enta et al. performed an ARUPS experiment on the single domain surface of the c(4×2) structure[11]. The observed spectra are shown in Fig.2 of the article by Kono in this issue. The top region of the spectra consists of two branches whose dispersions are almost the same as the room temperature ones with much narrower widths. The observed dispersion agrees well with the calculated one of the surface band of the c(4×2) structure[9]. The U-branch is due to the band which appears in c(4×2) structure as a result of the interaction with the reciprocal lattice vector of c(4×2). Persistence of the dispersion to room temperature can be explained by the existence of the locally c(4×2) ordered structure.

The purpose of the present article is to confirm the above picture by investigating the intensity of ARUPS spectra for low and high temperature phases[12,13]. We shall first calculate the intensity for SM of the c(4×2) structure and find that the intensity of the U-branch vanishes by symmetry for the electrons emitted along the dimer row. Then we shall try several kinds of displacements of the surface atoms from SM to explain observed magnitudes of the intensity. We also calculate electronic structure and ARUPS spectra for a disordered array of the dimer row. We show that the spectra consist of the N- and U-branches with dispersion similar to c(4×2) phase, as long as the dimer rows are moderately correlated.

§2. Models and Method of the Calculation

§2.1 Atomic Structure

We assume that the buckled dimer is ordered antiferromagnetically along the row. We choose the y-axis parallel to the row and the x-axis perpendicular to the row. The z-axis is the outward normal of the surface. Hence, the system is always periodic along the y-direction with the period 2b, where b is the distance between the nearest neighbouring atoms on the ideal surface. Two possibilities of the phase of the sense of the buckling is described by an index S. The row with S=-1 differs from the row S=1 by the translation b along the y-axis (see Fig.1). The p(2×2) and c(4×2) structures are denoted by $S_n=1$ and $S_n=(-1)^n$, respectively. The lattice parameter along the x-axis is denoted by a=2b. Later, we shall introduce several modifications of the atomic configuration within the row.

In the disordered phase, S_n randomly takes the values 1 and -1 with short-range correlation. To simulate the structure, we assumed the interaction of the type JS_nS_{n+1} and performed a Monte Carlo simulation for the array of 502 rows with the periodic boundary condition. The calculation was done for a few values of the parameter $\lambda = J/kT$. A snapshot is employed as a model of the disordered array of the row.

§2.2 Electronic State

The LCAO method is employed to describe the surface electronic states, because we are to consider them for several arrangements of the dimers. The wave function of the surface bands is expanded in terms of the π-like dangling bond orbital of each atom. The parameters are chosen to reproduce the results of the pseudopotential calculation[9]. The off-diagonal parameters for c(4×2) structure and their values in 0.01eV are shown in Fig.1. The diagonal energies of up and down atoms are 4.41 and 4.55eV, respectively. In the periodic structure the state is classified by the surface wave vector $\vec{k}_s = (k_x, k_y)$. We obtain two bands for each $\vec{k}_s$ in the Brillouin zone of the (2×1) structure, one occupied and the other unoccupied for a semiconducting case. We also simulate the results for the p(2×2) structure as a reference for the room temperature calculation. It turns out that the surface band structure for the c(4×2) and p(2×2) structures are quite similar, when they are plotted in the

(2×1) Brillouin zone. In fact the values of the intra-row parameters are practically the same. Slight differences of the bands are due to the difference in the inter-row parameters and the elementary reciprocal vectors.

In the disordered phase, the electronic states are calculated for a snapshot of random array of the 502 dimer rows with the periodic boundary condition. The inter-row parameters are chosen for $S_nS_{n+1}>0$ and <0 as those of p(2×2) and c(4×2) structures, respectively. Since the system is periodic along the y-axis, the state is still classified by k_y, while k_x is no longer a quantum number and the spectrum is spreading over the region of the k_x-dispersion of the periodic structures. The actual spectra will be seen in section 4 from the calculated ARUPS spectra.

§2.3 Calculation of ARUPS Spectra

The intensity of the ARUPS is calculated by a one electron approximation. The spectral density is proportional to the following expression:

$$I = \sum_i |\langle\vec{p}'|\vec{A}\vec{P}|\phi_i\rangle|^2 W\sqrt{\varepsilon_f}\,\delta(\varepsilon_i+h\nu-\varepsilon_f). \qquad (1)$$

In the above, i denotes the initial state with energy ε_i and wave function ϕ_i, $\vec{A}$ is the vector potential of the incident light, $h\nu$ is its energy, and $\vec{P}$ is the momentum operator. The calculation is done under the experimental condition of Enta et al.[7,11] that unpolarized light with $h\nu$ = 21.2eV is incident with the angle 45 degrees and photoelectrons are detected in the plane of incidence. $|\vec{p}'\rangle$ is the single plane wave of the final state inside the solid with the energy ε_f. The latter transmits into the vacuum with the probability W and is detected as a plane wave $|\vec{p}\rangle$. We assume a step function for the surface barrier potential so that only p_z' differs from p_z. We assume the height of the barrier as 2.65eV.

The matrix element of eq.(1) is proportional to the factor

$$\sum_j c_j \exp[-i\vec{p}'\vec{Q}_j], \qquad (2)$$

where j denotes the lattice point with the lattice vector $\vec{Q}_j$ and c_j is determined by LCAO coefficient and the Fourier transform of the dangling bond orbital. The selection rule comes out from the factor (2) and $\vec{p}'\vec{A}$.

§3. ARUPS Spectra of c(4×2) Structure

Since the initial state is classified by $\vec{k}_s$ in the Brillouin zone of (2×1) structure, the selection rule for the photoelectorn $\vec{p}$ is given by

$$k_x = p_x + 2m\pi/a, \qquad k_y = p_y + 2n\pi/b \tag{3}$$

$$k_x = p_x + (2m\pm1)\pi/a, \quad k_y = p_y + (2n\pm1)\pi/b. \tag{4}$$

Thus, two branches are observed for a given $\vec{p}$: eq.(3) and eq.(4) correspond to the N-and U-branches, respectively.

SM c(4×2) structure is invariant under the reflection with respect to the plane bisecting the dimer row followed by the translation b along the row. We call the operation the RT operation. When $p_x = 0$, i.e., the photoelectron is observed along the $\bar{\Gamma}$-$\bar{J}'$ direction, the relevant electronic state has a parity with respect to RT-operation apart from the exponential factor $\exp[ip_y b]$. The parity of the plane wave $|\vec{p}'\rangle$ is even. The parity of the initial states with $(2m\pi/a, p_y+2n\pi/b)$ and $((2m\pm1)\pi/a,\ p_y+(2n\pm1)\pi/b)$ are even and odd, respectively. Since only the p-polarization part of $\vec{p}'\vec{A}$ with even parity contributes for $p_x = 0$, the transition from the latter initial state (U-branch) is strictly forbidden by symmetry.

Since the U-branch is actually observed, the symmetry of the arrangement of the dimers must be lower than the SM structure. We consider several modifications of the atomic configuration of the dimer row to the SM structure and calculate the ARUPS spectra. We introduce the displacements of the surface atoms along the y and z directions with the magnitudes Δy and Δz, respectively: these cover the modulation of the angle of the buckling, the height of the dimer as a whole, the distance between the dimers along the row, and the rotation of the dimer around the z-axis. We have examined 729 cases of this kind with $\Delta y = 0.05b$ and $\Delta z = 0.045b$. We neglect the change of the electronic structure and consider the effect only through the exponential factor (2). Results of the calculation on the SM and model structures illustrated in Fig.2 are shown in Table 1. We find that, for a reasonable magnitude of the displacements, a sizable intensity of the U-branch can be obtained only when the height of the dimer is modulated alternately along the row (e.g, models C and D). The model D reproduces the p_y-dependence of the observed intensity of the U-branch much better. We call the structure "alternate modulation of the height and distance" (AMHD) structure, as the distance between the dimers is also modulated alternately along the row.

Table 1.

Intensity of the N-branch and U-branch at several points in the surface Brillouin zone for the SM structure and model configurations of the buckled dimers illustrated in Fig.2. The intensity of the N-branch at the $\bar{\Gamma}$ point for the SM structure is assumed to be 1. The unit of p_x is π/a and p_y is π/b.

model		p_x ($\bar{\Gamma}$-$\bar{J}$)				p_y ($\bar{\Gamma}$-$\bar{J}'$)		
		0.0	1.0	2.0	3.0	0.5	1.0	1.5
SM	N	1	0.95	0.57	0.024	1.4	1.7	2.0
	U	0	0.052	0.18	0.20	0	0	0
A	N	1	0.95	0.57	0.024	1.4	1.6	1.9
	U	0	0.052	0.18	0.20	0.001	0.011	0.014
B	N	1	0.95	0.57	0.024	1.4	1.6	1.9
	U	0	0.052	0.18	0.20	0.009	0.038	0.11
C	N	0.86	0.63	0.38	0.053	1.2	1.5	1.8
	U	0.14	0.27	0.33	0.17	0.19	0.19	0.21
D	N	0.86	0.63	0.38	0.053	1.1	1.3	1.5
	U	0.14	0.27	0.33	0.17	0.27	0.38	0.57

Calculated ARUPS spectra along the $\bar{\Gamma}$-$\bar{J}'$ direction for the AMHD c(4×2) structure is shown in Fig.3a by a vertical solid line. The height of the line is proportional to the intensity. The energy is measured from that of the N-branch at $\bar{\Gamma}$. p_y is in the unit of $250\pi/251b$. The result reproduces the observed trend of the intensity of the U-branch: it is rather small at $\bar{\Gamma}$, increases with p_y, and becomes discernible near $\bar{J}'$.

The calculated spectra for the AMHD c(4×2) structure along the $\bar{\Gamma}$-$\bar{J}$ direction is shown in Fig.3b by a vertical solid line. p_x is in the unit of $250\pi/251a$. In this case the intensity of the N-branch decreases with p_x, while that of the U-branch gradually increases and reaches a maximum. When p_x is larger than about $2\pi/a$, the intensity of the U-branch becomes larger than that of the N-branch. Such a trend is observed in the experiment. In the experiment, the intensity of the U-branch grows to be comparable to that of the N-branch at $\bar{\Gamma}$, while the calculated intensity is not so large. The discrepancy would be overcome , if we introduced displacements of the dimer along the x-axis. For $p_x \neq 0$, the intensity of the U-branch does not vanish even for the SM structure. However, the discrepancy of the p_x-dependence of the spectra with the experiment is larger than the AMHD structure.

§4. ARUPS Spectra of Disordered "(2×1)" Structure

Since the system is still periodic along the y-axis in the present model, selection rule

$$k_y = p_y + M\pi/b \qquad (5)$$

holds. The cases of even and odd M correspond to the N- and U-branches, respectively. The RT-symmetry is broken by the disorderd array of the dimer row even when we assume the SM structure for each row. Strictly speaking, the intensity of the U-branch does not vanish for $p_x = 0$ in that case. The calculated value is, however, very small: less than 0.1% of the intensity of the N-branch at $\bar{\Gamma}$. Therefore, we assume the AMHD structure for each row and calculate ARUPS spectra for a random array of them.

Calculated spectra for the case $\lambda = 1.0$ are shown in Fig.3a ($\bar{\Gamma}$-$\bar{J}'$) and Fig.3b($\bar{\Gamma}$-$\bar{J}$). In this case the phase

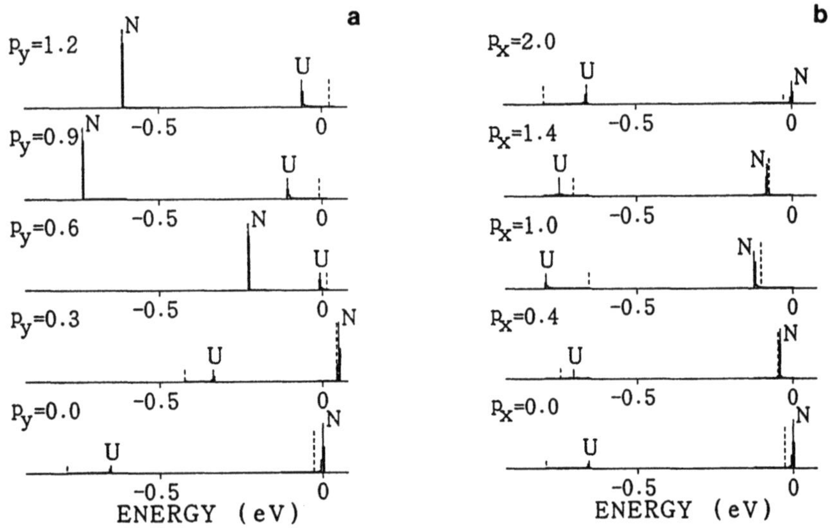

Fig.3

Calculated ARUPS spectra along (a) the $\bar{\Gamma}$-$\bar{J}'$ direction and (b) the $\bar{\Gamma}$-$\bar{J}$ direction for moderate short-range correlation ($\lambda = 1.0$).
The N- and U-branches are shown by histograms. The unit of p_y is $250\pi/251b$. Vertical solid and broken lines are the spectra for c(4×2) and p(2×2) structures, respectively. The energy is measured from the position of the N-branch of c(4×2) structure at the $\bar{\Gamma}$ point.

of the row is moderately correlated antiferromagnetically along the x-axis over a few rows. We also calculated the spectra for the p(2×2) structure with the same AMHD modulation and show them in Figs. 3a and 3b by vertical dashed lines for reference. For $p_x=0$, both the N- and U-branches of the random system have sharp peaks with asymmetric tails. The position of the peaks practically coincides with that of the c(4×2) structure and has a similar dispersion as a function of p_y. The p_y-dependence of the intensity is also similar to that of the c(4×2) structure. These characteristics can be explained as follows. For a k_y given by eq.(5), the energy levels of the random sysytem are spreading over the region of the corresponding k_x-band of c(4×2), because the system contains many locally ordered parts of a finite length. When we observe the photoelectron with p_x, we pick up mostly the components $k_x = p_x$ and $k_x = p_x+\pi/a$ for the N- and U-branches, respectively, and the intensity has a peak at the energy of the corresponding band of c(4×2). Since the position coincides with the edge of the k_x-band for $p_x = 0$, the peak becomes very sharp due to the edge singularity of the density of states. For $p_y = 0$($\bar{\Gamma}$-$\bar{J}$ direction), the position of the peaks of both the N- and U-branches also coincide with that of the c(4×2) structure. The spectra are, however, much broadened compared with the case $p_x = 0$. Particularly, the peak of the U-branch is not sharp except for the case $p_x = M\pi/a$ (M is an integer). For a general value of p_x, the corresponding k_x is not at the edge of the k_x-band of the c(4×2) structure so that the enhancement due to the singularity is absent. The peak would be more pronounced, if we improved the agreement with the intensity of the U-branch of the c(4×2) structure by further modulating the atomic position along the x-axis.

When temperature is further raised, the inter-row correlation becomes weaker and the p(2×2) configuration also appears locally. When $\lambda = 0$, the system has no correlation and c(4×2) and p(2×2) structures appear locally with equal probability. Calculated spectra for such a case are shown in Fig.4a ($\bar{\Gamma}$-$\bar{J}'$) and Fig.4b ($\bar{\Gamma}$-$\bar{J}$). The spectra of the N-branch still have sharp peaks with the dispersion similar to the one at low temperature. This is because the peaks of the c(4×2) and p(2×2) structures are located pretty close to each other. On the other hand, the spectra of the U-branch are very broad and the peaks are too weak to be observed. The peaks of the U-branch of the c(4×2) and p(2×2) structures are separated so that two peaks appear at the corresponding positions in the disordered structure.

Thus, ARUPS spectra, particularly the intensity and profile of the U-branch, reflect the degree of the

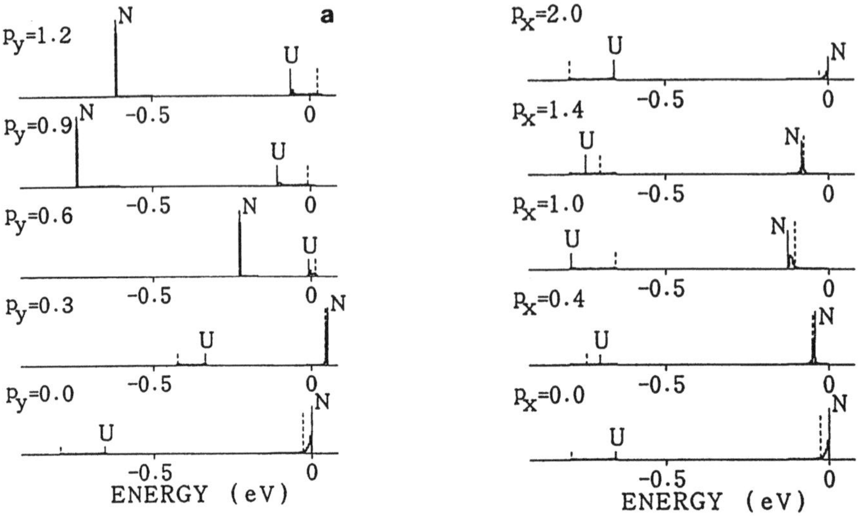

Fig.4

Calculated ARUPS spectra along (a) the $\bar{\Gamma}$-$\bar{J}'$ direction and (b) the $\bar{\Gamma}$-$\bar{J}$ direction for completely disordered structure (λ = 0).
The N- and U-branches are shown by histograms. The unit of p_y is $250\pi/251b$. Vertical solid and broken lines are the spectra for c(4×2) and p(2×2) structures, respectively. The energy is measured from the position of the N-branch of c(4×2) structure at the $\bar{\Gamma}$ point.

correlation of the disordered structure. The experimental fact that the spectra at room temperature are similar to that of the c(4×2) structure can be understood by the presence of a moderate short-range order among the dimer rows.

§5. Concluding Remarks

We have seen that the analysis of ARUPS spectra provides us with detailed information on the atomic ordering of the Si(001) surface. In the low temperature phase of c(4×2) structure, the existence of the U-branch suggests that both the height and distance of the buckled dimers are modulated alternately along the row. In the room temperature phase, "(2×1)" structure is likely to be a disordered array of the dimer row with a modulated arrangement within the row.

In the present calculation the final state is approximated by a single plane wave. The corresponding Bloch wave has the same parity as the plane wave, if the system has the RT symmetry. For the photoelectron Bloch state, emitted along the $\bar{\Gamma}$-$\bar{J}'$ direction, the N- and

U-branches are coupled to p- and s- polarizations, respectively. The coupling to the s-polarized light is not zero, because the Bloch wave contains plane waves whose p_x is not zero. Recently, Johansson et al. investigated the polarization dependence of ARUPS spectra on the single-domain "(2×1)" surface[14]. They observed that, along the $\bar{\Gamma}$-$\bar{J}'$ direction, N- and U-branches are suppressed for s- and p-polarizations, respectively. More calculations and experiments with polarized light are necessary to examine the effect of the Bloch state.

More detailed studies from both the experimental and theoretical sides are needed to see whether the modulated structure is realized or not. The Madelung energy is lowered by the modulation. However, the amount is so small that the change of the energy due to other mechanisms should be carefully investigated. Zhu et al. showed that total energy of the c(4×2) structure is lowered by modifying the position of the atoms from the SM structure[9]. Though the modification breaks the RT-symmetry, the calculated energy band does not agree with the observed dispersion for both the branches. Therefore, the modified structure proposed by them does not explain the ARUPS spectra.

Acknowledgements

This article is written on the basis of joint work with K.Inoue and H.Kawai[12,13]. The author would like to thank Prof.S.Kono of Tohoku University for informing him of experimental results prior to publication. The work was partly supported by Grant-in-Aids on the priority areas of the Ministry of Education, Science, and Culture of Japan.

REFERENCES

[1] For a review on the semiconductor surface, see D.Haneman: Rep. Prog. Phys., 50(1987)1045.
[2] For a review on the calculation of the surface structure, see J.Ihm: Rep. Prog. Phys.,51(1988)105.
[3] D.J.Chadi: Phys. Rev. Letters, 43(1979)43.
[4] M.T.Yin and M.L.Cohen: Phys. Rev., B24(1981)2303.
[5] J.Ihm, M.L.Cohen, and D.J.Chadi: Phys. Rev.,B21 (1980)4592.
[6] For a review on ARUPS study of semiconductor surface, see G.V. Hansson and R.I.G.Uhrberg: Surface Sci. Rep., 9(1988)197.
[7] Y.Enta, S.Suzuki, S.Kono, and T.Sakamoto: J. Phys. Soc. Japan, 59(1990)657.

[8] T.Tabata, T.Aruga, and Y.Murata: Surf. Sci.,179 (1987)L63.
[9] Z.Zhu, N.Shima, and M.Tsukada: Phys. Rev.,B40(1989) 11868.
[10] K.Kakitani and A.Yoshimori: to be published.
[11] Y.Enta, S.Suzuki, and S.Kono: Proc. Yamada Conference XXVI "Surface as a New Material",1990. Surface Sci., 242(1991)277; Phys. Rev. Letters, 65(1990)2704.
[12] K.Inoue, H.Kawai, and M.Nakayama: Proc. Yamada Conference XXVI "Surface as a New Material",1990. Surface Sci.,242(1991)244.
[13] K.Inoue, H.Kawai, and M.Nakayama: to be published in Surface Sci.
[14] L.S.O.Johansson, R.I.G.Uhrberg, P.Martensson, and G.V.Hansson: Phys. Rev., B42(1990)1305.

Aspects of Core-Level Photoemission from Semiconductor Surfaces: Resonant Emission and Final State Interaction

K. Cho

Faculty of Engineering Science, Osaka University,
Toyonaka 560, Japan

Abstract. Two cases of core level photoemission are considered that have close relationships with surface reconstruction. On GaAs(110), the intensity ratio of $d_{5/2}$- and $d_{3/2}$-resonances is discussed with emphasis on the difference from that in optical absorption. On Si(100), the effect of final state screening is shown to lead to a possibility of interpreting the observed data of $2p$ core emission in terms of the buckled dimer model.

1. Introduction

Atomic displacements such as relaxation and reconstruction are very commonly seen on semiconductor surfaces, and they are accompanied by specific surface electronic states, which reflect the symmetry and the magnitude of the displacements. Therefore, the study of these surface states can contribute very much to understanding the ordering of the surface. Photoemission, especially an angle-resolved measurement, is a powerful method to study these electronic states. Besides its role as a useful method, the photoemission process is interesting by itself, especially when core electrons are involved, in view of the dynamical effects arising from electron-electron interaction. From these two points of view, the study of core-level photoemission is an attractive theme on semiconductor surfaces. Here we present theoretical studies for two different cases, i.e., the resonant emission from the surface core excitons on GaAs(110) and the effect of final state screening on the direct emission of $2p$ core electron on Si(100).

2. Resonant Photoemission from GaAs(110)

When photon energy is resonant with the formation energy of a surface core exciton (SCE) on this surface, there opens an additional channel of photoemission for valence electrons, since the decay of SCE due to the electron-electron interaction can produce a photoelectron and a valence hole. The existence of SCE is closely connected with the relaxation of this surface, namely the raising (lowering) of surface As (Ga) atoms. These atomic displacements change the characters of chemical bonds of the surface atoms, and hence those of surface dangling bonds (DB's). The DB of a surface Ga (As) atom becomes *p*-like (*s*-like), which causes the transfer of the DB electron from Ga to As. Then, the empty p-like DB orbital of a surface Ga can be used to accommodate an electron excited from the $3d$ core of the same atom, which produces an atom-like, well-localized SCE state. Since the SCE is possible only on the surface, we can expect a very surface-sensitive signal from the photoemission process resonant with this SCE. Generally, the decay of a core exciton can proceed in two channels, i.e., direct recombination (DR) and Auger recombination (AR). In DR (AR), there remains, in addition to a photoelectron, a valence hole (two valence holes and an electron), which can be distinguished by the energy of the photoelectron. In the case of GaAs(110), it is shown that the DR process dominates, which indicates the well-localized character of SCE [1].

One of the peculiar features of this resonant emission is the strong polarization dependence: The emission is much stronger for the light polarization parallel to the Ga DB orbital than for the other cases [1], which cannot be ascribed to the difference in the optical absorption probability [2]. This puzzle has been solved by a careful consideration of various anisotropies involved in this photoemission process [2], namely [**a**] the anisotropic angular distribution of photoelectrons inside the crystal, [**b**] the polarization dependence of its anisotropy, due to the fixed direction of the Ga DB orbital, [**c**] surface refraction of photoelectrons which favors the emission of electrons running perpendicularly to the surface, and [**d**] the partially angle integrated measurement using a CMA (cylindrical mirror analyzer) with its aperture oriented in a particular direction.

A $3d$ core hole can take either $j = 3/2$ or $j = 5/2$ states, and hence the resonance occurs at the two core exciton energies originating from these core hole states [1]. The intensity ratio of these resonances is quite different, or even reversed, from the ratio of the statistical weights. A similar problem was encountered many years ago in the optical absorption of valence excitons, and was interpreted as due to the interplay of spin-orbit interaction (H_{so}) and electron-hole exchange interaction (H_{ex}) [3]. The application of this theory to our problem shows [4] that the mechanism of the intensity reversal works "more than once" in the case of photoemission, while it works just once in the absorption. Since this aspect is rather unknown, we will discuss it explicitly below.

Let us consider the case of p-polarization, for example. Among the $(3d)^9(4p_z)^1$ states, the following two are active for this polarization:

$$|z\rangle_{5/2} = \{cY_2^0\alpha + dY_2^1\beta\}Y_1^0\alpha + \{dY_2^{-1}\alpha + cY_2^0\beta\}Y_1^0\beta \ , \qquad (2.1)$$

$$|z\rangle_{3/2} = \{dY_2^0\alpha - cY_2^1\beta\}Y_1^0\alpha - \{cY_2^{-1}\alpha - dY_2^0\beta\}Y_1^0\beta \ , \qquad (2.2)$$

where $\{\varphi\}\chi$ stands for a Slater determinant with an electron excited from φ to χ, $\{Y_\ell^m\}$ are the spherical harmonics, and (α, β) denote spin states. The radial part of the orbitals are omitted, and $(c, d) = (\sqrt{3}, \sqrt{2})/\sqrt{10}$. These two states diagonalize H_{so}. For an optical transition from the ground state to $\{\varphi\}\chi$ to be allowed, φ and χ must have the same spin part. Thus, the ratio of the optical absorption to these states is $I_{5/2}/I_{3/2} = c^2/d^2 = 3/2$. When we further consider H_{ex}, the above two states mix with each other, and the resulting eigenstates take the forms

$$|p_1\rangle = a_1\{\alpha\}\alpha + b_1\{\beta\}\alpha + b_1\{\alpha\}\beta + a_1\{\beta\}\beta \ , \tag{2.3}$$

$$|p_2\rangle = a_2\{\alpha\}\alpha + b_2\{\beta\}\alpha + b_2\{\alpha\}\beta + a_2\{\beta\}\beta \ , \tag{2.4}$$

where only the spin parts are explicitly written. In this case, the intensity ratio of optical absorption is $I_1/I_2 = |a_1/a_2|^2$. Onodera and Toyozawa [3] showed the variation of this ratio as a function of the relative strength of H_{so} and H_{ex}, which explains well many of the observed spectra where the spin-orbit splitting is not much changed but their intensity ratio (I_1/I_2) is quite different, or even reversed, from the value 3/2.

For the resonant emission via these exciton states, we need to evaluate another matrix element, too. Suppose the exciton $\{\varphi\}\chi$ decays into a photoelectron and a valence hole, i.e., $\{\psi_v\}\psi_k$. The matrix element of this direct recombination is written as

$$\begin{aligned}\langle\{\chi\}\varphi|H_{DR}|\{\psi_v\}\psi_k\rangle &= \langle\varphi(1)\psi_v(2)|(e^2/r_{12})|\chi(1)\psi_k(2)\rangle \\ &\quad -\langle\varphi(1)\psi_v(2)|(e^2/r_{12})|\psi_k(1)\chi(2)\rangle \ ,\end{aligned} \tag{2.5}$$

where the integration contains the spin inner product. In the first term, the spin states must be the same for each pair of (φ, χ) and (ψ_v, ψ_k). Namely there appear the same factors (a_1 and a_2) as in the optical absorption from this term. On the other hand, the second term requires that the spin states of each pair of (φ, ψ_k) and (χ, ψ_v) should be equal. Thus, both factors a_j and b_j $(j = 1, 2)$ contribute to this term. The probability of the resonant emission process via a particular exciton state is proportional to $|\langle\{\psi_v\}\psi_k|H_{DR}|\{\chi\}\varphi\rangle\langle\{\chi\}\varphi|H_{eR}|0\rangle|^2$ (H_{eR} : electron radiation interaction). Therefore, if the first term alone in (2.5) were to contribute, it would lead to the intensity ratio $|a_1/a_2|^4$ for the resonant emissions via $|p_1\rangle$ and $|p_2\rangle$. This would mean that the intensity reversal, if it exists, should be further enhanced. In reality, however, there is a contribution from the second term of (2.5). Therefore, one should expect the intensity ratio between $|a_1/a_2|^2$ and $|a_1/a_2|^4$ in realistic cases.

In addition to the above calculations, we have studied the angle and spin resolved emission spectra for this resonant process, by considering the interference effect between the direct and resonant emissions of the valence electrons [4]. The results are compared with available experimental data on the angle resolved spectra. For certain emission angles, the agreement is reasonable, but for others not so much. It seems to indicate the inadequacy of neglecting the effect of surface diffraction and the band effect for the photoelectrons. There is no spin resolved measurement to be compared. The details of these studies will be published elsewhere [5].

3. Final State Screening by DB Electrons

Core electron spectroscopy is often used for the chemical analysis of the ground electronic state, since the energy of the emitted electron sensitively reflects the charge distribution in the ground state. Though the redistribution of the remaining charges in the photoionized state should also contribute to the energy of the photoelectron, it is usually considered to be of secondary importance (or to be not so sensitive to the chemical environment). As the second subject of this note, we want to propose that this effect of charge redistribution in the excited state might play an essential role in the interpretation of the $2p$ core level emission from Si(100). The ideal surface of Si(100) consists of a square lattice of topmost Si atoms, each of them having two DB's sticking out in the $\langle 11 \rangle$ and $\langle \bar{1}\bar{1} \rangle$ directions. It is widely accepted that, on the real surface, one of the DB's is used to form a dimer with the neighboring atom, which stabilizes the surface. Various arrangements of the dimers are thought to correspond to the observed superstructures such as 2×1, $p(2 \times 2)$, $c(2 \times 4)$ etc. As to the behavior of the remaining DB, it is controversial whether the buckling of the dimer further stabilizes the surface or not. The existence of the phase transition between 2×1 and $c(2 \times 4)$ phases at low temperature [6] seems to be favorable to the buckled dimer model. However, the occurrence of a single surface emission peak in the direct emission of $2p$ core electrons [7] seems to support the symmetric dimer model: If the dimer is buckled, the two Si atoms are not equivalent, and therefore one would expect two emission peaks corresponding to the processes which leave the core hole on either the raised or the lowered atom. It is about this argument that we have reservations, because we can show the possibility of a different interpretation of the experimental result. The point is the consideration of the final state interaction between the core hole and DB electrons: Since the DB electrons are more mobile than the electrons in chemical bonds, they would easily adjust their distribution to the core hole. Therefore, the consideration of this redistribution of DB electrons would possibly change the energy of photoelectron and thus the physical picture of core electron emission.

In order to test this idea, we have performed a Hartree-Fock (HF) calculation of the ground and excited states of a dimer as a function of its tilt angle, which represents the degree of the buckling [8, 9] We have used a simple cluster, Si_2H_4, and the basis set STO-3G [10] for Si orbitals, which might be rather crude to get a quantitatively accurate result. But, for a qualitative argument to look for the possibility of the new interpretation mentioned above, it is sufficient. For the core ionized states, we made UHF (unrestricted HF) calculations by keeping a $2p$ core state (of a given spin) vacant on one of the Si atoms of the dimer in each loop of the SC (self-consistent) procedure. We did not consider the spin-orbit interaction, and the $2p$ hole orbital was chosen perpendicular to the dimer direction. A remarkable result is that the state ($|A\rangle$) with a $2p$ core hole and (most of the) two DB electrons on the raised atom can be obtained as a SC solution, but not the state ($|B\rangle$) with a $2p$ core hole on the lowered atom and (most of the) two DB electrons on the raised atom. These two states should correspond to the two final states starting from the ground state of a buckled dimer according to the naive picture of a core emission. One can take the state $|B\rangle$ as an initial configuration of the SC calculation, but it converges, as a SC solution, to the state ($|C\rangle$) where the two DB electrons are separated on different Si atoms. (On the atom with the core hole, the spin of DB electron is opposite to that of the missing core electron.) This means that the state $|B\rangle$ is unstable against the deformation of the electronic orbitals. This character will be preserved even if one employs a better basis set for the electronic orbitals. The electric dipole matrix elements between the ground state of a buckled dimer and the excited states $|B\rangle$ and $|C\rangle$ (with a photoelectron in a plane wave state) are very different from one another. For the state $|C\rangle$, the change in the DB orbitals gives a small overlap integral with those in the ground state, leading to a considerable reduction of the dipole matrix element. The calculation of the emission intensities for all the SC solutions of this model actually shows that, even for a buckled dimer, one may expect a surface core emission spectrum consisting almost of a single peak [9]. A more refined calculation will be necessary to draw a more definite conclusion, but we have shown the possibility of reconciling the conflicting result of ref.7 with the buckled dimer model on Si(100) by considering the effect of final state interaction in the presence of a core hole.

Acknowledgments

I acknowledge the contributions of Mr. Yoshiyuki Miyamoto, Mr. Yasuhiko Takeda, and Mr. Tatsuji Yamazaki in their collaboration in the work mentioned in this note. I am grateful to Professor Akio Yoshimori for useful comments and for his support during this work.

References

[1] G. J. Lapeyre and J. Anderson: Phys. Rev. Letters **35** 117 (1975).

[2] K. Cho and Y. Miyamoto: Surf. Sci. **192** L835 (1987).

[3] Y. Onodera and Y. Toyozawa: J. Phys. Soc. Japan **22** 833 (1967).

[4] Y. Takeda: Master Thesis (Osaka University, 1989).

[5] Y. Takeda, Y. Miyamoto, and K. Cho: in preparation.

[6] T. Tabata, T. Aruga, and Y. Murata: Surf. Sci. **179** L63 (1986).

[7] D. H. Rich, T. Miller, and T.-C. Chiang: Phys. Rev. B**37**, 3124 (1988).

[8] T. Yamazaki: Master Thesis (Osaka University, 1990).

[9] T. Yamazaki and K. Cho: *Surface as a New Material* (Proc. 26th Yamada Conf., Osaka, 1990) to be published in Surf. Sci.

[10] R. Ditchfield, W. J. Hehre, and J. A. Pople: J. Chem. Phys. **54** 724 (1971).

Molecular-Dynamics Simulation of the Structural Phase Transition on the Si(100) Surface

S. Sawada

Fundamental Research Laboratories, NEC Corporation,
34 Miyukigaoka, Tsukuba, Ibaraki 305, Japan

Abstract. The phase transition of the Si(100) surface from the c(4×2) structure to the (2×1) disordered structure is simulated by molecular dynamics using a transferable semi-empirical tight-binding model, which reproduces very well the cohesive energies of various structured crystals and a Si_2 molecule obtained by ab initio calculations and the energy and the structure of the reconstructed (100) surface at zero temperature.

1. Introduction

There has been considerable interest in the structure of the Si (100) surface. Recent low-energy electron diffraction (LEED) experiments have shown c(4×2) patterns at temperatures lower than about 200 K and (2×1) patterns at higher temperatures [1]. This phase transition has been predicted theoretically by using the effective spin model before experimental observation [2,3]. In recent scanning tunneling microscopy (STM) experiments of Si(100) at room temperature, only symmetric dimers have been observed far from surface defects and edges [4]. This can be understood as follows: Only the average of the thermal motion of the dimers can be observed as symmetric dimers.

In this paper, we present molecular dynamics simulations of this phase transition using a transferable semi-empirical tight-binding model for Si, which reproduces very well the energies of a Si_2 molecule and various crystal structures obtained by ab initio calculations and the energy and the structure of the (100) reconstructed surface. The simulations show the phase transition from the c(4×2) structure to the disordered (2×1) structure around 130 K.

2. Model Hamiltonian

The total energy of the system is given as a sum of a band-structure energy and a repulsive energy. The band-structure energy is obtained by diagonalizing the tight-binding Hamiltonian

$$H_{tb} = \sum_{i,\alpha} \varepsilon^0_\alpha a^\dagger_{i\alpha} a_{i\alpha} + \sum_{i,\alpha} \sum_{j,\beta} t_{i\alpha,j\beta} (a^\dagger_{i\alpha} a_{j\beta} + \text{h.c.}), \quad (2.1)$$

where ε^0_α are energy levels of isolated atoms and $t_{i\alpha,j\beta}$ parametrized hopping integrals. The latter are assumed to have the following form:

$$t_{i\alpha,j\beta} = \eta_{\alpha\beta} \cdot S(r_{ij}) \cdot r_{ij}^{-\nu} \quad (2.2)$$

with

$$S(r) = 1 / (1 + \exp(-\mu(r-R_c)), \tag{2.3}$$

where r_{ij} is the distance between the i-th and j-th atoms and the parameters ν, μ and R_c are dependent on the orbitals α and β.

The repulsive energy is assumed to have the form

$$E_{rep} = \frac{1}{2} \cdot \sum_i \sum_j A_{ij} \cdot S(r_{ij}) \cdot r^{-\nu}_{ij}, \tag{2.4}$$

where S is given by (2.3). The coefficient A_{ij} is assumed to be linearly dependent on the effective coordination numbers of both the i-th and j-th atoms, Z_i and Z_j:

$$A_{ij} = a_0 - a_1 (Z_i + Z_j)/2, \tag{2.5}$$

where a_0 and a_1 are positive constants. The effective coordination number of the i-th atom Z_i is defined by

$$Z_i = \sum_j \exp(-\lambda_1 (r_{ij} - R_i)^2) \tag{2.6}$$

with

$$R_i = \sum_j r_{ij} \cdot \exp(-\lambda_2 r_{ij}) / \sum_j \exp(-\lambda_2 r_{ij}), \tag{2.7}$$

where λ_1 and λ_2 are constants. In the limit of $\lambda_2 \to \infty$, R_i tends to the minimum interatomic distance between the i-th atom and its neighbors. A not infinite but sufficiently large value of λ_2 is chosen so that derivatives of potential, i.e. forces, are continuous as atoms move. It should be noted that each term in (2.4) includes not only two-body interactions but also multi-body interactions because of the coefficients dependent on the effective coordination numbers. It is not unreasonable since the repulsive energy part can be considered to contain core-core interactions minus doubly counted electron-electron Coulomb interactions plus exchange interactions and so on, which are multi-body interactions in general. This idea is similar to that of Tomanek and Schluter [5].

The values of the parameters in this model are chosen to reproduce the properties of silicon known from experiments and ab initio calculations [6,7]. We fit, not energies, but rather those differences among the various structures in the bulk since the calculated values of the latter must be more accurate than those of the former. (The value of the cohesive energy of the diamond structure obtained by ab initio calculation is 4.84 eV [6] while its experimental value is 4.63 eV [8].) The resultant values of the parameters are summarized in Table I. The choice of the values for the exponents of the hopping integrals ν has been suggested by Robertson's work [9], in which the pressure-dependent band structure calculated by the pseudopotential method was compared with those calculated by the sp^3s^* tight-binding model [10].

Table I. The values of the parameters

	Hopping integrals				Repulsive
	ssσ	spσ	ppσ	ppπ	potential
ν	4	3	2	2	5
μ (Å^{-1})	5.96	5.96	2.55	2.55	2.55
R_c(Å)	3.17	3.17	3.83	3.83	3.83
η (eV·Å$^{\nu}$)	-63.9	27.7	13.1	-2.94	--

λ_1 = 1.086 Å^{-2} a_0 = 314.5 eV·Å^5 a_1 = 10.42 eV·Å^5

λ_2 = 8.511 Å^{-1}

The calculated binding curve for a Si_2 molecule is shown (the solid line) with the ab initio result (the broken line) [7] in Fig.1 (a). The cohesive energies of the bulk in the various structures calculated by the recursion method [11] are shown in Fig.1 (b) as a function of normalized atomic volume. The axial ratio in ß-tin (Tin) structure is assumed to be 0.5516. The interactions are taken into account up to the second neighbors in diamond, FCC, BCC and SC structures. On the other hand, the third and forth neighbor interactions are included in Tin structure. The result for the bulk modulus is 8.35×10^{11} erg·cm^{-3}. (The experimental value is 9.88×10^{11} erg·cm^{-3} [8]). The values of the hopping integrals for the first and second neighbor are almost equal to those given by Pandey and Phillips [12], which reproduce well the valence band and the bottom of

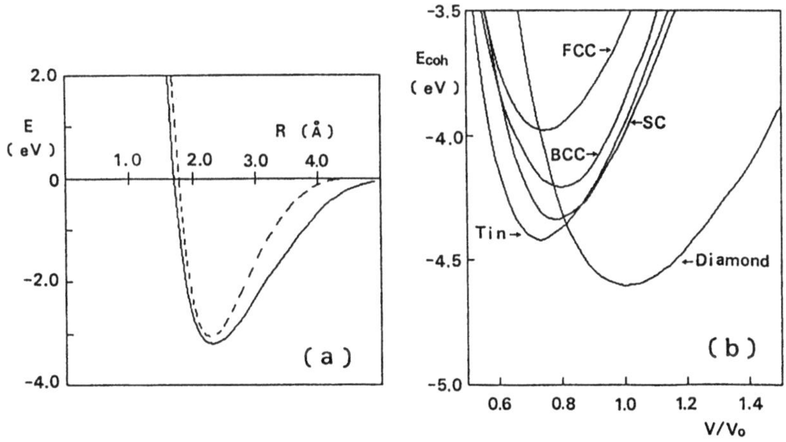

Fig.1. (a) Binding curves (energy in eV vs.interatomic distance in Å) for Si_2: The present result (solid line) and the ab initio result (broken line) after Refs. 5 and 7. (b) Cohesive energies (in eV) of the bulk in the various structures as functions of normalized atomic volume.

the conduction band calculated by the empirical pseudopotential method [13].

3. Simulations of the Phase Transition

For simplicity, we deal instead of with semi-infinite crystals with clusters which are parts of semi-infinite crystals. We use the following trick to suppress edge effects: For edge atoms, only the sp^3-hybrid orbitals toward other atoms in a cluster,

$$|h\rangle = (|s\rangle + \sqrt{3}\ |p\rangle) \ / \ 2, \qquad (3.1)$$

are used as basis orbitals while the hybrid orbitals toward the rest of the atoms of the semi-infinite crystal are eliminated. We solve the following Newton equations for atomic cores:

$$m\cdot\partial\, v_i/\partial t = -\ 2\cdot \sum_{\alpha}^{occ} \sum_{j,\beta}^{occ} P_{i\alpha,j\beta}\cdot\partial t_{i\alpha,j\beta}/\partial r_i \ -\ \partial E_{rep}/\partial r_i, \qquad (3.2)$$

where the first terms on the right-hand side denote the Hellmann-Feynman forces and $P_{i\alpha,j\beta}$ is a bond order defined by

$$P_{i\alpha,j\beta} = \langle\ a^{\dagger}_{i\alpha}\ a_{j\beta}\ \rangle. \qquad (3.3)$$

The second term on the right-hand side of (3.2) denotes a repulsive force. Values of the bond orders are calculated by diagonalizing the tight-binding Hamiltonian (2.1) at each time step. While the edge atoms are fixed, for their nearest neighbor atoms the Langevin equations,

$$m\cdot\partial\, v_i/\partial t = -\ 2\cdot \sum_{\alpha}^{occ} \sum_{j,\beta}^{occ} P_{i\alpha,j\beta}\cdot\partial t_{i\alpha,j\beta}/\partial r_i \ -\ \partial E_{rep}/\partial r_i$$

$$-\gamma\cdot v_i + F_i, \qquad (3.4)$$

are solved in order to include the effect of the rest of the crystal as a heat reservoir. The friction coefficient γ is related to the Debye frequency ω_D as $\gamma = \omega_D/6$ [14,15]. The F_i is a random force acting on the i-th atom obeying the following relationship obtained from the fluctuation-dissipation theorem:

$$\langle F_i^{\alpha}(t)\ F_j^{\beta}(t+\tau)\rangle = 2\ \gamma\ k_B\ T\ m\ \delta(\tau)\ \delta_{ij}\ \delta_{\alpha\beta}, \qquad (3.5)$$

where α and β refer to the x, y and z components of the random force and i, j refer to two different atoms. The k_B is the Boltzmann constant and T is the temperature.

In numerical calculations, we deal with a cluster consisting of 4 layers (24, 30, 25 and 30 atoms), 61 movable atoms and 48 fixed edge atoms. First, to obtain the surface structures corresponding to local or absolute minima of the potential energy, we anneal the system with adequate initial conditions for atomic co-ordinates and velocities, i.e. we solve eq. (3.2)-(3.4) with T = 0. The edge atoms are assumed to be fixed at the bulk equilibrium position.

We have found four different reconstruction configurations, the (2×1), p(2×2), c(4×2) and (4×1) structures, to correspond to

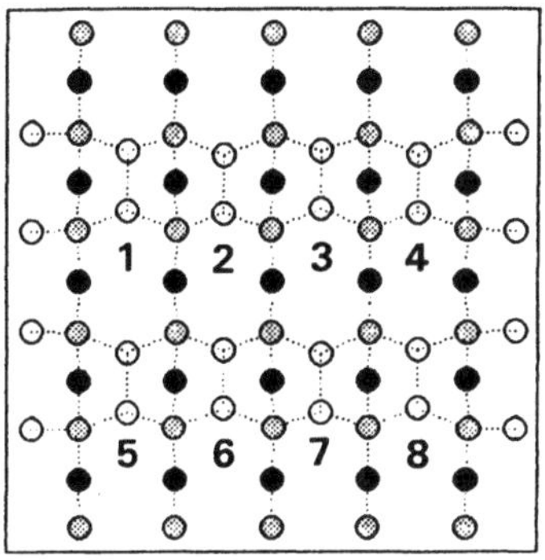

Fig.2. The c(4×2) reconstructed structure of the 109-atom cluster. White, gray and black circles show 1st-, 2nd- and 3rd-layer atoms, respectively. 4th layer atoms are not shown. The dimers are numbered for reference in Fig.3.

local energy minima having lower energies by 2.171, 2.186, 2.186 and 2.172 eV per dimer, respectively, than the ideal (1×1) structure. These values are comparable to the ab initio result of the energy gain of the (2×1) reconstructed surface, 1.7 eV per dimer [16] though the energy differences among the different structures are somewhat smaller than the ab initio results [17]. In these structures, all dimers are asymmetric. We have found that symmetric dimers are not stable. Only the c(4×2) structure is illustrated in Fig.2. In this structure, the averaged values of atomic displacements in the dimers from the ideal bulk positions are $\delta y_1 = 0.86$, $\delta z_1 = -0.16$, $\delta y_2 = -0.70$ and $\delta z_2 = -0.52$ (in Å), where the subscripts 1 and 2 denote up and down atoms, respectively. These results are qualitatively in good agreement with the ab initio results [16,17].

Next we simulate the phase transition from c(4×2) to (2×1). Starting with the c(4×2) structure, we solve the eq. (3.2)-(3.5) with finite values of T. In these calculations, the edge atoms in the surface are displaced from the ideal bulk positions by the same amount as the above calculated displacements of the atoms in the dimers and then fixed, so that all the surface atoms are arranged in the c(4×2) structure at T = 0. We show in Fig.3 (a), (b) and (c) results for the time evolutions of the differences in the z-coordinates between the up and down atoms in each dimer, Δ. The value of Δ changes its sign when a dimer flips. At T = 100 K, no dimers flip, as shown in Fig.3 (a), so the surface is in the c(4×2) ordered phase. At T = 130K, the dimers begin to flip as shown in Fig.3 (b). It has been found that as the temperature increases above 130 K, the dimers flip more frequently, where the surface is in the (2×1) disordered phase. The time evolutions of Δ for T = 200 K are shown in Fig.3 (c). Consequently, the transition temperature is estimated to be about 130 K. The periods of the flipping of the dimers are too short to be observed in the STM experiment so that only the thermal averages can be observed.

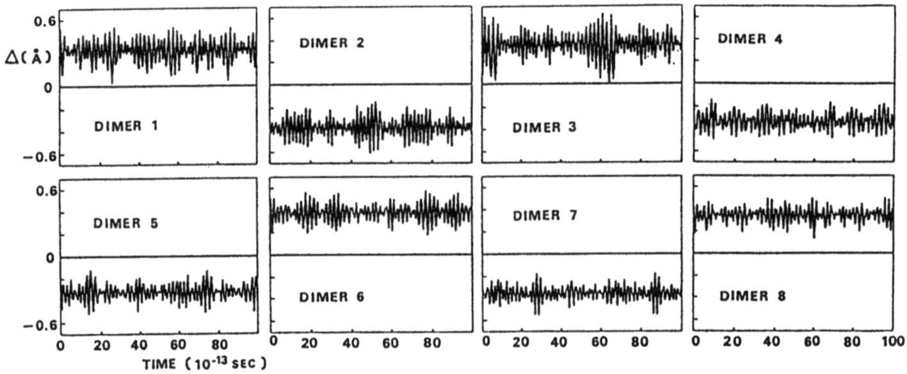

Fig.3. (a) The time evolutions of the differences in the z-coordinates between the up and down atoms in each dimer at the temperature T = 100 K; Time in 10^{-13}sec and distance in Å.

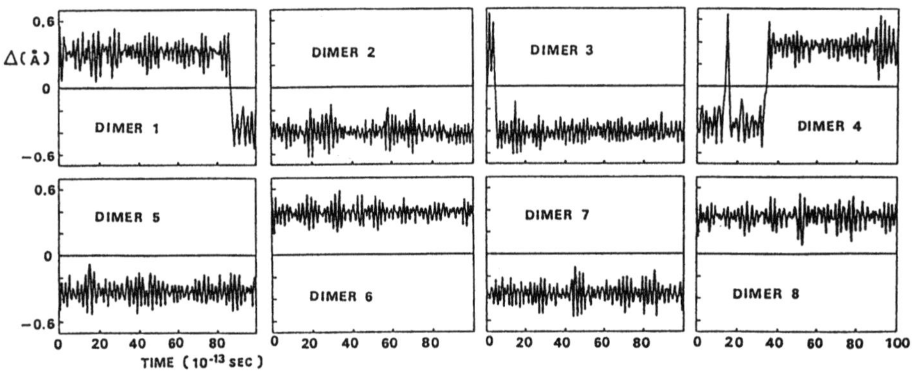

Fig.3. (b) The same as Fig.3(a) except the temperature T = 130K.

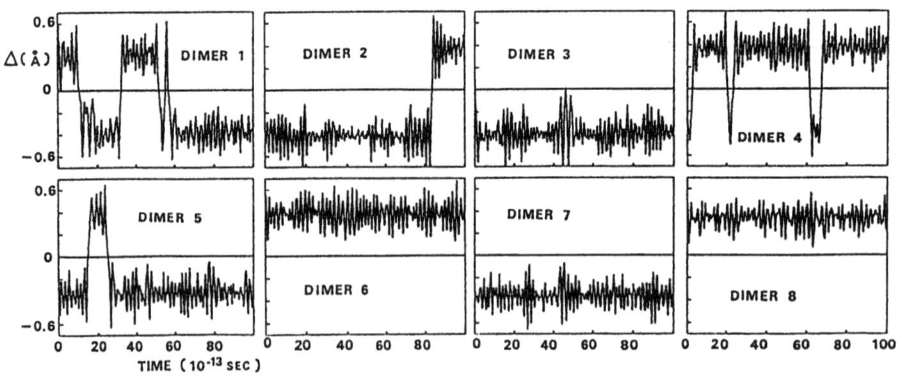

Fig.3. (c) The same as Fig.3(a) except the temperature T = 200K.

4. Concluding Remarks

In this paper, we have presented simulations of the phase transition of the Si(100) surface from the c(4×2) structure to the (2×1) disordered structure by molecular dynamics using a transferable semi-empirical tight-binding model. The estimated transition temperature is about 130 K, which is qualitatively in good agreement with the LEED experiment [1]. Simulations of larger systems and longer runs are needed to define clearly the transition temperature and to examine how the motions of dimers correlate.

In the present model, the main source of the stabilization of the c(4×2) structure is the repulsive interaction since distances between atoms in one of the dimers and those in its neighboring dimer are larger in the (4×2) structure than in the (2×1) and (4×1) structures. There is another factor of the stabilization, i.e. a dipole-dipole interaction between dimers since dimers are polarized due to buckling. Kakitani and Yoshimori have taken it into account in their effective spin model to simulate this phase transition by the Monte Carlo method [18]. This effect can be included by incorporating the intra- and inter-atomic electron Coulomb interactions into the present tight-binding model. In that model, we can expect a higher transition temperature, which may be closer to the experimental value. This work is now in progress.

Acknowledgments

The author would like to acknowledge Dr. S. Saito for his valuable and helpful discussions and comments.

References

1. T. Tabata, T. Aruga and Y. Murata, Surf. Sci. 179, L63 (1986).
2. J. Ihm, D. H. Lee, J. D. Joannopoulos and J. J. Xiong, Phys. Rev. Lett. 51, 1872 (1983).
3. A. Saxena, E. T. Gawlinski and J. D. Gunton, Surf. Sci. 160, 618 (1985).
4. R.J. Hamers, R.M. Tromp and J.E. Demuth, Phys. Rev. B 34, 5343 (1986)
5. D.Tománek and M.A. Schlüter, Phy. Rev. Lett. 56, 1055 (1986); Phys. Rev. B 36, 1208 (1987);
6. M.T. Yin and M.L. Cohen, Phys. Rev. Lett. 45, 1004 (1980), Phys. Rev. B 26, 5668 (1982).
7. K. Raghavachari, J. Chem. Phys. 84, 5672 (1986).
8. C. Kittel, Introduction to Solid State Physics, 6th ed., Wiley, New York (1986).
9. J. Robertson, Phil. Mag. B 47, L33 (1983)
10. P. Vogl, H.P. Hjalmarson and J.D. Dow, J. Phys. Chem. Solids 44, 365 (1983).
11. R. Haydok, Solid State Physics 35, p. 215. Academic, New York (1980).
12. K.C. Pandey and J.C. Phillips, Solid State Commun. 14, 439 (1974); Phys. Rev. Lett. 32, 1433 (1974); Phys. Rev. B. 13, 750 (1976).
13. J.R. Chelikowsky and M.L. Cohen, Phys. Rev. B 14, 556 (1976)

14. S.A. Adelman and J.D. Doll, J. Chem. Phys. 64, 2375 (1975); Acc. Chem. Res. 10, 378 (1977), S.A. Adelman and B.J. Garrison, J. Chem. Phys. 65, 3751 (1976); B.J. Garrison and S.A. Adelman, Surf. Sci. 66, 253 (1977).
15. M. Shugard, J.C. Tully and A. Nitzan, J. Chem. Phys. 71, 1630 (1979); J.C. Tully, ibid. 73, 1975 (1980).
16. M.T. Yin and M.L. Cohen, Phys. Rev. B 24, 2303 (1981).
17. Z. Zhu, N. Shima and M. Tsukada, Phys. Rev. B40, 11868 (1989).
18. K. Kakitani and A. Yoshimori, in this issue.

Short Range Order in the Phase Transition of the Si(100) Surface Reconstruction

K. Kakitani and A. Yoshimori

Department of Material Physics, Faculty of Engineering Science, Osaka University, Toyonaka, Osaka 560, Japan

Abstract. Analysis of the phase transition of the Si(100) clean-surface reconstruction is made by computer simulation. The asymmetric dimer is described by an Ising spin and interactions short ranged as well as dipolar are assumed between them. The computer simulation is prepared to incorporate the longer ranged dipolar interactions. The anisotropy in the short range order above the transition point, which is observed experimentally, is examined in detail in the above mentioned model.

1. Introduction

The Si(100) surface undergoes the phase transition of surface reconstruction at about 200K [1] from the low temperature c(4×2) phase to the high temperature (2×1) one [2]. It is now believed [3] that the low temperature c(4×2) structure consists of the ordered asymmetric dimers, whose ordered structure has been proposed by Yin and Cohen [4]. Each asymmetric dimer has two equivalent configurations, which is specified by an Ising spin ($S = \pm 1$), so that the system can be described by an interacting Ising system [5].

The short range order effect is one of the focuses of research on this surface. According to LEED by Kubota and Murata [6], at 400k, this surface has already dominant short range order in each dimer row without any correlation of order between dimer rows. This situation continues down to 300K, and, at some temperature between 300K and 200K, the short range order between dimer rows sets in toward Yin-Cohen's structure below 200K. This short range order affects strongly the electronic structure of the surface dangling bond state, which is observed by ARUPS [7] and discussed theoretically [8].

In this paper, this phase transition is analyzed by computer simulation for the Ising spin system from the asymmetric dimer model, with particular emphasis put on the short range order. The two-spin interactions, short ranged as well as dipolar, are assumed. Since the asymmetric dimer has an electric dipole moment and most of the low total energy configurations of this reconstructed surface have semiconductive surface electronic states [3] as well as the bulk states of the semiconductor, the electric dipolar interaction is not screened in a metallic form.

This paper is organized as follows. In the next section, our model is introduced and the inclusion of dipolar interaction in our computer simulation is mentioned. With the two sets of parameters, by which the observed short range order can be explained, detailed analyses are made of the two spin correlation function in section 3. Discussions are given in section 4.

2. Model and Computer Simulation

On the Si(100) reconstructed surface, the asymmetric dimers form a rectangular lattice, whose lattice constants are 2a and a. As is mentioned in the introduction, the two equivalent configurations of the ℓ-th asymmetric dimer are expressed by an Ising spin $S_\ell (= \pm 1)$.

The two-spin interactions are assumed and the Hamiltonian, H, is given by

$$H = \sum_{\ell,\ell'} J_{\ell\ell'} S_\ell S_{\ell'}, \tag{1}$$

where $J_{\ell\ell'}$ consist of the two parts, that is, the short range ones, $J^s_{\ell\ell'}$ and the dipolar one, $J^d_{\ell\ell'}$. For the short range one, the two kinds of interaction, J_x and J_y are assumed in this paper as shown in Fig.1.

The x-axis is taken perpendicular to the dimer row. The dipolar one, $J^d_{\ell\ell'}$, is simply assumed to be given by

$$J^d_{\ell\ell'} = \frac{\mu^2}{\mid \mathbf{R} \mid^3_{\ell\ell'}} \left[1 - \frac{3X^2_{\ell\ell'}}{\mid \mathbf{R}_{\ell\ell'} \mid^2} \right], \tag{2}$$

where μ is the effective electric dipole moment of the asymmetric dimer, $\mathbf{R}_{\ell\ell'}$ the distance vector between the ℓ and ℓ'-th sites and $X_{\ell\ell'}$ its x-component.

The dipolar interaction itself brings a very strong anisotropy in the short range order in this system. The Fourier transform of $J^d_{\ell\ell'}$, $J^d(\mathbf{k})$ ($\mathbf{k}$ is wavenumber), is calculated and its values at wavenumbers of $\bar{\Gamma}$,

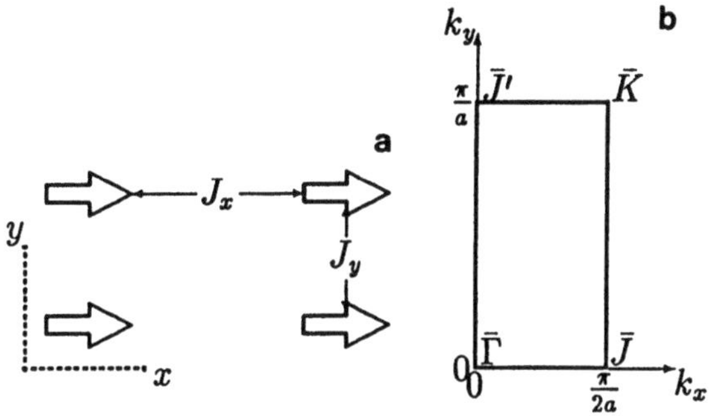

Fig.1 (a) The two kinds of interaction in this model (b) The surface Brillouin zone of Si(100)2 × 1

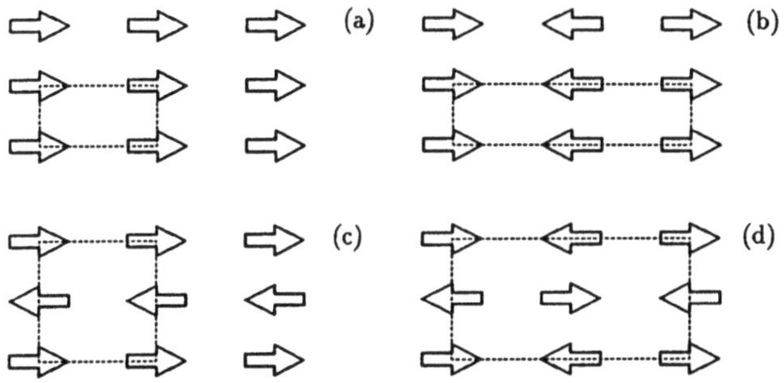

Fig.2 (a) p(2 × 1), (b) p(4 × 1), (c) p(2 × 2), (d) c(4 × 2)

$\bar{J}$, $\bar{J}'$ and $\bar{K}$ in the surface Brillouin zone (Fig.1), corresponding to the energies of the 2×1, 4×1, 2×2 and c(4×2) structure, are given by 1, 4.25, −2.48 and −2.26 in unit of $J^d(0)$, respectively. Those four structures are shown in Fig.2. From these values, we see very strong antiferromagnetic interaction in the dimer row, which favors the $+-+-$ structure in the row, and the interaction between the antiferromagnetic dimer rows is weak but favors the 2×2 structure.

When the short range interactions dominate, the c(4×2) structure becomes stable for $J_y, J_x < 0$. Where the dipolar interaction is appreciable, we have to have, at least, a $J_x(< 0)$ sufficient to make the c(4×2) structure more stable than the 2×2 structure.

The computer simulation is made by the Metropolis sampling. When the dipolar interaction is involved, an energy calculation needs some caution, because of the longer ranged nature of the interaction. With the periodic boundary condition, we take into account enough n×n systems surrounding the central n×n system in order to have the dipolar energies convergent. Most of the analyses are made in the system of 20×20.

3. Analyses of the Two Spin Correlation Function

The two spin correlation function in real space and its Fourier transform are analyzed above the transition temperature, T_t, in this section. This would reflect on the diffuse streak in LEED, though the actual analyses of the LEED intensity of the diffuse streak will be difficult because of the necessity of the dynamical LEED theory. Here qualitative features obtained from the diffuse LEED results [6] are tried, to find from the computer simulation that the antiferromagnetic correlation within the dimer rows remains beyond $T/T_t = 2$ and becomes very small at $T/T_t = 2.5$ and that the correlation between the dimer rows takes place below $T/T_t = 1.5$.

As mentioned in the previous section, we are going to examine mainly two sets of the interaction parameters in order to examine the effects due to the dipolar interaction in the two spin correlation function; one is the (s) set that $J_y = -1$, $J_x = -0.05$ and $\mu = 0$, and the other is the (d) set that $J_y = 0$, $J_x = -0.075$ and $\mu^2/a^3 = 1$, which is determined in such a way that

$$\frac{(J^{(d)}(\bar{\Gamma}) - J^{(d)}(\bar{K}))}{(J^{(s)}(\bar{\Gamma}) - J^{(s)}(\bar{K}))} = \frac{(J^{(d)}(\bar{\Gamma}) - J^{(d)}(\bar{J}'))}{(J^{(s)}(\bar{\Gamma}) - J^{(s)}(\bar{J}'))},$$

where $J^{(d)}(\bar{\Gamma})$ etc. are values of the Fourier transform of the interaction for the two set (s) and (d) at the symmetry points in the surface Brillouin zone. The values of T_t for those two sets are obtained from the specific heat peak. T_t are 0.76 for the (s) set and 0.64 for the (d) set in the unit of $k_B = 1$.

The two spin correlation function, $\langle S_\ell S_{\ell'} \rangle$, in real space and its Fourier transform, $F(\mathbf{k})$ are obtained at several temperatures above T_t by computer simulation for the (s) and (d) sets. The short range order parameters, σ_y and σ_x, which are defined by the absolute value of the two spin correlation function for the nearest neighbor pair along the dimer row and perpendicular to that, are obtained as functions of temperature.

When we have well developed correlation between spins, $\mathbf{k}$-dependence of $F(\mathbf{k})$ may be described approximately by the Lorentzian as

$$F(\mathbf{k}) = C/(\varepsilon q_x^2 + q_y^2 + \xi^2) \quad , \qquad \mathbf{q} = \mathbf{k} - \mathbf{k}_m, \qquad (3)$$

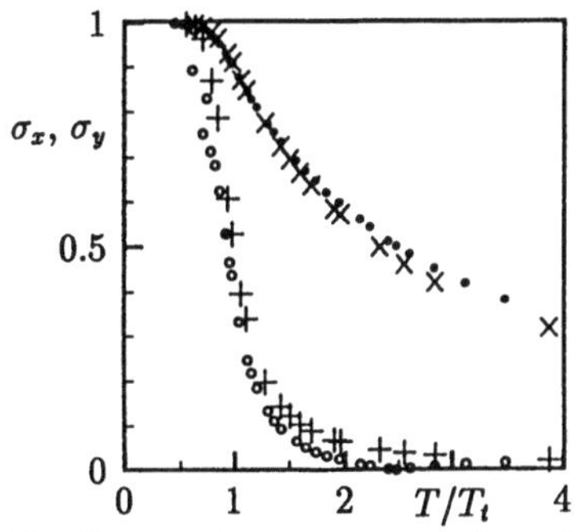

Fig.3

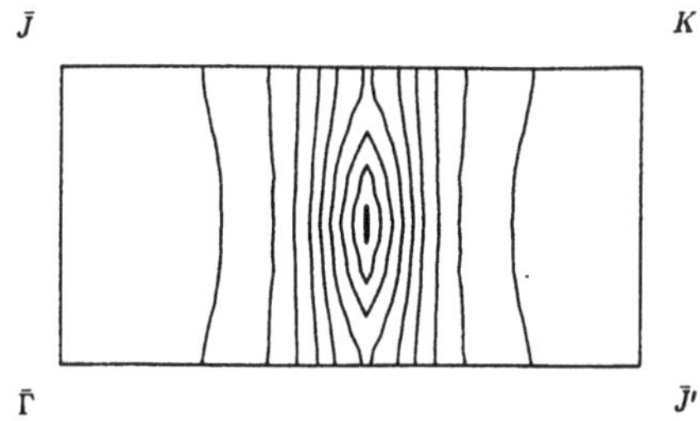

Fig.4

Fig.3 $\circ$ and $\bullet$ are σ_x and σ_y in (d) set and $+$ and $\times$ are those in (s) set.

Fig.4 The contour map of $F(\mathbf{k})$ at $T/T_t = 2$ ((d) set).

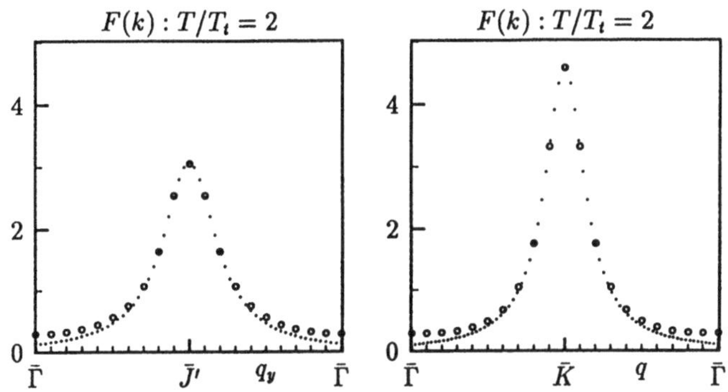

Fig.5 The lorentzian fitting for $F(\mathbf{k})$ at $T/T_t = 2.0$((d) set).

where principal axes of anisotropy are assumed to be along the x- and y-axes. In the present case, $\mathbf{k}_m = (\pi/(2a), \pi/a)$ and the principal axes of anisotropy are indeed along the x- and y-axes. From the expression, we can obtain the half width defined by the half peak value.

The temperature dependence of σ_x and σ_y is shown in Fig.3. The contour map of $F(\mathbf{k})$ and its Lorentzian fit are shown in Fig.4 and 5, respectively, and the half widths in $\mathbf{k}$-space along q_x and q_y directions are shown in Fig.6(a) as functions of temperature.

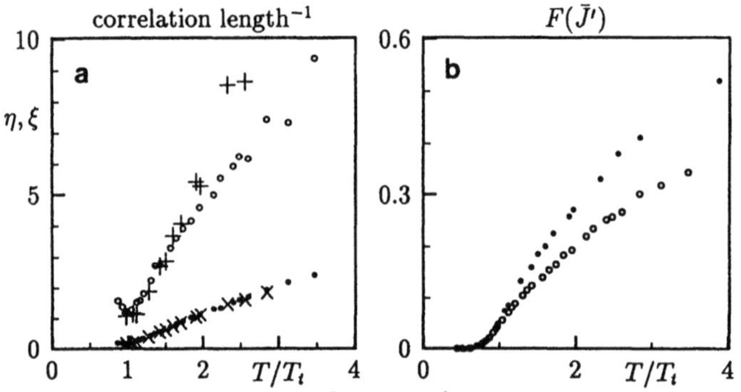

Fig.6 (a) $\circ$ and $\bullet$ are η^{-1} and ξ^{-1} in (d) set and $+$ and $\times$ are those in (s) set. (b) $\circ$ is F(**k**) on $\bar{J}$ in (d) set and $\bullet$ is one in (s) set.

4. Discussion

The T_t value for the (s) set can be estimated as 0.74 in the unit of $J_y = 1$ and $k_B = 1$ from the exact expression in terms of J_x and J_y, and it should be compared with the value, 0.76, from the specific heat peak by the computer simulation. From this we see the degree of approximation of our computer simulation. Generally speaking, the nature of the short range order is not very different between the two sets in the quantities calculated in the previous section, though some difference is seen. Another aspect to note in both sets examined is the degree of short range order in the dimer row. We have $\sigma_x \approx 0$ and $\sigma_y \approx 0.6$ at $T/T_t = 2$ in both sets. If one requires higher σ_y values at $T/T_t = 2$, then one should have higher anisotropy in the interaction. This means a further smaller ratio of J_x to J_y in the (s) set and a further smaller difference between $J^d(\bar{J}') - J^d(\bar{K})$ and $2J_x$. It would be difficult to say which would be likely, but the (d)-like case might be more likely.

The most prominent difference between the (s) and (d) sets is the very high energy of the 4×1 structure in the (d) set. It comes from the nature of the dipolar interaction as is seen from the value, 4.25, of $J^d(\bar{J})$ in the unit of $J^d(0)$. This difference does not seem to give a very large difference in the properties of the two spin correlation function, though some can be clearly attributed to it like the values of $F(\bar{J})$ shown in Fig.6(b)

Acknowledgments

This work was supported by a Grant in Aid for Scientific Research on Priority Areas of the Ministry of Education, Science and Culture of Japan. The authors would like to express many thanks to K. Cho, K. Makoshi, Y. Ohfuchi and other members of their group for discussions. Numerical calculations were done on the SX-2 system at the Computer Center, Osaka University.

Reference

[1] T.Tabata, T.Aruga and Y.Murata, Surf. Sci. Lett. **179**, L63(1987).

[2] R.E.Schlier and H.E.Farnsworth, J. Chem. Phys. **30**, 917(1959).

[3] Z.Zhu, N.Shima and M.Tsukada, Phys. Rev. B**40**, 11868(1989).

[4] M.T.Yin and M.L.Cohen, Phys. Rev. B**24**, 2303(1981).

[5] J.Ihm, D.H.Lee, J.D.Joannopoulos and J.J.Xiong, Phys. Rev. Lett. **51**, 1872(1983).

[6] M.Kubota and Y.Murata, Phys. Rev. B15 to be submitted.

[7] Y.Enta, S.Suzuki, S.Kono and T.Sakamoto, J. Phys. Soc. Jpn **59**, 657(1990).

[8] K.Inoue, M.Nakayama and H.Kawai, Surf. Sci. to be published.

Mechanism of Reconstruction of Si(100) Surfaces

M. Tsukada[1], *N. Shima*[1,*], *Z. Zhu*[1,**], *K. Kobayashi*[1], *H. Kageshima*[1], *and N. Isshiki*[2]

[1]Department of Physics, Faculty of Science,
University of Tokyo, Hongo 7-3-1, Bunkyo-ku, Tokyo 113, Japan
[2]Institute for Knowledge and Intelligence Science,
Kao Corporation, Bunka 2-1-3, Sumida-ku, Tokyo 131, Japan
*Present address: Department of Material Science, Faculty of Science,
Himeji Institute of Technology, Shosha 2167, Himeji 671-22, Japan
**Present address: Department of Physics, Fudan University,
Shanghai 200433, People's Republic of China

Electronic structures of Si(100) buckled 2×1, p(2×2) and c(4×2) surfaces are revealed by first-principles LDA calculation. The c(4×2) surface is found to be the most stable structure. The mechanism of the stabilization for this surface is different from the electrostatic dipole-dipole interaction of the asymmetric dimers. The dispersion of the surface band compares fairly well with the ARUPS observation. The theoretical simulations of STM image for Si(100) surface is described.

1. Introduction

Recently many experiments have been performed on the reconstructed Si(100) surfaces, but there are still controversial problems [1-12]. This comes from the fact that these surfaces include inherent defect structures and their concentration depends on the preparation conditions. That the essential structure ingredients of the Si(100) surface are the dimers as shown in Fig.1 had been proposed by the

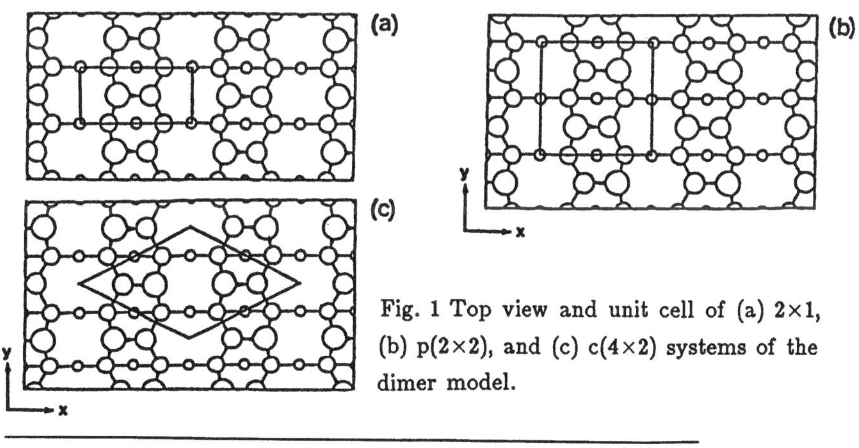

Fig. 1 Top view and unit cell of (a) 2×1, (b) p(2×2), and (c) c(4×2) systems of the dimer model.

pioneering work by Schlier and Farnsworth [1] as early as 1959. But the real establishment of this model was not until the observation of the STM image by Hamers *et al.* in 1985 [7,8]. Until now it has remained an open problem whether the most stable structure of this surface is the symmetric dimer structure observed by STM or the asymmetric dimer structure, which is supported by many other experiments like LEED, ARUPS [10-12], and so on. It is thus very important to clarify the stabilization mechanism and calculate the total energy of various reconstructed structures based on the first-principles electronic state theory. In the present paper these are discussed based on the calculations of the electronic states with a local density functional (LDA) approach by the present authors [13].

2. Models and the method of calculation

The configurations of the asymmetric dimers we have chosen for the calculations are the 2×1, p(2×2), c(4×2) structures as shown in Fig.1. Though the 2×1 structure has not been observed in the STM images, the electronic structure of this model is calculated to provide a reference for comparison with other models. First, the results of the calculation using the dimer structure as shown in Fig.2 are discussed in section 3 and section 4. In this model, the structure of the deeper part is fixed as in the bulk (this model is called the unoptimized model hereafter). We also make calculations for the completely optimized structures. This optimization can be performed by the potential energy minimization method. The electronic structure is calculated by the norm-conserving pseudopotential method by Bachelet *et al.* [14] The number of layers in a single repeated slab is ten including the dimer layers on the both sides. The cutoff energy of the plane wave basis functions is 5.8 Ry.

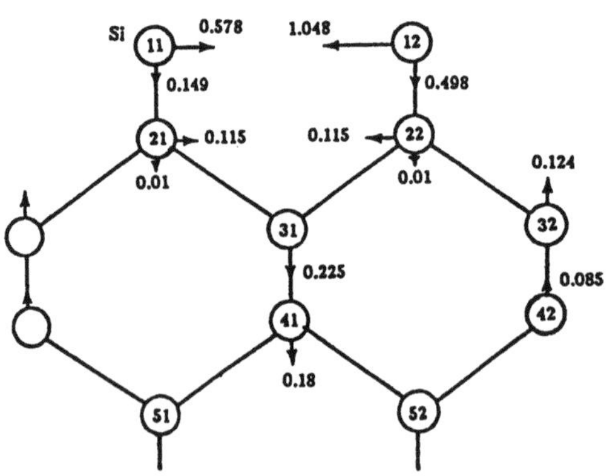

Fig. 2 Geometry of a single dimer used for comparing calculations of 2×1, p(2×2), and c(4×2) systems; units are Å.

3. The energy dispersions of the surface states

For the unoptimized models the surface band structures show semiconducting character only for the c(4×2) structure. A semimetallic feature is found for the 2×1 and p(2×2) structures which is not consistent with the UPS measurements. The band gap of the c(4×2) structure is very small. These features are not unexpected, because the energy gap of semiconductors tends to be calculated too small by the LDA formalism as compared with experiments. However, it should be remarked that the gap opens only for the case c(4×2). This fact is closely related to the unique stabilization mechanism of this surface as shown below.

How well does the theoretical surface band dispersions compare with ones observed by angle resolved ultraviolet photoemission spectroscopy (ARUPS)? There are two domains in the Si(100) surface; the dimer rows in one type of domain run perpendicular to the rows in the other domain. There remains some ambiguity when extracting a unique conclusion from the double domain surface. However, recently single domain surfaces were prepared and careful ARUPS experiments have been performed by Kono's group [10,11]. The experiment clearly showed the existence of the two branches of the surface state bands in the unit cell. This indicates the surface structure including two asymmetric dimers in the unit cell. The experimental dispersion of the Si(100) 2×1 surface plotted in Fig.3 compares fairly well with the calculated dispersions for the p(2×2) or c(4×2) surfaces. Therefore one might speculate the following model for the Si(100) 2×1 surface.

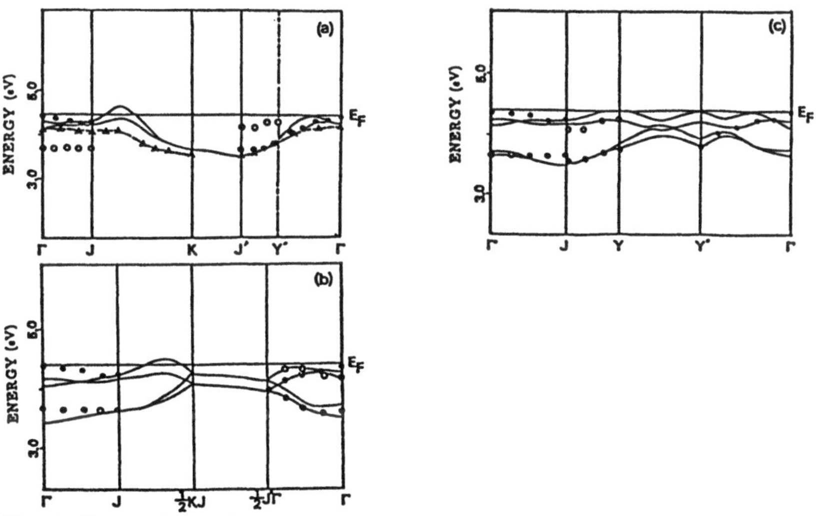

Fig. 3. Comparison of energy dispersion with ARUPS experiments (a) in a 2×1 BZ, (b) in a p(2×2) BZ, and (c) in a c(4×2) BZ. Solid triangles represent experimental data from Ref. 12. Solid and open circles represent the data from Ref. 11.

Namely, the local structure of this surface is nearly the same as those of p(2×2) or the c(4×2) structure, but perfect order is not developed in the long range. This speculation is supported by the Monte Carlo simulation using appropriate couplings between neighboring asymmetric dimers. Very recently the single domain c(4×2) surface was realized by Kono's group and an ARUPS experiment was performed on this surface [11]. The result of the dispersion showed an excellent agreement with the calculated one, i.e. that shown in Fig.3(c). In particular along the direction Y-Y′ , the dispersion is observed for the first time and showed good agreement with the theory.

4. The total energy and the stabilization mechanism

In the previous section the energy dispersions are calculated for the structures 2×1, p(2×2), c(4×2) which are formed by the fixed dimer geometry shown in Fig.2. The total energy and its decomposition are shown in Table I for the respective structures. The most stable structure among the three is the c(4×2) structure. The total energy per dimer of c(4×2) surface is lower by 0.023eV than the 2×1 structure. On the other hand, the total energy per dimer of the p(2×2) structure is lower by 0.013eV than that of the 2×1 structure, but higher than 0.01eV than that of the c(4×2) structure. It is found from Table I that the mechanism of the stabilization of the higher order arrangement of the asymmetric dimers is different between the structures p(2×2) and c(4×2). That is to say, the core-core interaction energy $E_{c\text{-}c}$ and the electron-electron Coulomb interaction energy $E_{e\text{-}c}$ decrease in the former structure and increase in the latter structure as compared with the 2×1 structure. On the other hand the opposite trend is found for the electron-core interaction energy $E_{e\text{-}c}$. The sum of $E_{c\text{-}c}$ and $E_{e\text{-}e}$ corresponds to the electro-static interaction energy among the dimers, while $E_{e\text{-}c}$ represents the energy gain by the band formation. Therefore the mechanism of the stabilization for the p(2×2) structure is the Coulomb interaction between the neighboring dimers: the

Table I. Differences of energies among asymmetric 2×1, p(2×2), and c(4×2) systems.

Energy component (eV/dimer)	$E_{p(2\times2)}-E_{(2\times1)}$	$E_{c(4\times2)}-E_{(2\times1)}$	$E_{c(4\times2)}-E_{p(2\times2)}$
$E'_{c\text{-}c}$	−0.616	0.204	0.820
$E'_{e\text{-}e}$	−0.933	0.398	1.331
$E'_{e\text{-}c}$	1.545	−0.671	−2.216
E_{kin}	0.013	0.064	0.051
Exchange	−0.022	−0.018	0.003
Total energy	−0.013	−0.023	−0.011

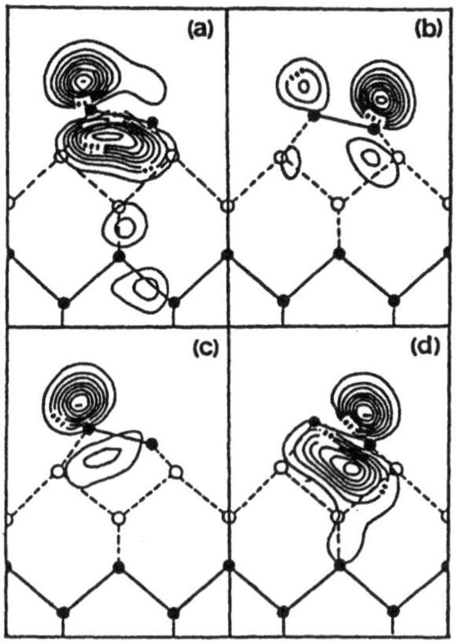

Fig. 4 Contour plots of the square of wave functions at point A in the c(4×2) BZ. (a),(b) Surface bonding and antibonding states of the 2×1 system. (c),(d) Surface bonding and antibonding states of the c(4×2) system.

asymmetric dimer has a significant dipole moment due to the partial charge transfer from the lowered to the raised atom. As seen in Fig.1 the distance between the raised atoms is kept large in the p(2×2) structure. On the other hand, in the c(4×2) structure, the dipole-dipole repulsive interaction among the asymmetric dimers between the adjacent dimer rows is increased as seen in Fig.1(c). Significant lowering of the band energy, which overcomes the increase of the dipole-dipole electrostatic energy, is found for the c(4×2) structure. This effect comes from the orbital overlapping between the raised dimer atoms in the adjacent dimer rows. The orbital shape of the occupied surface band changes to enhance this effect as shown in Fig.4.

We have also searched for the most stable structure by optimizing all the geometrical structure, i.e., dimer configuration, as well as the coordinates of the substrate atoms. The most stable structure is found also to be the c(4×2) structure. The total energy of the fully relaxed c(4×2) structure is 0.067eV per dimer lower than the fully relaxed 2×1 structure, and lower than the fully relaxed p(2×2) structure. The geometry of dimers in the fully relaxed c(4×2) structure showed a reasonable agreement between the one experimentally inferred. However, it should be remarked that the dispersion of the occupied surface state band of the fully relaxed surface does not show such a good agreement when compared to the one for the unrelaxed surface shown in Fig.3(c).

5. Theoretical simulation of the scanning tunneling microscopy image of Si(100) surface

Tsukada *et al.* [15] developed a first-principles theory of the STM image taking into account the atomic structure of the tip. By a number of examples of numerical simulation, it has been shown that for the tip with a single apex atom, the tunneling current is almost concentrated on this apex atom. The theory by Tersoff and Hamann [16], which claims the tunnel current is in proportion to the surface local DOS at the center of the curvature of the tip cannot be applied for the case of finite bias region. Further the STM image significantly depends on the structure and electronic states of the tip apex, when the wave function of the sample surface includes a nodal structure.

We have calculated the contour map of the tunneling current for the Si(100) symmetric dimer surface and asymmetric 2×1 surface with the LCAO-DV-Xα method. Typical results of the Si(100) buckled 2×1 surface are shown in Fig.5 for the tip-surface distance 2.6A and the tip bias voltage 0.5V for Fig.5(a) and —0.5V for Fig.5(b), respectively. The tip is modeled by the cluster W_{10}[111] formed by 10 W atoms. As for the surface an isolated slab of 10 atomic layers is used and the electronic structure is calculated by Bardeen's perturbation approach. The details of the method of calculation has been published elsewhere [15].

In Fig.5(a), the maxima of the tunnel current are concentrated at the raised atom positions, which is consistent with the contour of the squared amplitude of the surface state as shown in Fig.6. On the other hand, local maxima of the tunnel current are also found around the down atom of the dimer for the case of the unoccupied state as seen in Fig.6(b). This feature is explained well by the charge distribution of the surface state of the dimer as shown in Fig.6. The amplitude is more concentrated in the down atom. The feature of the STM image of asymmetric dimers is well reproduced by the theoretical simulation. The simulated image of the symmetric dimer shows a good agreement with experimental observation for the positive tip bias region. However, we found the effect of the tip state is quite significant around the central region of the dimer. The tip model

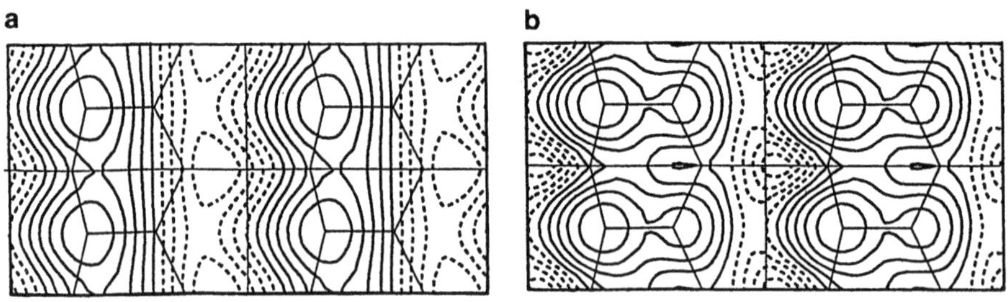

Fig. 5 Contour maps of STM image of Si(100) buckled 2×1 surface. The tip bias voltage is (a) 0.5 eV and (b) —0.5 eV.

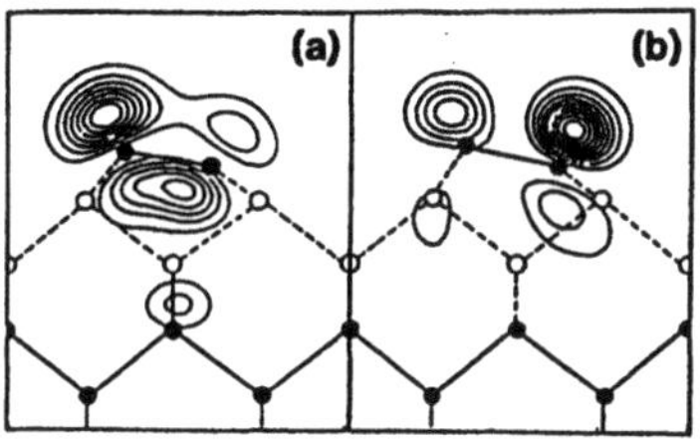

Fig. 6 Contour plots of the square of wave functions at the J point in the 2×1 BZ. (a) Surface bonding and (b) antibonding states for the asymmetric dimer model.

$W_{10}[111]$ does not reproduce a sharp dip of the current experimentally observed. On the other hand the local DOS of the unoccupied surface states shows a much sharper dip than the observation. Despite these delicate points, the STM image observed for the dominant regions of the Si(100) surface can be explained by the symmetric dimer arrangement.

6. Summary and discussions

Results of the first-principles calculation in this paper show that the most stable structure of the Si(100) reconstructed surfaces is the c(4×2) structure. Especially the agreement of the surface band dispersion calculated for this model with the ARUPS observation for the single domain c(4×2) surface is excellent. That the structure is most stable is also consistent with the LEED observation showing the spots corresponding to c(4×2) below about 200K [9].

However, there remain several problems for the structure of this surface. For example, not the asymmetric dimers but the symmetric dimers are observed in the STM image of wide terrace regions of good quality surface. The asymmetric dimers are observed in only local regions around missing dimer defects, or at the terrace edges. How is this feature reconciled with the above conclusion of the lowest energy of the c(4×2) structure? Some authors consider the effect of the flip flop oscillation of the asymmetric dimers between the bistable configuration. According to the present calculation the activation energy for this oscillation is rather high, which would preclude very frequent oscillation. Another reason for the observation of symmetric dimers in the STM might be the interaction between the tip and the dimers during the scanning. If the tip-surface distance is in the range of several Angstroms, the interaction is strong enough to cause the deformation of asymmetric dimers. To elucidate the reason for the discrepancies of the STM images with other experimental and theoretical conclusions, careful experiments at low temperature, changing the tip-surface distance, are required.

References

1. R.E.Schlier and H.E.Farnsworth: J. Chem. Phys. **30**, 917 (1959).
2. F.Jona, H.D.Shih, A.Ignatiev, D.W.Jepsen, and P.M.Marcus: J. Phys. **C10**, L67 (1977).
3. T.Aruga and Y.Murata: Phys. Rev. **B34**, 5654 (1986).
4. J.A.Appelbaum and D.R.Hamann: Surf. Sci. **74**, 21 (1978).
5. D.J.Chadi: Phys. Rev. Lett. 43, 43 (1979).
6. J.Ihm, M.L.Cohen, and D.J.Chadi: Phys. Rev. **B21**, 4592 (1980).
7. R.M.Tromp, R.J.Hamers, and J.E.Demuth: Phys. Rev. Lett. **55**, 1303 (1985).
8. R.J.Hamers, R.M.Tromp, and J.E.Demuth: Phys. Rev. **B34**, 5343 (1986).
9. T.Tabata, T.Aruga, and Y.Murata: Surf. Sci. **179**, L63 (1987).
10. Y.Enta, S.Suzuki, S.Kono, and T.Sakamoto: J. Phys. Soc. Jpn. **59**, 657 (1990).
11. Y.Enta, S.Suzuki, and S.Kono: Phys. Rev. Lett. **65**, 2704 (1990).
12. R. I. G. Uhrberg, G. V. Hansson, J. M. Nicholls, and S. A. Flodstrom: Phys. Rev. **B24**, 4684 (1981).
13. Z.Zhu, N.Shima, and M.Tsukada: Phys. Rev. **B40**, 11868 (1989).
14. G.B.Bachelet, D.R.Hamann, and M.Schluter: Phys. Rev. **B26**, 4199 (1982).
15. M.Tsukada, K.Kobayashi, and S.Ohnishi: J. Vac. Sci. Technol. **A8**, 160 (1990).
16. J.Tersoff and D.R.Hamann: Phys. Rev. **B31**, 805 (1985).

Scanning Tunneling Microscopy of a Metastable c(4×4) Structure Formed on a Si(100) Surface

T. Mizutani and T. Ide

Fundamental Research Laboratories, NEC Corporation,
34 Miyukigaoka, Tsukuba, Ibaraki 305, Japan

Abstract. A metastable c(4x4) structure is formed on a silicon (100) surface by hydrogen exposure of $2x10^{-5}$ Torr at 700°C for 30 minutes. Atomic images acquired simultaneously at negative and positive biases using an ultrahigh vacuum scanning tunneling microscope (UHV-STM) are inconsistent with the c(4x4) structural models previously proposed. The authors propose a new structural model, where the unit cell has a single topmost dimer.

1. Introduction

On the technologically important silicon (100) surface, surface structures of (2x1), c(4x2) [1-5], p(2x2) [4], and c(4x4) [6,7] on clean surfaces, and (2xn) [8-10] and c(8x8) [10] on metal contaminated surfaces, were observed by the low energy electron diffraction (LEED) technique, a He scattering technique and various other techniques. Though these techniques have been used to study surface structures in k-space, most of them, excluding the c(4x4) and c(8x8) have also been observed in real space using an ultrahigh vacuum scanning tunneling microscope (UHV-STM) [11,12].

Observation of a c(4x4) structure formed on a silicon (100) surface was first reported by Wang et al. [6]. They formed the surface by heat treatment. Based on their LEED observations, they proposed a structural model where the unit cell had three silicon dimers and one missing dimer. Kato et al. [7], formed a c(4x4) surface by hydrogen exposure at room temperature followed by heat treatment. They studied the surface by the LEED technique and obtained essentially the same result as Wang et al. They also reported that the structure was metastable based on their findings that the surface structure changed irreversibly from c(4x4) to (2x1) when the sample temperature was raised more than 700°C. On the other hand, Pandey [13] suggested that the c(4x4) structure with π-bonded defects was energetically favorable based on his pseudopotential calculation.

This paper describes the filled- and empty-state images of a c(4x4) structure simultaneously acquired by a UHV-STM at negative and positive biases. The images show that the unit cell of the c(4x4) structure has a single topmost silicon dimer.

2. Experimental

A UHV-STM equipped with LEED optics, a cylindrical mirror analyzer (CMA) for Auger measurements, a field ion microscope (FIM) for observation and cleaning of the probe tip [14], and a load lock chamber for fast sample exchange was developed. The STM unit was similar to that developed by Demuth et al. [15]. For vibration isolation, the microscope was placed on a metal stack, which was suspended by 25cm long springs. The resonant frequency of the spring system was about 1 Hz. The base pressure of the STM chamber was $8x10^{-11}$ Torr.

The samples were p-type doped with boron, oriented to within 0.5 degrees of the (100) direction. The resistivity was 0.01-0.02 Ω-cm. After the clean surface was produced by a well-known cleaning technique [16], the formation of the (2x1) surface was confirmed by STM observation. Then, the sample was exposed to hydrogen with the pressure of $2x10^{-5}$ Torr at 700°C for 30 minutes. The mole ratio of water in the hydrogen was less than 10 ppm. A B-A gauge was turned on to monitor the pressure during the hydrogen exposure. When the vacuum pressure was decreased to the 10^{-9} Torr range after the hydrogen exposure, the sample temperature was decreased. The formation of the c(4x4) structure was confirmed by LEED observations. The surface contamination was checked by Auger measurement, and no trace of metal contamination was observed.

To obtain filled- and empty-state STM images simultaneously from the same area, a technique similar to that developed by Stroscio et al. [17] was applied, where two x-scans were made at negative- and positive-biases at each y-position. The probe tip was made from a tungsten wire by electrochemical etching. The oxide formed on the tip was field evaporated just before the STM experiments by using an FIM unit until an atomic image of the tip emerged.

3. Results

Figure 1 is a typical LEED pattern of the surface produced by the above procedure. The fundamental spots are indicated by arrows. Fractional order spots of c(4x4) are clearly observed.

A filled- and an empty-state image of the c(4x4) structure are reproduced in Fig. 2(a) [-1.5V sample bias] and 2(b) [1.5V sample bias]. The tunneling current was 2 nA. The square indicates a unit cell positioned in the same place. Compared with an STM image of a (2x1) structure, the protrusions observed on the filled-state image were found to arrange with a c(4x4) symmetry. Overlapping the two images, two ball-like protrusions of the empty-state image were found to sit on both ends of the bean-like protrusion of the filled-state image.

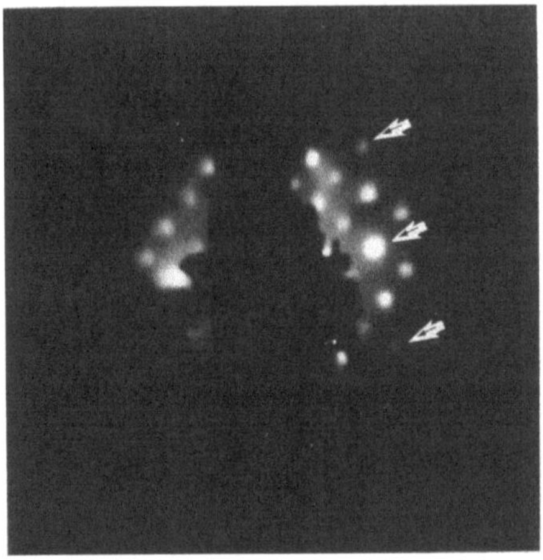

Fig.1 A typical LEED pattern of the c(4x4) structure. Fundamental spots are indicated by arrows.

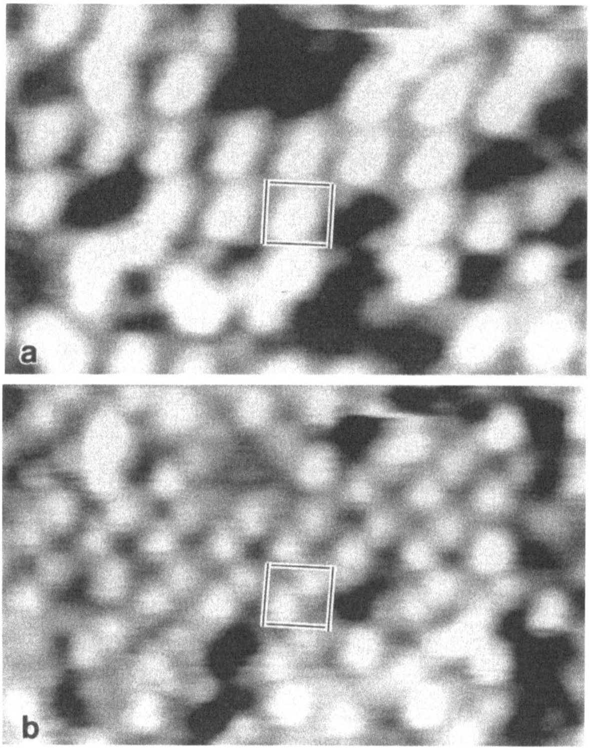

Fig.2 (a) A filled- and (b) an empty-state STM image of the c(4x4) structure simultaneously acquired on an identical area. Bias voltages are -1.5V and 1.5V, respectively. The square indicates a unit cell positioned in the same place.

4. Discussions

The structural models previously proposed for the c(4x4) by Pandey [13], Wang et al. [6] and Kato et al. [7] are schematically illustrated in Fig. 3(a) and 3(b), respectively. However, these models are inconsistent with the present STM observations. Before we discuss the c(4x4) structure in detail, let us remember how a silicon dimer was observed by STM. In Fig. 4(a) and 4(b), we show filled- and empty-state STM images of a (2x1) structure. The bias voltages were -1.2 and 1.2V, respectively. A silicon dimer is indicated by a rectangle in both figures. Although each dimer contacts with adjacent dimers in the filled-state image, a dimer is observed as a bean-like protrusion [Fig. 4(a)]. It is observed as two ball-like protrusions in the empty state image [Fig. 4(b)]. Overlapping the two images, a ball-like empty-state protrusion sits on both ends of the bean-like filled-state protrusion. This is qualitatively the same as the STM image of the c(4x4) structure except the size of the protrusions.

Here, we would like to propose a new model for a c(4x4) structure formed on a silicon (100) surface. It is schematically illustrated in Fig. 5. This model has a single topmost dimer in a unit cell of the c(4x4) structure. This structure explains why the apparent dimer size observed on the c(4x4) surface by STM is larger than that observed on the (2x1) surface.

Looking at the empty-state image of the (2x1) surface, the distance between the ball-like protrusions that correspond to the dimer in Fig. 4(b) is longer than that between the

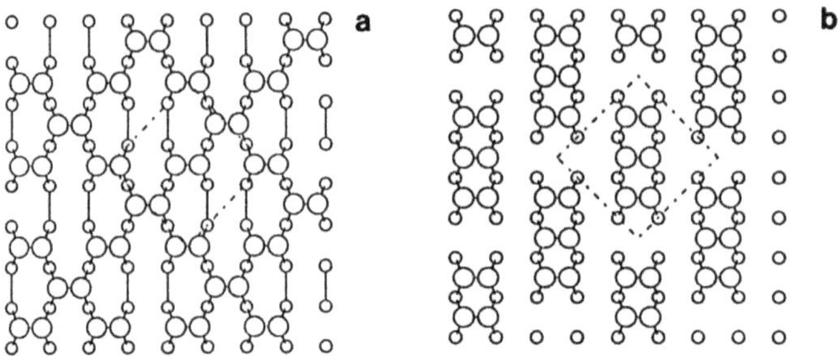

Fig.3 Schematic illustration of the proposed model for the c(4x4) structure so far; (a) the model proposed by Pandey, and (b) the model proposed by Wang et al. and Kato et al.

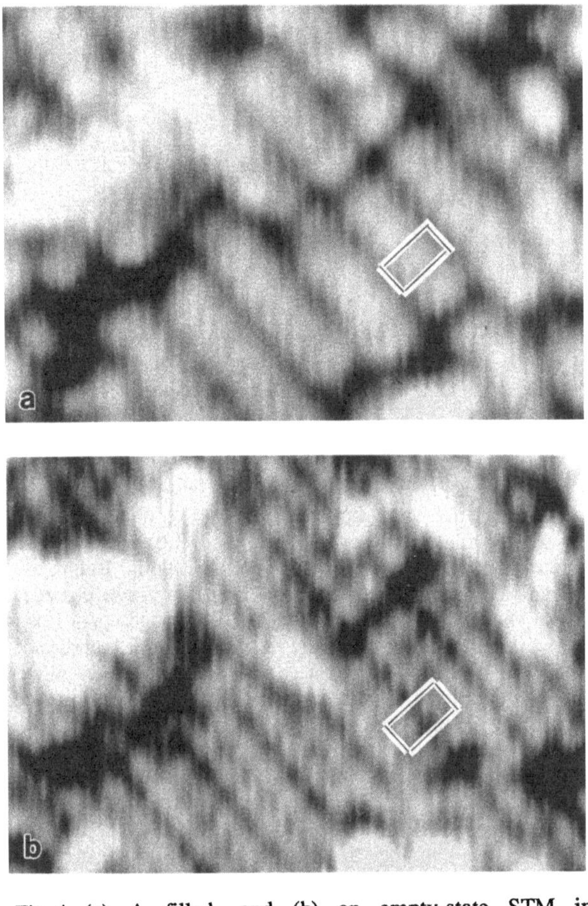

Fig 4 (a) A filled- and (b) an empty-state STM image of the (2x1) structure simultaneously acquired on an identical area. Bias voltages are -1.2V and 1.2V, respectively. The rectangle indicates a dimer.

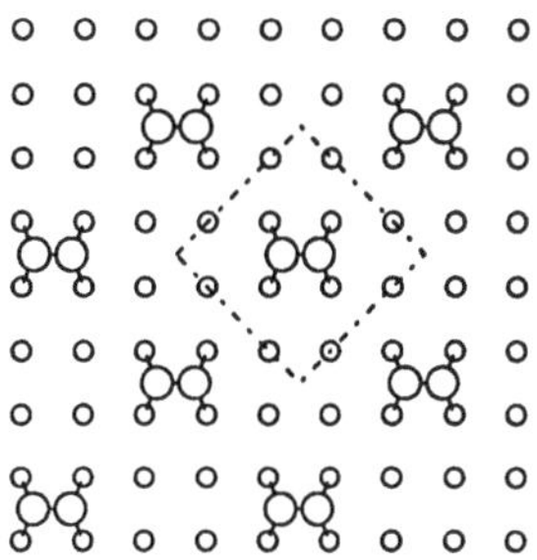

Fig.5 A structural model for the c(4x4).

neighboring dimers. In other words, the dimer is observed larger than the real size by STM even on the (2x1) surface. A cluster calculation on the electronic states of a silicon dimer shows that the distance between the maxima of the local density of states at 5 A above the surface is about 0.38 nm [18], which is longer than the dimer bond length (0.23 nm). This result is consistent with the STM images of the (2x1) surface. However, on the (2x1) surface, the enlargement of the apparent dimer size is limited by the existence of the adjacent dimers. On the contrary, the absence of adjacent dimers on the c(4x4) surface permits the surface wave functions to have long tails which spread not only vertically to the surface but also in lateral directions. Hence, the dimers are observed larger on the c(4x4) surface than that on the (2x1) surface.

This structure will be reconstructed further to reduce the surface energy. There are two possible reconstructions. One is obtained by making the subsurface dimer bonds so that the topmost dimers sit on the subsurface dimer rows. The other is obtained by making the subsurface dimer bonds so that the topmost dimers sit between the subsurface dimer rows. It is not possible to determine solely from the STM images which reconstruction occurred, because we could not "see" the subsurface atoms in the present STM images (see Fig. 2).

The photoemission study [19] and the low electron energy loss study [7] revealed that no hydrogen remains on the hydrogen treated silicon (100) surface after the sample temperature was elevated by more than 500°C. Therefore, it can be ruled out that the surface hydrogen remaining forms a c(4x4) structure. It is reported that boron atoms segregate at the subsurface to form a $\sqrt{3}$x$\sqrt{3}$ structure on a silicon (111) surface [20]. However, on a silicon (111) surface with the same boron doping concentration, we observed a well-ordered (7x7) structure. Moreover, we confirmed the formation of a (2x1) structure on a clean silicon (100) surface before the formation of the c(4x4) structure. Therefore, we can rule out the possibility that the dimers observed in the present STM experiment are boron atoms. Metal contamination induced surface reconstructions [8-10] are known to occur on silicon surfaces. However, no metal traces were observed on the present c(4x4) surface. Therefore, we can also rule out the possibility that the c(4x4) structure was produced by metal contamination.

On the STM images of the present c(4x4) structure, there exist irregularly imaged parts most likely due to oxide or carbide. This is probably because the partial pressure of H_2O, CO, CO_2, and hydrocarbons increased to the low 10^{-8} Torr range during the hydrogen exposure. However, Kato et al. [7] reported that exposure of a silicon (100) surface to H_2O, O_2, CO_2, and N_2 did not form the c(4x4) structure. Although the effects of CO and hydrocarbons have not yet been checked, it was usually found that cleaning silicon surfaces under bad vacuum quality produces similar irregularly imaged parts with the (2x1) structure on the remaining clean part of the surface.

Kato et al. suggested that the c(4x4) was produced through the ordering of missing dimers which were produced by vaporization of volatile silicon hydrides [7]. The present STM images require a change in the model for ordering mechanisms, suggesting that the c(4x4) structure was produced through the ordering of dimers. However, the ordering mechanism is an open question at present.

5. Conclusions

A metastable c(4x4) structure was formed on a silicon (100) surface through hydrogen exposure at 700°C for 30 minutes. The atomic images, simultaneously obtained from filled- and empty-states of an identical area using a UHV-STM, revealed that the unit cell of the c(4x4) structure has a single topmost dimer. A new structural model of the c(4x4) was proposed based on the STM images.

Acknowledgments

The authors would like to thank Prof. T. Sakurai and Prof. T. Hashizume for valuable discussions on the STM instruments, and Drs. J. Matsui, Y. Matsumoto, K. Ishida and M. Ogawa for their encouragement.

References.

[1] J. J. Lander and J. Morrison, J. Chem. Phys. 37 (1962) 729.
[2] T. D. Poppendieck, T. C. Ngoc and M. B. Webb, Surf. Sci. 75 (1978) 287.
[3] M. J. Cardillo and G. E. Becker, Phys. Lev. Lett. 40 (1978) 1148.
[4] M. J. Cardillo and G. E. Becker, Phys. Rev. B 21 (1980) 1497.
[5] T. Tabata, T. Aruga and Y. Murata, Surf. Sci. 179 (1987) L63.
[6] H. W. Wang, R. Lin and X. Wang, Phys. Rev. B 36 (1987) 7712.
[7] K. Kato, T. Ide, T. Nishimori and T. Ichinokawa, Surf. Sci. 207 (1988) 177.
[8] E. G. McRae, R. A. Malic, and D. A. Kapilow, Rev. Sci. Instrum. 56 (1985) 2077.
[9] K. Kato, T. Ide, S. Miura and T. Ichinokawa, Surf. Sci. 194 (1988) L87.
[10] L. Barbier and J. Lapujoulade, J. Vac. Sci. Technol. A8 (1990) 2662.
[11] R. J. Hamers, R. M. Tromp, and J. E. Demuth, Phys. Rev. B34 (1986) 5343.
[12] H. Niehus, U. K. Köhler, M. Copel and J. E. Demuth, J. Microscopy 152 (1988) 735.
[13] K. C. Pandey, in:Proc. 17th Intern. Conf. on the Phys. of Semiconductors, Eds. D. J. Chadi and W. A. Harrison (Springer, New York, 1984) p. 55.
[14] T. Sakurai, T. Hashizume, I. Kamiya, Y. Hasegawa, T. Ide, M.Miyao, I. Sumita, A. Sakai, and S. Hyodo, J. Vac. Sci. Technol. A7 (1989) 1684.
[15] J. E. Demuth, R. J. Hamers, R. M. Tromp and M. E. Welland, IBM J. Res. Develop. 30 (1986) 396.
[16] B. S. Swartzentruber, Y. -W. Mo, M. B. Webb and M. G. Lagally, J. Vac. Sci. Technol. A7 (1989) 2901.
[17] J. A. Stroscio, R. M. Feenstra and A. P. Fein, Phys. Rev. Lett. 57 (1986) 2579.
[18] S. Ohnishi, private communication
[19] T. Sakurai and H. D. Hangstrum, Phys. Rev. B 14 (1976) 1593.
[20] P. Bedrossian, R. D. Meade, K. Mortensen, D. M. Chen, J. A. Golovchenko, and D. Vanderbilt, Phys. Rev. Lett. 63 (1989) 1257.

Structural Phase Transition on the W(001) Surface

K. Terakura[1] *and N. Hamada*[2]

[1]Institute for Solid State Physics, University of Tokyo, Minato-ku, Tokyo 106, Japan
[2]NEC Fundamental Research Laboratories, 34 Miyukigaoka, Tsukuba 305, Japan

Among many examples of reconstructions on metal surfaces, the c(2 × 2) of W(001) has been studied most extensively and is still being actively studied. The nature of the phase transition of the W(001) surface was elucidated recently by experiments and some of the basic experimental data were analyzed theoretically with a phenomenological model Hamiltonian, the suitability of which will be discussed here.

1. Introduction

Since the discovery of the surface reconstruction of the (001) surface of Mo and W by Felter et al. [1], much experimental [2-6] and theoretical [7-16] work has been carried out to elucidate the basic mechanism of the reconstruction and the precise atomic arrangement. As for the W(001) surface, the ground state property is now fairly well understood except for some quantitative details. Most of the recent arguments have been concentrated on the nature of the phase transition and also on the nature of the high temperature phase. Whether the phase transition is an order-disorder type or an order-order type had been in debate up to a couple of years ago. Although support for the latter type of transition is still alive [4], some of the recent studies [6,15,16] seem to have given almost conclusive evidence for the former. The most direct observation supporting the order-disorder type of the phase transition, which was made by an x-ray diffraction by Robinson et al. [6], is the absence of change in the integrated intensity of the half-order spot across the transition temperature in contrast to an abrupt change in the peak height. The same experiment showed also that the half-order spot has an ellipsoidal shape. The temperature dependence of the intensity profile was used to estimate the temperature dependence of the anisotropic correlation length of the atomic displacement. These experimental data were analyzed fairly successfully by Han and Ying [16] by a Monte Carlo simulation with a model Hamiltonian. One of the purposes of the present work is to study the suitability of the model Hamiltonian itself.

2. Procedure of the Calculation

The tool we used in this work is exactly the same as the one in our previous work [7,8]. As for the electronic part we adopted the tight-binding d band model and the hopping integrals were adjusted to give the bulk d band structures at the high symmetry k points. The recursion method was used to solve the electronic structure [17,18]. The atomic d level in the bulk crystal is set to be zero. The Fermi energy is assumed to be zero also. Then the number of d electrons per atom in the bulk crystal is 4.4. The interatomic repulsion was treated by the simple Born-Mayer type potential whose parameters were determined so as to give the experimental equilibrium lattice constant and the bulk modulus. The actual values for all the parameters were tabulated in Tables 1 and 2 of Ref.7. It turned out that the shear elastic constants of the bulk Mo and W were reproduced very well by those parameters [8].

The parameters determined by the bulk properties were transferred to surfaces with no change except the atomic levels in the first two layers from the surface. They were tuned so that the d electron numbers in the surface region are the same as those in the bulk. The shift values are 0.015 Ry (surface) and 0.012 Ry (the 2nd layer) for Mo and the corresponding values for W are 0.017 Ry and 0.013 Ry. We first optimized the surface relaxation to be 4% (3%) of the surface interlayer spacing for Mo (W) and then introduced the lateral displacement of surface atoms. The shift values for the atomic d level were fixed even if these atomic displacements were introduced. This simplification produces a negligibly small effect on the energetics of the reconstruction [8].

In the following section, we will derive the values for the parameters in Han and Ying's model Hamiltonian [16] and compare them with those assumed by Han and Ying. Let us describe briefly the procedure of such calculations. The pair-interaction parameter is derived in the following way. We consider a case where the atoms at the site i and site j are displaced by $\mathbf{u}_i$ and $\mathbf{u}_j$ simultaneously. We can calculate the energy change associated with this configuration change both for the d-band energy part and the Born-Mayer potential part and express the net energy change as $E_{ij}^{(2)}(\mathbf{u}_i, \mathbf{u}_j)$. Similarly we calculate the energy change for the single-atom displacement, $E_i^{(1)}(\mathbf{u}_i)$ and $E_j^{(1)}(\mathbf{u}_j)$. The pair interaction energy is defined as

$$\Delta E_{ij}^{(2)} = \{E_{ij}^{(2)}(\mathbf{u}_i, \mathbf{u}_j) - E_i^{(1)}(\mathbf{u}_i) - E_j^{(1)}(\mathbf{u}_j)\}/\delta_i\delta_j, \tag{1}$$

where δ_i and δ_j are defined by $|\mathbf{u}_i| = a_0\delta_i$ and $|\mathbf{u}_j| = a_0\delta_j$ with a_0 the bulk lattice constant.

3. Results and Discussion

3.1. Energy Change for the c(2×2) Reconstruction

We express the total energy change per one surface atom associated with the reconstruction mode of Fig.1 as follows:

$$\Delta E(\delta, \theta) = \epsilon^{(2)}\delta^2 + (\epsilon^{(4)} + \epsilon^{(4,\theta)} \sin^2 2\theta)\delta^4, \tag{2}$$

where δ is the lateral displacement of a surface atom in units of a_0. The parameters in this equation were evaluated with the procedure described in the preceding section and are tabulated in Table 3 of Ref.7. The negative sign of $\epsilon^{(4,\theta)}$ implies that $\theta = \pi/4$ (the Debe-King model) is stabilized. The equilibrium displacements $a_0\delta$ are 0.236Å for Mo and 0.234Å for W. Experimentally, $a_o\delta$ is 0.24±0.025Å for W [5].

Although some quantitative differences are found in the parameters of eq.(2) between Mo and W, they should not be taken seriously. We admit that ambiguities were introduced inevitably to the *d*-band parameters and the Born-Mayer parameters when they were transferred from bulk to surface. In the following, therfore, we regard the set of parame-

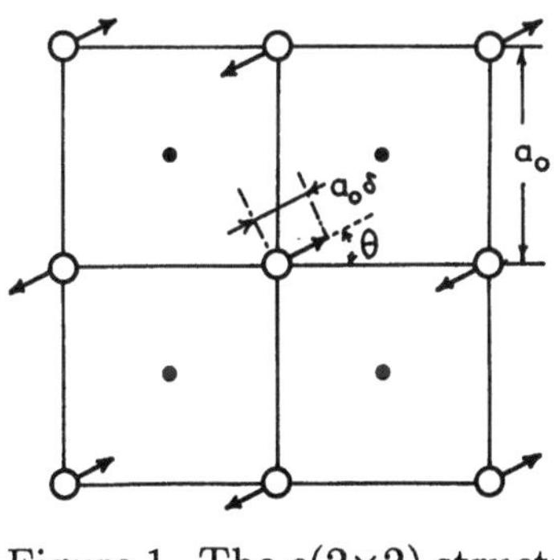

Figure 1. The c(2×2) structure with direction of the atomic displacement specified by θ and the magnitude of it by $a_0\delta$. a_0 is the equilibrium cubic lattice constant of the bulk crystal : 3.14Å for Mo and 3.16Å for W.

ters for Mo as representative both for Mo and W and use them even for the discussion on W. This will not affect the following arguments, which are rather qualitative.

3.2. The Present Approach versus Han and Ying's Model

As mentioned in the preceding chapter, various aspects of the experimental data by Robinson et al.[6] were fairly successfully explained by Han and Ying [16]. They carried out a Monte Carlo simulation based on the following effective Hamiltonian:

$$H = \sum_i [\frac{A}{2}\mathbf{u}_i^2 + \frac{B}{4}\mathbf{u}_i^4 + 8H_4 u_{ix}^2 u_{iy}^2 + c_1(\mathbf{u}_i \cdot \mathbf{u}_j + \mathbf{u}_i \cdot \mathbf{u}_k) + c_2\{u_{ix}(u_{j'y} - u_{k'y}) + u_{iy}(u_{j'x} - u_{k'x})\}], \qquad (3)$$

where the first three terms represent an energy change due to a single atom displacement, the 4-th term corresponds to a pair interaction between the first nearest neighbor (n.n.) sites and the 5-th term that for the second n.n. sites. We discuss these terms separately in the following.

Single Atom Term

The energy change due to a single atom displacement calculated by our approach is shown in Fig.2a both for the (110) and (100) displacements by a solid line and a broken line, respectively. The corresponding one

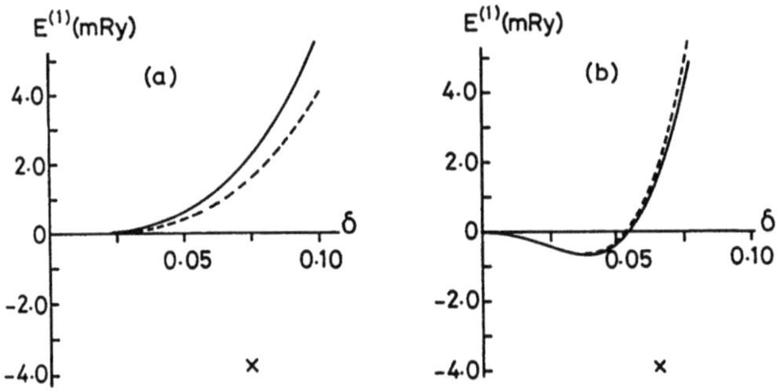

Figure 2. The energy change due to a single atom displacement with reference to the unrelaxed surface. (a) for the present result and (b) for the model II in Han and Ying's Hamiltonian [16]. The solid lines represent the case with the (110) displacement and the broken lines the (100) displacement. The crosses denote the energy minimum for the Debe-King mode.

obtained by the first three terms of eq.(3) for the model II of Han and Ying [16] is shown in Fig.2b. At a first glance one may notice some qualitative differences between Fig.2a and Fig.2b. For example, there is no energy minimum for a finite δ in our result (Fig.2a), whereas a shallow minimum exists at about δ=0.04 in Han and Ying's result (Fig.2b). However, this will not produce any qualitative difference. The most important aspect common to both cases is that the cooperative atomic displacement to form the Debe-King mode produces a deep energy minimum (as indicated by a cross in Fig.2a or 2b) at a fairly large value of δ. Therefore, whether a single atom displacement may produce a shallow minimum at a small but finite δ or not may not affect the qualitative nature of the phase transition. On the other hand, a large energy gain by the cooperative atomic displacement is a very important aspect and may be the origin of the observed large short range order above the transition temperature.

The second aspect concerns the anisotropy. In our approach the easy axis in the single atom term is in the (100) direction rather than in the (110) direction and the reversal of the easy axis in the c(2×2) pattern (eq.(2)) is a result of pair interactions as discussed below. On the other hand, Han and Ying assumed that the easy axis is in the (110) direction already in the single atom term.

First Nearest Neighbor Interaction

Han and Ying assumed a simple form of a scalar product of the displacement vectors for the pair interaction between the first n.n. atoms. We first checked whether this is a suitable assumption or not. Figures 3a and b show the calculated pair interaction parameters for two different pair displacements. Clearly, the present result (the numbers outside the parentheses) is consistent with Han and Ying's model (the numbers inside the parentheses). However, this is so only for the lowest order of the atomic displacement. In Fig.4, we show the pair-interaction contributions to the energy changes associated with the two different modes. There are two important aspects in this result. First, for the magnitude of the atomic displacement around the equilibrium value in the Debe-King mode, the first n.n. pair interaction favors the (110) displacement. Second, the energy minimum for the (110) pair displacement occurs almost exactly at the equilibrium value of the atomic displacement in the Debe-King mode. This strongly supports our previous view [7] that the basic origin of stabilization of the Debe-King mode is not the Fermi line nesting but the local bonding effect.

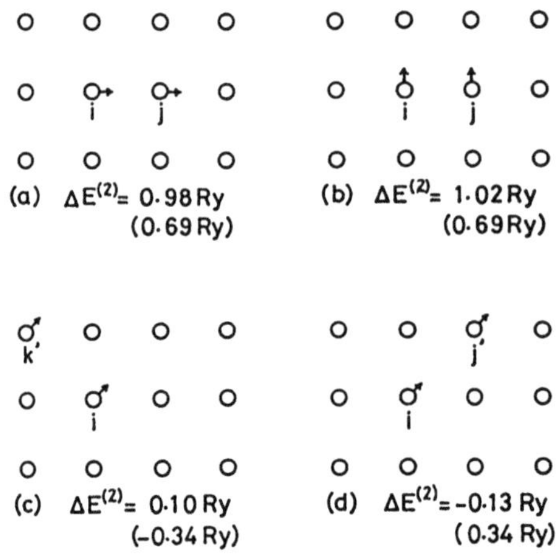

Figure 3. The pair interaction energies (eq.(1)) with reference to the unrelaxed surface. The numbers outside the parentheses are the present result and those inside the parentheses correspond to the model II of Han and Ying's Hamiltonian [16].

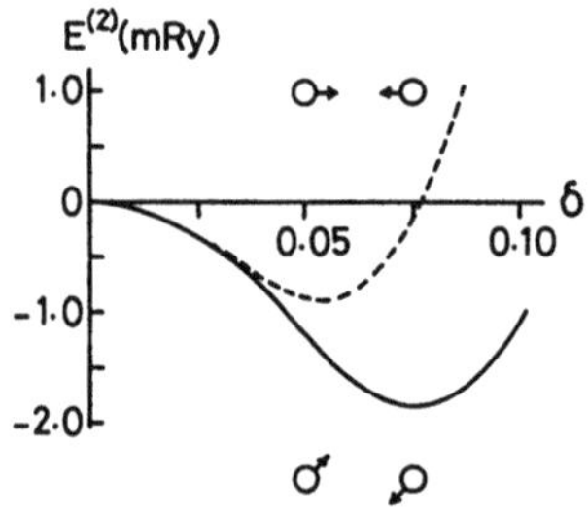

Figure 4. The first nearest neighbor pair interaction energies as defined by $\{E^{(2)}(\mathbf{u}_i, \mathbf{u}_j) - E^{(1)}(\mathbf{u}_i) - E^{(1)}(\mathbf{u}_j)\}$ for (110) (solid line) and (100) (broken line) displacements.

Second Nearest Neighbor Interaction

Han and Ying [16] introduced the second n.n. interaction (the 5th term in eq.(3)) in order to explain the anisotropic correlation length observed by Robinson et al. [6]. For the c(2×2) structure of Fig.1 with $\theta = \pi/4$, the Hamiltonian of eq.(3) assumes that the sign of the (i, k') pair interaction corresponding to the pattern of Fig.3c is negative while the one (i, j') of Fig.3d is positive. Taking the analogy with the magnetic interaction, we call the former a ferromagnetic (F) interaction and the latter an antiferromagnetic (AF) one. In this sense, the first n.n. interaction in Figs.3a and b is AF. The essence of Han and Ying's argument on the anisotropic

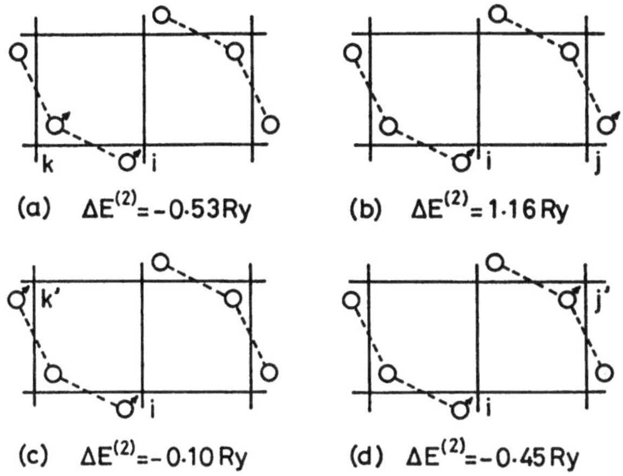

Figure 5. The pair interaction energies with reference to the Debe-King mode of the surface relaxation.

correlation length can be viewed in the following way. The indirect second n.n. interaction via the AF first n.n. interaction is F for both the (i, j') and (i, k') pairs. The indirect interaction enforces the direct F interaction for the (i, k') pair, while the indirect F and the direct AF interactions for the (i, j') pair cause a frustration. Therefore the correlation length is longer in the $(\bar{1}10)$ direction than in the (110) direction, being consistent with the experimental observation [6]. We found, however, that our approach gives opposite signs to the (i, j') and (i, k') pairs: the (i, j') coupling is F and the (i, k') one is AF. Does this mean that our approach gives a wrong result for the anisotropy in the correlation length? In the following, we will show that it is not so. So far the interaction parameters were derived by introducing appropriate atomic displacements with respect to the unreconstructed background. However, the anisotropy in the correlation length is concerned with the reconstructed domain even if we discuss the high temperature range above the transition temperature. We should consider the interaction energies with the reconstructed phase as a reference. Therefore, we introduced atomic displacements to the equilibrium c(2×2) phase and calculated the pair interactions in the same way as before.

The pair interaction energies for the four different types of the pair displacements are summarized in Fig.5. The broken lines connecting atoms demonstrate that the Debe-King mode is a chain structure. The coupling is AF for the (i, j) pair (the inter-chain first n.n.) and F for other pairs. Consequently, for the (i, k') pair (the intra-chain second n.n.), both the indirect interaction (via the first n.n. interactions) and the direct one are F, whereas for the (i, j') pair (the inter-chain second n.n.), the direct F interaction competes with the indirect AF interaction. Therefore, although the basic ingredients are different, the relative relation between the direct and indirect interactions for the second n.n. is quite similar between Han and Ying's model and the present approach.

4. Concluding Remarks

The present method based on the tight-binding *d* band and the Born-Mayer repulsive potential works fairly well for the bulk elastic properties of Mo and W and the c(2×2) surface reconstruction of the W(001) surface. On the other hand, Han and Ying [16] performed a Monte Carlo simulation based on a model Hamiltonian and reproduced nicely several experimental observations about the phase transition on the W(001) surface. Therefore we tried to elucidate the inter-relation between our approach and Han and Ying's model Hamiltonian. Most of the essential aspects of the model Hamiltonian can be derived from our approach. However, there are also some qualitative differences. It is certainly very interesting to see what aspects may be affected by such qualitative differences by performing a simulation based on the present approach. This is what we are aiming to do in the near future.

Acknowledgements

This work was partly supported by a Grant-in-Aid for Scientific Research from the Ministry of Education, Science and Culture.

References

1. T. E. Felter, R. A. Barker and P. J. Estrup: Phys. Rev. Lett. **38**, 1138(1977).
2. M. K. Debe and D. A. King: Phys. Rev. Lett. **30**, 708(1977) and Surf. Sci. **81**, 193 (1979).
3. D. A. King: Phys. Scr. **T4**, 34(1983), and references therein.
4. H. -J. Ernst, E. Hulpke and J. P. Toennies: Phys. Rev. Lett. **58**, 1941(1987).
5. M. S. Altman, P. J. Estrup and I. K. Robinson: Phys. Rev. B **38**, 5211(1988).

6. I. K. Robinson, A. A. MacDowell, M. S. Altman, P. J. Estrup, K. Evans-Lutterodt, J. D. Brock and B. J. Birgeneau: Phys. Rev. Lett. **62**, 1294(1989).
7. I. Terakura, K. Terakura and N. Hamada: Surf. Sci. **111**, 479(1981).
8. K. Masuda-Jindo, N. Hamada and K. Terakura: J. Phys. C **17**, L271(1984).
9. J. E. Inglesfield: Prog. Surf. Sci. **20**, 105(1985) and references therein.
10. D. Singh, Su-Huai Wei and H. Krakauer: Phys. Rev. Lett. **57**, 3292(1986).
11. X. W. Wang, C. T. Chan, K. M. Ho and W. Weber: Phys. Rev. Lett. **60**, 2066(1988).
12. C. Z. Wang, E. Tosatti and A. Fasolino: Phys. Rev. Lett. **60**, 2661(1988).
13. V. Heine and J. J. A. Shaw: Surf. Sci. **193**, 153(1988).
14. C. L. Fu and A. J. Freeman: Phys. Rev. B **37**, 2685(1988).
15. J. Pendry, K. Heinz, W. Oed, H. Landskrou, K. Müller and G. Schmidtlein: Surf. Sci. **193**, L1(1988).
16. W. K. Han and S. C. Ying: Phys. Rev. B **41**, 9163(1990).
17. R. Haydock, V. Heine and M. J. Kelly: J. Phys.C **8**, 2591(1975).
18. C. M. M. Nex: J. Phys. A **11**, 653(1978).

Ab-Initio Investigations of Surfaces and Grain Boundaries in Germanium

M.C. Payne[1], *G.P. Francis*[1], *M. Needels*[2], *E. Tarnow*[2], *P. Dallot*[2], *P.D. Bristowe*[2], *and J.D. Joannopoulos*[2]

[1]Cavendish Laboratory, Madingley Road, Cambridge, CB3 0HE, UK
[2]Massachusetts Institute of Technology, Cambridge, MA 02139, USA

Abstract. This paper describes the application of ab-initio total energy pseudopotential calculations to investigate surfaces and grain boundaries in germanium. Parameters have been calculated for an effective Hamiltonian that describes the system of dimers on the (001) surface and calculations have been performed to determine the mechanism that stabilises the Takayanagi reconstruction of the (111) surface. A study of the Σ=5 [001] twist boundary reveals that the boundary has many local energy minima. This behaviour is attributed to the frustration induced by the combination of directional bonding and the twist geometry of the boundary.

1. Introduction

In the total energy pseudopotential method calculations are performed using periodic supercells, the ionic potentials are replaced by pseudopotentials [1] and the electronic wavefunctions are expanded in terms of a plane wave basis set. Density functional theory [2] is used to map the system of interacting electrons onto a system of non-interacting electrons moving in an effective potential, V_{XC}, the exchange-correlation potential. The electronic groundstate is determined by solving the self-consistent Kohn-Sham equations for the electrons [3]. The total energy of the system, the forces on the ions and the stresses on the unit cell can all be computed once the groundstate electronic configuration has been located. Since the forces on the ions can be calculated the equilibrium geometry of the system can be determined as part of the calculation.

Total energy pseudopotential calculations have been used to calculate accurate equilibrium lattice constants, bulk moduli and phase transition pressures and temperatures [4]. The numerical methods used to locate the electronic groundstate have developed extremely rapidly in recent years following the introduction of the molecular dynamics method by Car and Parrinello [5]. These methods allow total energy pseudopotential calculations to be applied to much larger systems than previously possible so that calculations can now be routinely performed on systems containing up to hundred atoms in the unit cell. A conjugate gradients method has recently been developed which should allow calculations to be performed on systems containing hundreds of atoms in the unit cell [6]. In the future with the use of parallel computers it will be possible to study systems containing a thousand atoms in the unit cell thus allowing increasingly complex systems to be investigated using the total energy pseudopotential technique.

2. Dimer reconstruction of the (001) surface of germanium

When a crystal of silicon or germanium is cut along the (001) surface each atom in the unreconstructed surface has two dangling bonds. The surface energy is considerably reduced when pairs of atoms at the surface move towards each other and form 'dimer' bonds. In the case of the germanium surface it is found that the dimers do not remain parallel to the surface but tilt or buckle so that one atom in the dimer moves closer to the surface and the other atom moves further away from the surface [7]. An infinite number of reconstructions can be formed by varying the directions of tilting of the dimers. There is no point attempting to determine the lowest energy configuration simply by calculating the energies of different configurations of buckled dimers. Progress can be made by writing the energy of the surface in terms of the interactions between the dimers. There are two degrees of freedom associated with the buckling of each dimer and it possible to represent the system of buckled dimers by a system of spins as illustrated schematically in figure 1. The spin that represents a particular dimer points in the direction of the raised atom of the dimer. The energies of different configurations of dimers can be expressed by an effective spin Hamiltonian, $\mathcal{H}$, as follows

$$-\mathcal{H} = V\sum_{ij} s_{i,j}s_{i,j+1} + H\sum_{ij} s_{i,j}s_{i+1,j} + D\sum_{ij} s_{i,j}s_{i+1,j+1}$$
$$+\text{higher order terms} \qquad (1)$$

The parameters V, H and D represent the interactions between the dimers shown in figure 1. Since the interactions between the dimers are dominated by the elastic strain energy in the system the values of the interaction parameters will decrease as the distance between the dimers increases. Hence it is expected that truncating the effective Hamiltonian after the three terms shown in (1) should provide an accurate description of the system. However it is essential that the values of the parameters V, H and D in the effective Hamiltonian are given the values that correctly describe the interactions between the buckled dimers on the surface. These parameters can be determined by calculating the energies of a number of dimer configurations. Such calculations have been carried out [8] and the values of the coefficients V, H and D deduced from these calculations are shown in table I. It can be seen the strength of the interaction decreases rapidly with distance between the dimers. This suggests that the truncation of the effective Hamiltonian to include only a small number of interactions is justified. The strongest interaction is the interaction between nearest

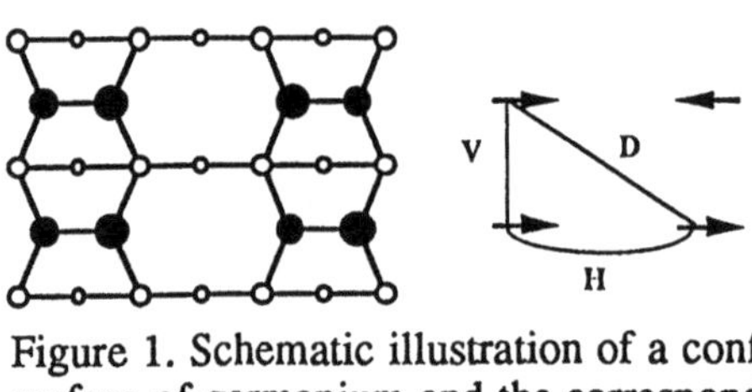

Figure 1. Schematic illustration of a configuration of buckled dimers on the (001) surface of germanium and the corresponding representation as a set of spins. V, H and D represent interactions between the dimers.

Table I. Values of parameters in effective spin Hamiltonian that describes the system of buckled dimers on the (001) surface of germanium.

Interaction	Interaction energy (meV)
V	- 43
H	10
D	4

neighbour dimers. This interaction favours an alteration in direction of the tilting from dimer to dimer which is indeed observed in experiments [7].

The effective Hamiltonian can be used to study the thermodynamic behaviour of the surface. A renormalisation group analysis of the effective Hamiltonian predicts a value of 380 ± 100 K for the critical temperature for the order-disorder transition on the surface [8] in reasonable agreement with experimental results [9,10].

3. The Takayanagi reconstruction

The (111) surfaces of silicon and strained germanium form the Takayanagi structure [11] on annealing. The Takayanagi structure is extremely complex and consists of three distinct structural elements: dimers, adatoms and stacking faults. Hence the structure is often referred to as the DAS structure. A family of Takayanagi reconstructions can be formed in (2n+1)x(2n+1) unit cells, where n is an integer. The lowest member of the family of reconstructions is the 3x3 reconstruction which is illustrated schematically in figure 2(a). The surprising feature of the Takayanagi reconstruction is that it does not minimise the density of dangling bonds or the density of defects on the (111) surfaces of silicon and strained germanium yet it is the lowest energy reconstruction that can be formed on these surfaces. This can only

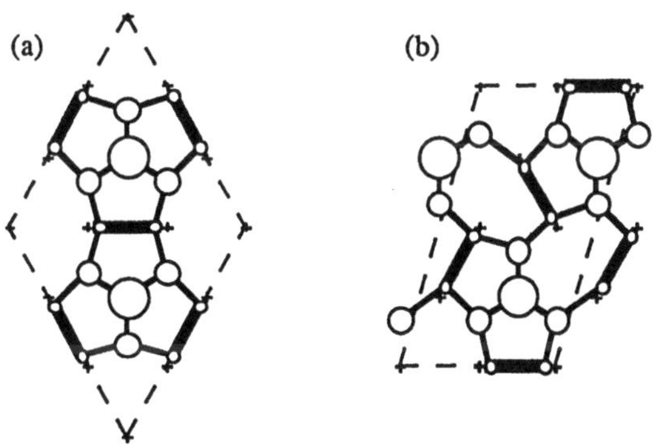

Figure 2. (a) 3x3 Takayanagi structure and (b) c(2x8) Takayanagi structure. The small circles show second layer atoms, the medium circles first layer atoms and the large circles show the adatoms. The crosses show the positions of the second layer atoms in the unreconstructed surface and the thick lines show the dimer bonds.

be the case if an interaction between the elements of the Takayanagi structure lowers the energy of the structures. This interaction stabilises the Takayanagi structure. In order to determine the nature of this interaction calculations have been performed to determine the energies of the 3x3 Takayanagi structure, the c(2x8) Takayanagi structure [12], which is illustrated schematically in figure 2(b), and a simple 2x2 adatom reconstruction of the (111) surface. The c(2x8) Takayanagi structure is also formed from the DAS structural elements but the structural elements are arranged differently in the c(2x8) and 3x3 reconstructions. The energies of the Takayanagi structures without the adatoms have also been calculated in order to provide more detailed information about these reconstructions. The Takayanagi structures without adatoms will be referred to as 'Takayanagi subsurfaces'.

The energies of the reconstructions are given in table II. The c(2x8) Takayanagi subsurface is slightly higher in energy than the 3x3 Takayanagi subsurface. This shows that the arrangement of the dimers and stacking faults in the 3x3 subsurface is energetically favourable. However, the difference in energy between the two Takayanagi subsurfaces is relatively small. The difference between the 3x3 and c(2x8) reconstructions is most apparent in the full Takayanagi structures. When adatoms are added to the unreconstructed surface to form the 2x2 structure the surface energy is reduced by 0.84 eV per adatom. However, when adatoms are added to the 3x3 Takayanagi subsurface the surface energy is reduced by 1.04 eV per adatom and when adatoms are added to the c(2x8) Takayanagi subsurface the surface energy is lowered by only 0.63 eV per adatom.

The results for the c(2x8) Takayanagi reconstruction show that an arbitrary arrangement of the DAS elements will not, in general, generate a low energy structure. The larger than expected lowering of the surface energy on adding adatoms to form the 3x3 Takayanagi structure can be attributed to the interaction between the adatoms and the dimers. Both of these elements favour a local contraction of the surface. When all the dimers lie alongside adatoms, as they do in the (2n+1)x(2n+1) Takayanagi structures, there is a favourable interaction between these two structural elements which arises from the increased local contraction of the surface. In the case of the c(2x8) structure not all of the adatoms lie alongside dimers and so some of the local contractions induced by these structural elements oppose each other, thus increasing the surface energy. It can be concluded that the (2n+1)x(2n+1) Takayanagi structures are stabilised by a direct strain interaction between the adatoms and the dimers.

Table II. Energies of reconstructions on the (111) surface of germanium.

	Energy per unit cell (eV)	Energy per surface atom (eV)
1x1 Unreconstructed surface	1.20	1.20
2x2 Adatom reconstructed surface	3.96	0.99
3x3 Takayanagi subsurface	10.00	1.11
3x3 Takayanagi structure	7.93	0.88
c(2x8) Takayanagi subsurface	9.28	1.16
c(2x8) Takayanagi structure	8.02	1.00

4. The Σ=5 [001] twist boundary in germanium

The Σ=5 [001] twist boundary in germanium is formed by rotating one half of an infinite crystal bounded by an (001) plane by either 36.9° or -53.1° around an [001] axis. For convenience the boundary formed by the 36.9° rotation will be referred to as the Σ=5 boundary and boundary formed by the -53.1° rotation as the $\Sigma=5^*$ boundary. Figure 3 shows the positions of the atoms in the two layers above and below an unrelaxed Σ=5 boundary formed by rotating the upper crystal by 36.9° around a rotation axis that passes through the shaded atom. It can be seen that the structure is periodic in the plane of the boundary. The dashed lines show a unit cell which contains five atoms per layer. The shaded atom is in a lattice site that belongs to both crystals hence this boundary is commonly referred to as the coincident site geometry. In a real boundary the two crystals are free to translate relative to each other and the ions will relax to their equilibrium positions. The boundary will adopt the translation state that has the relaxed structure of lowest total energy. The arrows to the left of figure 3 show the DSC vectors which are 1/5th the length of the unit cell vectors. When either of the crystals forming the boundary is displaced by any combination of integer multiples of the DSC vectors the structure of the unrelaxed boundary is unchanged. However the position of the coincident atom will, in general, change on translating the crystals.

The symmetry under displacements by DSC vectors implies that only the range of translations lying within the DSC cell has to be explored in order to locate the lowest energy translation state. To determine the equilibrium structure of the boundary a series of calculations have been performed in which the translation state of the boundary is kept fixed while the ions are relaxed to their equilibrium positions and the width of the boundary is allowed to vary [13]. Each calculation therefore generates a relaxed grain boundary structure and total energy for a particular translation state. By performing calculations for a series of translation states within the DSC zone the translation state that has the lowest boundary energy should be located. It is found that only translations in 1/4 of the DSC zone generate distinct boundary structures and so only 1/4 of the DSC zone has to be explored. The results of the calculations are shown in figure 4.

It can be seen that the energies of the relaxed boundaries lie within a narrow range. The relaxed boundaries are, however, found to have radically different structures.

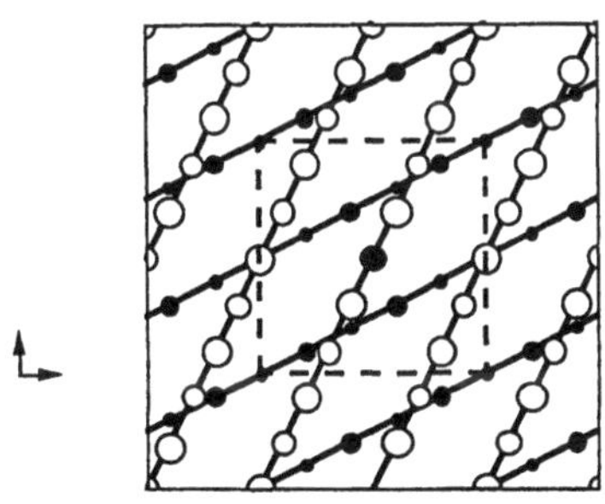

Figure 3. Structure of the unrelaxed Σ=5 [001] twist boundary in germanium. The black circles show atoms in the two layers below the boundary, the open circles show atoms in the two layers above the boundary. The boundary has been formed by rotation around the shaded atom.

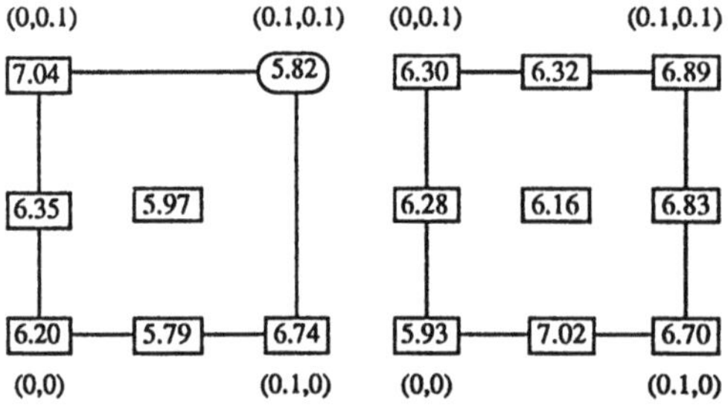

Figure 4. Total energy as a function of translation state for the $\Sigma=5$ and $\Sigma=5^*$ [001] twist boundaries in germanium. The numbers in brackets are the coordinates in terms of the unit cell vectors. The number in the ellipse gives the energy of the (0.05,-0.05) translation state of the $\Sigma=5$ boundary,. For convenience this is shown at the position of the (0.1,0.1) translation state.

Therefore, the near degeneracy cannot be attributed to a formation of a single low energy structure with variations in the boundary energy arising solely from the strain energy associated with a relative translation of the two crystals forming the boundary. There are instead many structures which are local energy minima in the energy surface, all of these structures have very similar energies. It is found that if the atoms above or below a relaxed boundary are translated by a DSC vector a completely new structure is obtained when the ions are relaxed to equilibrium. The new structure typically has an energy within the range of energies shown in figure 4. This behaviour is extremely surprising because the structure of the unrelaxed boundaries is invariant under a relative displacement of the crystals forming the boundary by a DSC vector. However, the loss of symmetry that occurs when the ions relax to equilibrium breaks the invariance of the boundary under displacement through a DSC vector. If there were no energy barriers between the new structure generated by displacement through a DSC vector and the original structure then the new structure would spontaneously relax to the original structure. However, it is found that in germanium there are energy barriers between the new and old structures and so relaxing the new structure generates a distinct equilibrium structure corresponding to the same translation state in the DSC zone as the original boundary. Hence there are many local energy minima at each point in the DSC zone rather than the single one predicted by simple theory. Although only one of these is the true lowest energy configuration for that translation state energy barriers separate the structures and prevent the metastable structures from spontaneously relaxing to the lowest energy configuration.

The behaviour found in the $\Sigma=5$ and $\Sigma=5^*$ twist boundaries is typical of the behaviour of a frustrated system. The frustration in these systems is caused by the combination of the strong directional bonding and the twist geometry. This behaviour is not observed in the $\Sigma=5$ tilt boundary in germanium. In this case the system is found to relax to the structure observed experimentally irrespective of the initial translation state of the boundary [14].

References

[1] V. Heine and M.L. Cohen, Solid State Physics Vol. **24** (1970).
[2] P. Hohenberg and W. Kohn, Phys. Rev. **136** B864 (1964).
[3] W. Kohn and L.J. Sham, Phys.Rev. **140** A1133 (1965).
[4] For reviews see J.D. Joannopoulos, in 'Physics of Disordered Materials', Plenum Publishing, p. 19 (1985); and W. Pickett, Computer Physics Reports **9** 115 (1985).
[5] R. Car and M. Parrinello, Phys. Rev. Lett. **55** 2471 (1985).
[6] M.P. Teter, M.C. Payne and D.C. Allan, Phys. Rev. B**40** 12255 (1989).
[7] J.A. Kubby, J.E. Griffith, R.S. Becker and J.S. Vickers, Phys. Rev. B**36** 6079 (1987).
[8] M. Needels, J.D. Joannopoulos and M.C. Payne, Phys. Rev. B**38** 5543 (1988).
[9] S.D. Kevan, Phys. Rev. B**32** 2344 (1985).
[10] R.J. Cuthbertson, Y. Kuk and L.C. Feldman, Surf. Sci. **167** 127 (1986).
[11] K. Takayanagi, Y. Tanashiro, S. Takahashi and M. Takahashi, Surf. Sci. **164** 367 (1985).
[12] K. Takayanagi and Y. Tanashiro, Phys. Rev. B**34** 1034 (1986).
[13] E. Tarnow, P. Dallot, P.D. Bristowe, J.D. Joannopoulos, G.P Francis and M.C. Payne, Phys.Rev. B**42** 3644 (1990).
[14] T. Arias and J.D. Joannopoulos, private communication.

Structure Ordering at the Si(111) Surface

A. Ichimiya

Department of Applied Physics, School of Engineering, Nagoya University, Nagoya 464-01, Japan

Abstract. On the basis of results of surface structure analyses by reflection high-energy electron diffraction (RHEED), we discuss that reconstruction of stacking faults of 7×7 structure into the normal stacking proceeds by homoepitaxial growth or adsorption of extra atoms on the Si(111) surface. During RHEED intensity oscillation with the homoepitaxial growth, a mixed phase of 5×5 and 7×7 structures is observed. This mixed structure at substrate temperatures higher than about 400 °C is determined as the dimer–adatom–stacking–fault structure by RHEED intensity rocking curve analysis. At about 300 ° C, however, it is found by rocking curve analysis that new 5×5 and/or 7×7 structures with pyramidal clusters are formed on the dimer-stacking-fault (DSF) framework. It is suggested that the formation of such metastable structure promotes successive epitaxial growth accompanied by stacking fault dissolution at the DSF framework. Formation of $\sqrt{3}\times\sqrt{3}$ structures on Si(111) by adsorption of metal atoms is also discussed based upon the results of the formation of the metastable structure. We also discuss the order of the phase transition from 7×7 to "1×1", the energy difference between T_4 and H_3 sites, and bonding processes during adsorption of atoms.

1. Introduction

The dimer–adatom–stacking fault (DAS) model of clean Si(111)7×7 surface structure [1] has been widely accepted by several experimental and theoretical studies [2–12]. For adsorption of several metal atoms on Si(111)7×7 at temperatures higher than 300 °C, the 7×7 structure changes into $\sqrt{3}\times\sqrt{3}$–R30 ° structures [13]. Since $\sqrt{3}\times\sqrt{3}$ structures include no stacking fault in most cases, the stacking fault layer is reconstructed into the normal stacking layer at the transition from the 7×7 structure to a $\sqrt{3}\times\sqrt{3}$ one. For the clean surface of Si(111), however, the 7×7 structure changes into "1×1" at about 850 °C [14], while reconstructions of adsorbed surfaces occur at lower temperatures. During

Eds.: A. Yoshimori · T. Shinjo · H. Watanabe

homoepotaxial growth on an Si(111)7×7 surface, very stable RHEED intensity oscillation is observed at the lower temperature [15–17]. It seems that defects and stacking fault layers scarcely remain in the grown layers during homoepitaxial growth on Si(111). Therefore the reconstruction from stacking fault layer to the normal stacking one is also performed by homoepitaxial growth at the low temperature. This difference of the reconstruction temperatures between clean and adsorbed surfaces may be discussed with the instability of surface structures due to adsorbed atoms.

In the present paper a mechanism of the stacking fault dissolution of the DAS structure by adsorbed atoms is discussed on the basis of results of structure analyses by RHEED for a high temperature Si(111)"1×1" surface, adsorbed surfaces and structures during homoepitaxial growth. From these RHEED intensity analyses, we also discussed the order of the phase transition from 7×7 to "1×1", energy differences between T_4 and H_3 sites, and bonding processes during adsorption of atoms.

For structure determination by RHEED intensity analysis an intensity rocking curve is measured under the one–beam condition [6] at first, in which the incident electron direction is chosen at a certain azimuthal angle with respect to a crystal zone axis. Under the one–beam condition, the RHEED intensity is a function of interlayer distances and atomic densities of surface parallel layers. The surface normal components of the atomic positions are determined by analysis of the one–beam rocking curve using RHEED dynamical calculations [18]. Then using the results for the one–beam condition, lateral positions of surface atoms are determined from analysis of rocking curves under many–beam conditions, where the direction of the incident beam is chosen along a certain crystal zone axis.

2. Structure of Si(111)"1×1" at high temperatures

In a RHEED pattern from the Si(111)"1×1" surface, diffuse $\sqrt{3}\times\sqrt{3}$ and 2×1 streaks are observed [19,20]. Therefore this surface includes disorder structures such as a randomly relaxed bulk–like structure (bulk–like structure), a relaxed bulk-like structure with random vacancies (vacancy structure), a relaxed bulk–like structure with random adatoms (adatom structure) and a random DAS structure. Fig. 1a shows an experimental rocking curve of the RHEED intensity under the one–beam condition for "1×1" at 900 °C [21]. Under the one–beam condition the azimuthal angle of the incident beam direction was set at 7.7 ° from [11$\bar{2}$] in order to avoid strong simultaneous reflections. The best fit curves obtained by RHEED dynamical calculations for the above four structure models are

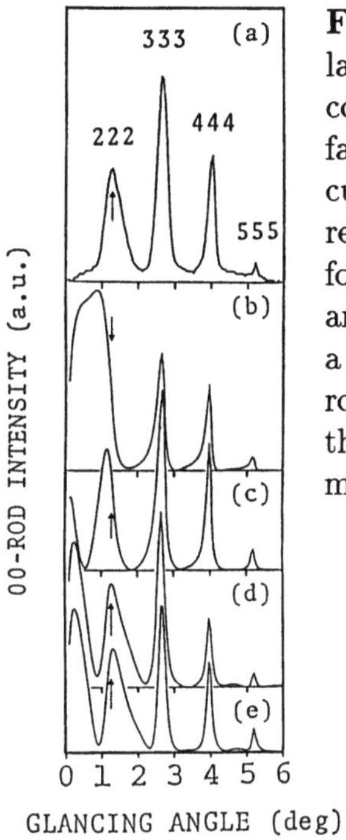

Fig. 1. Experimental and calculated rocking curves at one–beam condition from the Si(111) surface at 900 °C. (a) Experimental curve, (b) calculated curve for a relaxed bulk–like model, (c) that for a vacancy model, (d) that for an adatom model and (e) that for a random DAS model. The arrows indicate the peak position of the 222 reflection in the experimental curve (a).

shown in Fig. 1 with the experimental curve. In the calculations interlayer distances of surface parallel layers and atomic densities of the layers were widely changed. In Fig. 1, it can be seen that the curves for the adatom structure and for the random DAS structure model are in very good agreement with the experiment curve. For the vacancy structure model, the peak position at the 222 reflection of the rocking curve shifts about 0.2° to a lower angle compared with the position of the experimental curve indicated by arrows in Fig. 1. For the bulk–like structure model, the curve does not agree with the experimental one.

By the structure analysis under the one–beam condition, we can not determine either the adatom model or the random DAS model. In order to distinguish between the adatom model and the random DAS model, a 00-rod rocking curve at $[01\bar{1}]$ incidence was analyzed by a 7-beam dynamical calculation for the above two models. For the adatom model, calculated curves were obtained for the two adatom sites such as T_4 and H_3 sites. Fig. 2 shows experimental and calculated 00–rod rocking curves for $[01\bar{1}]$ incidence. A calculated curve for the vacancy model is also shown. In the calculated curves, curve (*e*) for the T_4 site is in better agreement with the experimental one than curve (*d*) for the H_3 site. The curves for the vacancy and the random DAS models do

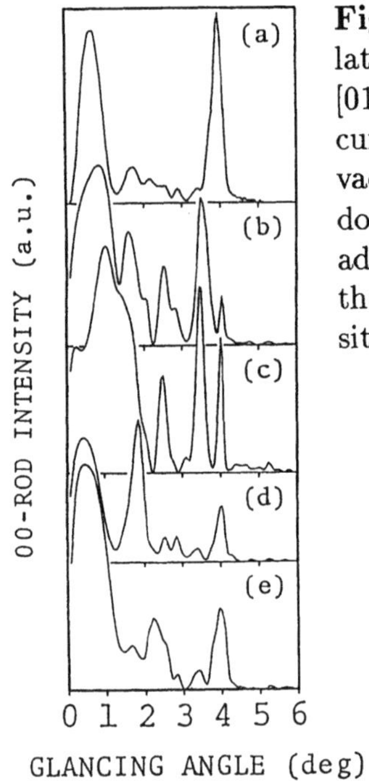

Fig. 2. Experimental and calculated (00)–rod rocking curves for [01$\bar{1}$] incidence. (a) Experimental curve, (b) calculated curve for a vacancy model, (c) that for a random DAS model, (d) that for an adatom model at H_3 site and (e) that for an adatom model at T_4 site.

not agree with the experimental one. At the Si(111)"1×1" structure, it is considered that the adatoms distribute randomly on both sites. By calculations for several mixing ratios of H_3 and T_4, H_3/T_4, from 1/6 to 1/1, a rocking curve for a mixing ratio of 1/4 is in the best agreement with the experimental curve [21]. The final structure determined by RHEED analysis is shown in Fig. 3 with several parameters. In this structure adatom coverage is 0.25 monolayer (ML), a value which is nearly equal to the adatom coverage of the DAS structure. In this structure the atoms just below the adatoms are pushed down, and the remaining atoms not bonding with the adatoms are pushed up from the bulk position. These features are very similar to those of the DAS structure [6,7].

From the above analysis, it is clarified that the Si(111)"1×1" structure includes no stacking fault [22,23]. Therefore the phase transition from the 7×7 DAS structure to the "1×1" adatom structure occurs by breaking the dimer–bonds and substantial dissolution of the stacking fault. Since the entropy is discontinuous at the transition, this transition process is the first–order one, but not a simple order–disorder process.

In the "1×1" phase at high temperatures, adatoms at H_3 and T_4 sites coexist on the (111) surface. It is considered that the mixing of the adatoms at both sites lowers the surface free energy because of the

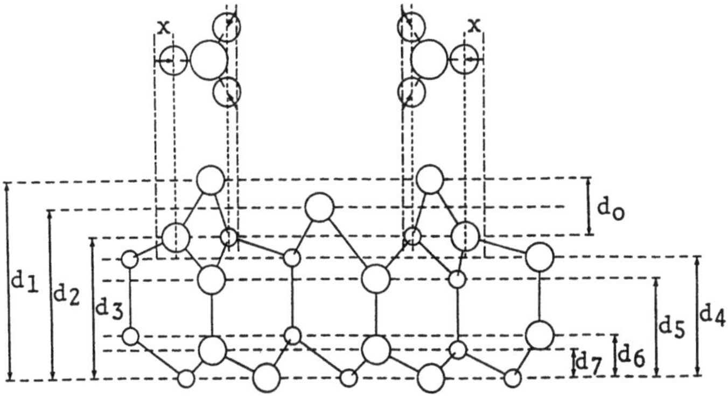

Fig. 3. $[01\bar{1}]$ side view of the best fit model for the "1×1" structure. Atomic positions perpendicular to the surface are as follows: d_0 is 1.31Å, d_1 5.08Å, d_2 4.11Å, d_3 3.77Å, d_4 3.24Å, d_5 2.92Å, d_6 0.92Å and d_7 0.78Å. The value of x is 0.26Å, which is the displacement parallel to the surface of the first layer atoms bonded to the adatoms.

large entropy term at the expense of energetical disadvantage of the H_3 configuration [24]. The energy difference between the T_4 and the H_3 geometry can be obtained by minimization of the surface free energy for the mixing site adatoms. For very rough estimation of the entropy, ΔE is obtained as

$$\Delta E = kT \ln [(N - n_H)n_T/n_H(N - n_T)],$$

where k, T, N, n_H and n_T are the Boltzmann constant, surface temperature, total number of sites at T_4 or H_3, number of adatoms at H_3 and that at the T_4 site, respectively. Since the adatom coverage, $(n_H + n_T)/N$, is 0.25 and the mixing ratio of $H_3/T_4 = n_H/n_T = 1/4$, ΔE becomes 0.16eV at 900 °C, at which kT is 0.1eV. Although the present evaluation of the entropy term is obtained in a very rough approximation, it seems that the present value of the logarithmic term does not so greatly deviate from the true value. Therefore it is expected that ΔE is nearly 0.2eV. This value is very small in comparison with 0.6eV obtained theoretically by Northrup [24].

3. Adsorption processes on the Si(111) surface

By adsorption of H, Li, Na, Ag, and Si atoms on the Si(111)7×7 surface at room temperature, the DAS structure changes into another 7×7 structure named δ7×7 [25–29]. Fig. 4 shows RHEED patterns from several

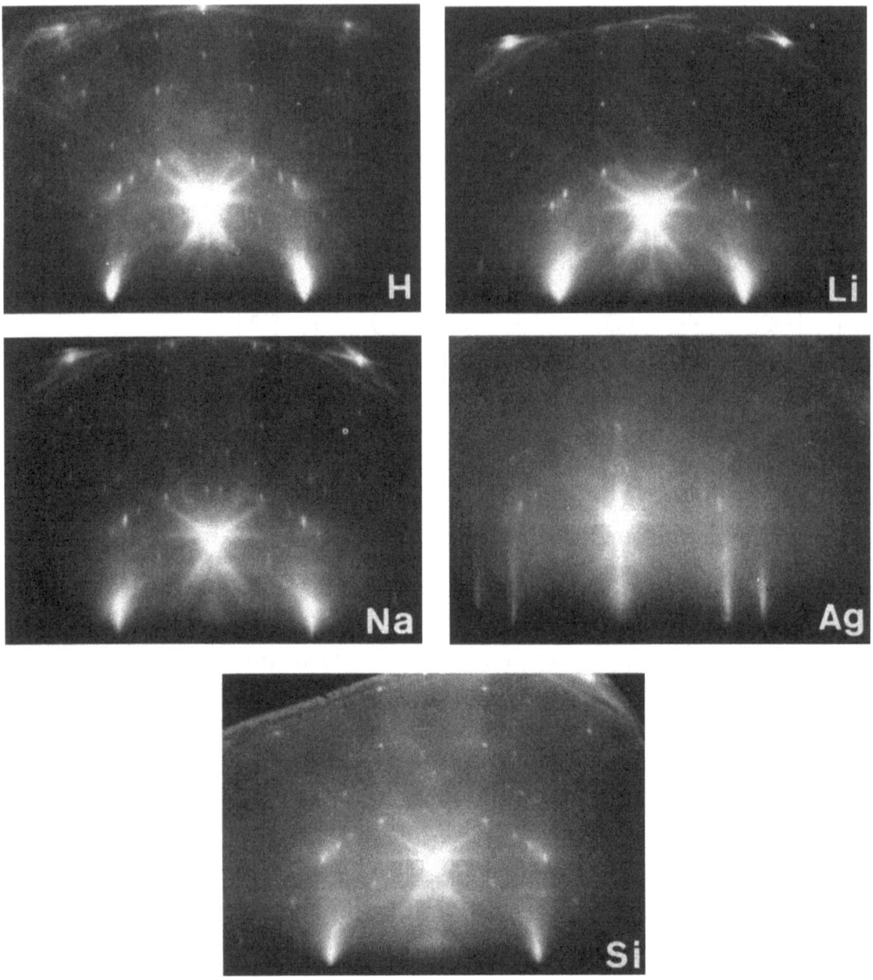

Fig. 4. RHEED patterns of the $\delta 7\times 7$ from Si(111) surfaces adsorbed with H, Li, Na, Ag and Si atoms at room temperature.

adsorbed surfaces with $\delta 7\times 7$ at room temperature. From rocking curve analyses [17,25,29,30] it has been concluded that the dimer-stacking-fault (DSF) framework is preserved for the adsorption processes with the $\delta 7\times 7$ reconstruction from the DAS structure. From analysis of rocking curves under the one–beam condition for several coverages of Li atoms shown in Fig. 5 [30], it has been found that back bonds of more than 80% (0.2ML) of Si adatoms in the DAS structure are cut off by adsorption of nearly the same number (0.2ML) of Li atoms. A similar process was also observed for Si deposition on an Si(111) surface at room temperature [29]. In this case the RHEED pattern changed completely into $\delta 7\times 7$ at

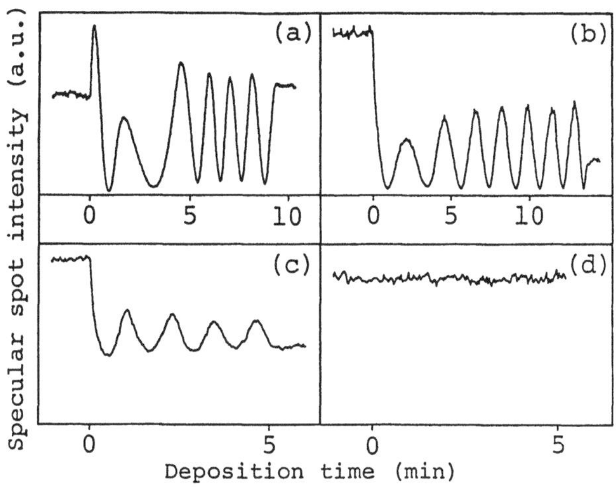

Fig. 5. RHEED (00)–rod intensity oscillations during molecular beam epitaxial growth of silicon; (a) at 280 °C, (b) at 400 °C, (c) at 600 °C and at 700 °C. These oscillations were taken at a glancing angle of 1.2 °.

0.5ML coverage of adsorbed Si atoms, and the intensity of the (3/7 3/7) rod, mainly due to periodicity of adatoms, decreased quickly at 0.25ML deposition. From rocking curve analysis during Si deposition, it is concluded that back bonds of adatoms are broken by adsorbed atoms and randomly distributed pyramidal clusters are formed subsequently on the DSF framework [17]. For further deposition of Si, an amorphous Si layer is formed on the $\delta 7\times 7$ structure.

For Li and Ag adsorbed $\delta 7\times 7$ surfaces, the surface structure changes into $\sqrt{3}\times\sqrt{3}$ by annealing at about 300 °C. For a Si deposited $\delta 7\times 7$ surface, annealing at above 500 °C results in solid phase epitaxy. On these structural changes from $\delta 7\times 7$ to $\sqrt{3}\times\sqrt{3}$ or under solid phase epitaxy, the stacking fault at the DSF layer is reconstructed into the normal stacking. Such stacking fault dissolution also occurs on homoepitaxial growth on Si(111)7×7. At temperatures higher than 300 °C, RHEED intensity oscillation during the homoepitaxial growth is very stable as shown in Fig. 5. For the beginning 5ML deposition up to 500 °C, the oscillation amplitudes are irregular, but after that the oscillations become stable and periodic. At this stable region, profiles of rocking curves changed periodically with the coverage [16]. Therefore it is concluded that no defects and stacking faults remain in the growing layers. At the growing surface, a mixed phase of 5×5 and 7×7 structures was observed. At the maximum intensity of the oscillation, the one–beam rocking curves

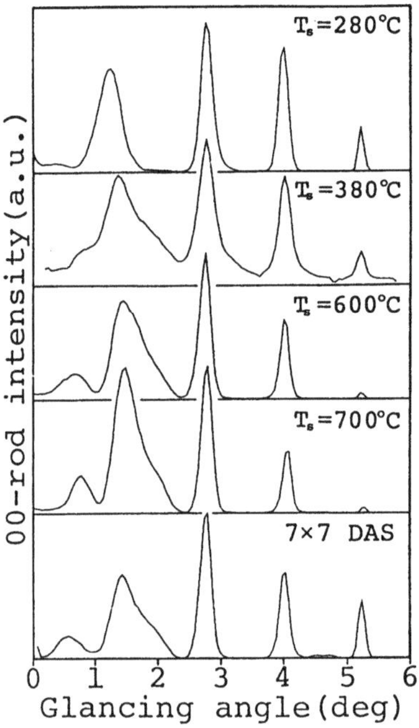

Fig. 6. RHEED intensity rocking curves at one–beam condition during homoepitaxial growth at several temperatures. The rocking curve from the 7×7 DAS structure is also shown in the figure.

at higher than 400 °C are very similar to the curve from the Si(111)7×7 DAS structure as shown in Fig. 6. Therefore the structures of 5×5 and 7×7 at the surface are nearly the same as the usual DAS structure.

Reconstruction of the stacking fault layer into the normal stacking by deposition of silicon atoms is assisted via breaking backbonds of adatoms and dimer bonds by coexistence of the 5×5 and 7×7 phases. At 280 °C, the one–beam rocking curve is very different from the rocking curve of the DAS structure at higher temperatures as shown in Fig. 6. The rocking curve is very similar to the curve from the δ7×7 structure as shown in Fig. 7. Since RHEED intensities on near the half order spots are stronger than those of other spots in the pattern from growing surface at 280 °C, the surface structure is a little different from the δ7×7 one, and $2a$ (a=3.84Å) periodicity remains on the surface such as a pyramidal cluster-like structure shown in Fig. 8. Since RHEED intensity oscillation is very stable at this temperature, there are no defects remaining in the growth layer. It seems that the stacking fault region is dissolved into the normal stacking due to the metastable structure with stimulation by further deposited atoms. Successive epitaxial growth at this temperature is assisted by formation of such metastable structure. At higher temperatures it is promoted that the metastable structure is formed at an intermediate stage of the layer growth.

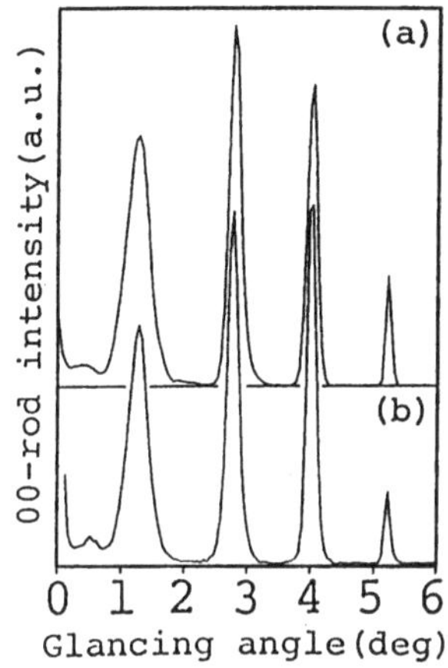

Fig. 7. (a) A rocking curve during silicon growth on Si(111) at 280 °C. (b) A rocking curve after 0.7 bilayer deposition of silicon at room temperature.

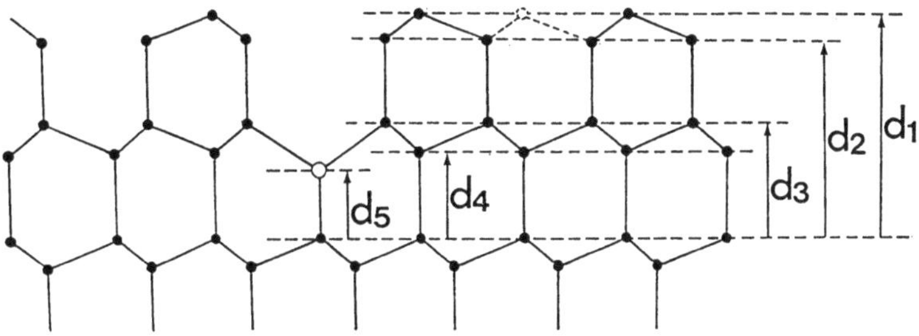

Fig. 8. [01$\bar{1}$] side view of the pyramidal cluster model. Atomic positions shown by broken lines are added for the $\delta 7\times 7$ structure. The values of d's are d_1=6.1Å, d_2=5.43Å, d_3=3.135Å, d_4=2.34Å and d_5=2.25Å.

In conclusion, the Si(111)7×7 DAS structure changes into $\delta 7\times 7$ by adsorption at room temperature. The structural change is due to breaking back bonds of adatoms of the DAS structure. It is considered that the $\delta 7\times 7$ structure is more unstable than the 7×7 DAS structure. Therefore such instability of the $\delta 7\times 7$ structure induces reconstruction of the stacking fault into the normal stacking at about 300 °C, while the stacking fault is dissolved at 850 °C for the clean Si(111)7×7 structure.

Acknowledgment

This work was carried out with the support of a Grant–in–Aid for Scientific Research on Priority Areas from the Ministry of Education, Science and Culture.

References

1 K. Takayanagi, Y. Tanishiro, M. Takahashi, S. Takahashi: Surf. Sci. **164**, 367 (1987).
2 R. M. Tromp: Surf. Sci. **155**, 432 (1985).
3 R. M. Tromp, E. J. van Loenen: Surf. Sci. **155**, 441 (1985).
4 I. K. Robinson, W. K. Waskiewicz, P. H. Fuoss, J. B. Stark, P. A. Bennett: Phys. Rev. **B33**, 7013 (1986).
5 G. Binnig, H. Rohrer, Ch. Gerber, E. Weibel: Phys. Rev. Letters **50**, 120 (1983).
6 A. Ichimiya: Surf. Sci. **192**, L893 (1987).
7 S. Y. Tong, H. Huang, C. M. Wei, W. E. Packard, F. K. Men, G. Glanden, M. B. Webb: J. Vacuum Sci. Technol. **A6**, 615 (1988).
8 Y. Horio, A. Ichimiya: Surf. Sci. **219**, 128 (1989).
9 Guo–Xia Qian, D. J. Chadi: Phys. Rev. **35**, 1288 (1987).
10 T. Yamaguchi: Phys. Rev. **B34**, 1085 (1986).
11 M. Fujita, H. Nagayoshi, A. Yoshimori: Surf. Sci. **208**, 155 (1989).
12 J. Yanagisawa, A. Yoshimori: Surf. Sci. **231**, 297 (1990).
13 H. Ohtani C.–T. Kao, M. A. van Hove, G. A. Somorjai: Prog. Surf. Sci. **23**, 155 (1986).
14 J. J. Lander: Surf. Sci. **1**, 125 (1964).
15 T. Sakamoto, N. J. Kawai, T. Nakagawa, K. Ohta, T.Kojima: Appl. Phys. Letters **47**, 617 (1985).
16 H. Nakahara, A. Ichimiya: J. Crystal Growth **95**, 472 (1989).
17 H. Nakahara, A. Ichimiya:Surf. Sci. **241**, 124 (1991).
18 A. Ichimiya: Jpn. J. Appl. Phys. **22**, 176 (1983) ; **24**, 1365 (1985).
19 S. Ino: Jpn. J. Appl. Phys. **16**, 891 (1977).
20 H. Iwasaki, S. Hasegawa, M. Akizuki, S.–T. Li, S. Nakamura, J. Kanamori: J. Phys. Soc. Jpn. **56**, 3425 (1987).
21 S. Kohmoto, A. Ichimiya: Surf. Sci. **223**, 400 (1989).
22 Y. Sakamoto, J. Kanamori: J. Phys. Soc. Jpn. **58**, 1512 (1989).
23 J. Kanamori, Y. Sakamoto: Surf. Sci. in press.
24 J. E. Northrup: Phys. Rev. Letters **57**, 154 (1984).
25 H. Daimon, S. Ino: Surf. Sci. **164**, 320 (1985).
26 A. Ichimiya, S. Mizuno: Surf. Sci. **191**, L765 (1987).
27 S. Mizuno, A. Ichimiya: Appl. Surf. Sci. **33/34**, 38 (1988).
28 S. Kohmoto, A. Ichimiya: Appl. Surf. Sci. **33/34**, 45 (1988).
29 H. Nakahara, A. Ichimiya: J. Crystal Growth **99**, 514 (1990).
30 S. Kohmoto, S. Mizuno, A. Ichimiya: Appl. Surf. Sci. **41/42**, 107 (1989).

Theoretical Modeling of Structure and Dynamics of Metal Surfaces

K.W. Jacobsen and P. Stoltze

Laboratory of Applied Physics, Technical University of Denmark, DK-2800 Lyngby, Denmark

The structure and dynamics of metal surfaces can be modeled using the effective medium theory to describe the interatomic interactions. In this paper we briefly outline the physical picture behind the effective medium theory and apply the method to the Cu(110)-surface. We determine the energies associated with different kinds of defects on the clean surface and discuss reconstruction energies, step-edge dynamics, diffusion, and premelting for this surface. We also comment on the changes in surface structure induced by chemisorption of alkali or oxygen atoms.

Introduction

The modeling of structure and dynamics of surfaces on an atomic scale requires some knowledge about the adiabatic potential energy surface or, in other words, the interaction potentials between the particles. Many methods ranging from *ab initio* local density functional calculations over tight binding models to empirically fitted pair potentials have been used to calculate interatomic potentials. The method used in this paper where we concentrate on the description of metallic systems is the effective medium theory[1]. The virtues of this approach is that there is a clear physical picture behind the construction of the interatomic potentials and, furthermore, that the calculation of the binding energy of a metallic system is fast enough to be used in Monte Carlo or Molecular Dynamics simulations with thousands of atoms.

In the following section we give a brief outline of the the effective medium theory. We then present the applications on the clean Cu(110) surface including calculations of defect energies and a discussion of step-edge fluctuations and premelting of this surface.

1 The Effective Medium Theory

The effective medium theory[1] is derived from density functional theory with the local density approximation for the exchange and correlation energy. An ansatz for the electron density is constructed by superposing densities from each of the atoms in the system. The atomic electron density for atom i is chosen as the density induced by the atom, when it is embedded in a homogeneous

Springer Series in Materials Science, Vol. 17 Ordering at Surfaces and Interfaces
Eds.: A. Yoshimori · T. Shinjo · H. Watanabe

electron gas with a density $\bar{n}_i$. The value for this embedding density is obtained by performing a spatial average over a sphere around atom i of the density tails sticking into the sphere from the surrounding atoms. The radius of the sphere is chosen so that the sphere is neutral, *i.e.*, the positive charge of the nucleus is exactly cancelled by the electrons inside the sphere.

The result of this construction is an approximate expression for the binding energy E_B of the system that consists of three terms.

$$E_B = \sum_i E_{c,Z_i}(\overline{n}_i) + E_{\mathrm{AS}} + E_{\mathrm{1el}}. \tag{1}$$

The first term is a sum over all the atoms in the system of the so-called *cohesive functions* $E_{c,Z_i}(\overline{n}_i)$. For a given atomic number Z the cohesive function $E_{c,Z}$ only depends on the embedding density $\overline{n}_i$ of the atom in question and it is closely related to the embedding energy of the atom in a homogeneous electron gas of density $\overline{n}_i$[1]. For most atoms the cohesive function exhibits a single minimum at a certain density indicating that the atom likes to be positioned in an environment with this electron density. The curvature of the cohesive function is denoted E_2 and is determined mainly by the kinetic energy of the electrons. It is proportional to the bulk modulus of the metal.

The second term E_{AS} is called the atomic-sphere-correction energy or overlap energy and it is an electrostatic energy. As mentioned above we associate a neutral sphere with each atom. The electrostatic interactions *within* such a sphere are already accounted for in the cohesive function for that atom. That means that we have included all the electrostatics in the system in the so-called atomic sphere approximation (ASA). This is known to be a very good approximation for bulk metals with a perfect close-packed lattice where the atomic spheres approximately cover all of the space and where the region of space where several spheres overlap is small and can be neglected. In situations where the ASA is not appropriate it is necessary to correct for the double counting of the electrostatic energy in the regions of space that falls within more than a single sphere. This is exactly what the overlap energy E_{AS} does.

The strength of the atomic-sphere-correction energy is determined by an electrostatic parameter α which is proportional to the elastic constant C_{44} for the metal (for details see ref. [1]).

The third term E_{1el} in Eq. (1) is the so-called one-electron energy. When an atom is embedded in a metallic system a change of the band-energy is induced. The one-electron energy E_{1el} is the difference in this change when the embedding of the atom in the real metallic system is compared to the embedding in the homogeneous electron gas. The one-electron energy can usually be neglected for simple metals or noble metals and this is what we shall do in the following. However, we shall return to the one-electron energy when discussing the oxygen-induced reconstructions.

The effective medium expression for the energy is computationally very efficient. It requires approximately the same computational effort as models with pairwise interactions. The effective medium theory does, however, contain many-body interactions which are essential for a correct description of the energetics of metallic systems.

2 Energetics of the Cu(110) surface

The structure and dynamics of surfaces are at low temperatures completely governed by the energetics and in this section we shall discuss some results on typical energies for reconstructions and defects on the Cu(110) surface.

Particularly simple estimates of defect energies can be obtained if we neglect relaxation effects *i.e.*, if we keep all interatomic distances at their bulk values. In that case the effective medium expression for the energy can be written as a sum over energies of all the atoms where the energy $E(N_1, N_2)$ of a given atom depends only on the number of nearest neighbors (N_1) and second nearest neighbors (N_2) of that atom. If we expand the energy to second order in N_1 and to first order in N_2 around the bulk values ($N_1 = 12$, $N_2 = 6$) we get

$$E(N_1, N_2) = \epsilon_1'(12 - N_1) + \frac{1}{2}\epsilon_1''(12 - N_1)^2 + \epsilon_2'(6 - N_2), \tag{2}$$

where the coefficients are given in terms of effective medium parameters as

$$\begin{aligned}
\epsilon_1' &= \frac{\alpha n_0}{12}\left(\frac{1}{\gamma_2} - \frac{1}{\gamma_1}\frac{\eta}{\eta + \eta_1}\right) \\
\epsilon_1'' &= \frac{1}{144}\frac{1}{\gamma_1^2}\left(\frac{\eta}{\eta + \eta_1}\right)^2\left(2E_2 - \alpha n_0\frac{\eta_1}{\eta}\right) \\
\epsilon_2' &= \frac{\alpha n_0}{6}\left(\left(1 - \frac{1}{\gamma_2}\right) - \left(1 - \frac{1}{\gamma_1}\right)\frac{\eta}{\eta + \eta_1}\right).
\end{aligned}$$

The parameters for Cu and Al can be found in ref. [1]. In the case of Cu the coefficients are calculated to be $\epsilon_1' = 0.116$ eV, $\epsilon_1'' = 0.005$ eV, and $\epsilon_2' = 0.029$ eV.

The physical interpretation of ϵ_1', ϵ_1'', and ϵ_2' is as follows. The coefficients ϵ_1' and ϵ_2' of the terms linear in the coordination numbers are half the energy cost of breaking a bond between nearest neighbor and second nearest neighbor atoms, respectively, in a simple model with *effective* pairwise interactions. It should be noted that the effective pair energy, ϵ_1', which is useful when considering moderate changes in coordination number away from the bulk value is much smaller than the pair energy determined by fitting pair potentials to bulk properties of the metal. In the case of Cu, $\epsilon_1' = 0.116$ eV, while the cohesive energy divided by 12 is of the order 0.3 eV. The coefficient ϵ_2' of the quadratic term is a typical energy associated with three-atom interactions. The magnitudes of ϵ_1' and ϵ_1'' are determined by the electrostatic parameter α which comes from the overlap energy while the three-atom term is seen to be the result of

Table 1: Various energy formulas and numerical values for Cu. E_s is the surface energy per area, $E_{rec}^{(1x2)}$ is the energy change for the (1x2) missing row reconstruction of the (110) surface, $E_{rec}^{(2x1)}$ is the energy change for the (2x1) missing row reconstruction of the (110) surface, E_{step} is the energy to create a step on the (110) surface for three different step directions (the listed direction is parallel to the step), E_{kink} is the energy required to create a *pair of kinks* on the steps, and finally E_{a-v} is the energy required to create an adatom vacancy pair on the three different surfaces. a denotes the lattice constant. In the calculation of the energies for Cu we have used the calculated value $a = 6.60$ a_0.

		Formula	Value for Cu (meV)
	(111)	$2\sqrt{3}(2\epsilon_1' + 3\epsilon_1'' + 2\epsilon_2')/a^2$	24.2 meV/a_0^2
E_s	(100)	$2(4\epsilon_1' + 8\epsilon_1'' + \epsilon_2')/a^2$	24.5 meV/a_0^2
	(110)	$\sqrt{2}(6\epsilon_1' + 13\epsilon_1'' + 4\epsilon_2')/a^2$	28.4 meV/a_0^2
$E_{rec}^{(1x2)}$	(110)	$-\sqrt{2}(2\epsilon_1'' - \epsilon_2')/a^2$	0.6 meV/a_0^2
$E_{rec}^{(2x1)}$	(110)	$\sqrt{2}(\epsilon_1' + 4\epsilon_1'')/a^2$	4.4 meV/a_0^2
	$[1\bar{1}0]$	$-\sqrt{2}(2\epsilon_1'' - \epsilon_2')/a$	3.9 meV/a_0
E_{step}	[001]	$(2\epsilon_1' + 7\epsilon_1'')/(2a)$	20.3 meV/a_0
	$[1\bar{1}2]$	$(2\epsilon_1' + 5\epsilon_1'' + 2\epsilon_2')/(\sqrt{6}a)$	19.4 meV/a_0
	$[1\bar{1}0]$	$2\epsilon_1' + 9\epsilon_1''$	277 meV
E_{kink}	[001]	$-2(\epsilon_1'' - \epsilon_2')$	47 meV
	$[1\bar{1}2]$	$-2\epsilon_1''$	-10 meV[a]
	(111)	$3(4\epsilon_1' + 17\epsilon_1'')$	1649 meV
E_{a-v}	(100)	$2(4\epsilon_1' + 15\epsilon_1'' + 4\epsilon_2')$	1307 meV
	(110)	$4\epsilon_1' + 11\epsilon_1'' + 4\epsilon_2'$	633 meV

[a] The value becomes positive (and still very small) if third nearest neighbor interactions are included.

a balance between a contribution from the cohesive function ($2E_2$) and the overlap energy ($\alpha n_0 \eta_1/\eta$).

In Table 1 a number of structural energies are listed. The table shows the analytic expressions for the energies in terms of the above defined coefficients and also the actual values obtained for Cu. The calculated surface energies are seen to be in good agreement with the experimental value[2] 31.9 meV/a_0^2.

A very small energy change is involved in the so-called (1x2) missing row reconstruction of the (110)-surface of fcc metals. This is due to the fact that the reconstruction does not lead to any net breaking of nearest neighbor bonds. This is reflected in the formula for $E_{rec}^{(1x2)}$ in the table because ϵ_1' does not appear. The driving force behind the reconstruction comes from the term in ϵ_1'' proportional to E_2. As mentioned above E_2 is proportional to the bulk modulus of the metal and mainly determined by the kinetic energy of the electrons. The reconstruction is opposed by the electrostatic terms involving the parameter α which is proportional to the elastic constant C_{44} of the metal. Therefore the higher the value for B/C_{44} for a given metal the greater the tendency for the metal to reconstruct should be[3]. This is in rough agreement with the experimental results which show that only Au, Pt, and Ir exhibit a reconstruction of the clean surface.

For other fcc metals the (1x2) missing row reconstruction can be induced by chemisorption of small amount of alkali atoms on the surface. The reason for this is that the alkali atoms bind more strongly in the valleys on the missing row reconstructed surface than on the unreconstructed one due to the higher effective coordination number of metal atoms here[4]. This additional chemisorption energy is of the order 0.2 eV for K on Cu[4] and even a very low coverage of alkali atoms is therefore sufficient to overcome the small reconstruction energy for the clean surface.

The energy cost for the (2x1) missing row reconstruction of the Cu(110) is much higher than for the (1x2) reconstruction as can be seen in the table. The (2x1) structure is formed by removing every second atom in the close packed rows and involves a net breaking of nearest neighbor bonds. Experimentally[5] oxygen is known to induce the (2x1) reconstruction and this has also been modelled using the effective medium theory[6]. The hybridization between the oxygen *p*-levels and the copper *d*-levels is so strong that antibonding states can be shifted up above the Fermi level. It is therefore necessary in this situation to include the one-electron energy E_{1el} in the description. The driving force between the reconstruction is a shift up of the surface d-band when the surface reconstructs leading to an additional binding because more antibonding O-*p*–Cu-*d* states are shifted above the Fermi level.

The energy required to create steps for three different orientations on the (110)-surface is also shown in the table. The energy for the step along the $[1\bar{1}0]$ direction is much smaller than for the other two directions shown because there is no net change in the number of nearest neighbor bonds when the step is formed.

The kink-pair energy E_{kink} is defined as the energy required to form a kink-antikink pair on the step. It can be seen from the table that the energy required to create a kink-pair on the step parallel to the $[1\bar{1}2]$ direction is much smaller than the energy necessary on the $[1\bar{1}0]$ or [001] steps. This has consequences for the dynamics of the step edges as will be discussed in the next section.

The last entry in the table, $E_{\text{a-v}}$, is the energy required to form an adatom vacancy pair on three different surfaces. It is seen that the energy for the (110) surface is considerably smaller than for the (111) and (100) surfaces. This has consequences for the onset of surface premelting which will also be discussed in the following section.

3 Dynamics of the Cu(110) surface

In this section we discuss some aspects of the dynamics of the Cu(110) surface. We start out by noting that the differences in energy between the (110) and the other low index planes of Cu determine the equilibrium shape of the crystallites observed in simulations at low temperatures[7]. These simulations show that the (100) and in particular the (111) surfaces are more stable than the (110) surface. The ease of formation of defects near the edges are reflected in the rounding of the crystallite shapes at higher temperature.

Experimentally the dynamics of steps on the Cu(110) surface has been studied by scanning tunneling microscopy (STM)[8]. STM pictures recorded at near atomic resolution on Cu(110) surfaces indicate that the edges of steps parallel to the $[1\bar{1}0]$ or [001] directions are quite stable at room temperature while step edges in a third direction exhibit strong fluctuations. The elementary excitations responsible for these fluctuations must be the formation of kinks. The kink-pair energy for the $[1\bar{1}0]$ or [001] steps are quite large compared with the temperature while for the step parallel to the $[1\bar{1}2]$ direction the kink-pair energy is small because there is no net breaking of nearest or second nearest neighbor bonds. It is therefore plausible that the observed fluctuations are around this direction. Stochastic simulations based on the calculated kink formation energies resemble at least qualitatively the experimental STM pictures.

A number of metal surfaces show a gradual loss of order well below the bulk melting point for the crystal. This phenomenon, known as surface premelting, is interesting for the understanding of the role of the surface in the bulk melting process as well as for the understanding of the structure and thermodynamics of surfaces at high temperatures.

The structure of the premelted surface has been studied by a number of experimental techniques[9] and computer simulations have been performed for Al, Cu, and Ni[11]. Recently it has been shown that the experimental results using MEIS and the simulations using effective medium theory are in excellent agreement. In both experiments and simulations it has been found that premelting is only observed on the more open surfaces. Simulations[11] indicate that the disordering at the premelted surface is caused by the formation of adatom-vacancy pairs. As the temperature is increased the formation of these defects will gradually cause a loss of the structure parallel to the surface as well as the layer structure in the surface region. The results in Table 1 show that the formation-energy E_{a-v} of adatom-vacancy pairs is lower on the (110) surface than on the more close-packed (111) and (100) surfaces in agreement with the fact that only the open surfaces can exhibit premelting.

The structure factor as measured in a diffraction experiment or calculated in the simulations may be used to calculate order parameters in a characterization of the disordering process. From the simulations it was found that the disordering process appears anisotropic only for some choices of order parameter. One might say that the apparent anisotropy found in diffraction experiments is caused by anisotropy in the *detection* of the disorder and is not an intrinsic property of the surface itself.

The dynamics of atoms at a premelted surface has been studied by He atom scattering[10]. In these experiments the diffraction peaks were found to be broadened and this broadening was interpreted as a huge diffusivity of atoms at the premelted surface. Interestingly the experimental data could not be reproduced by simple models for diffusion on a lattice.

For simulations of the Al(110) and Cu(110) surfaces we find a large diffusivity at the premelted surface. The diffusivity is moderately anisotropic. The results of the simulations show that the dominant diffusion mechanism at the onset of premelting is a vacancy mechanism, where atoms from the surface layer jump up into the troughs and then skate along in the very shallow potential along the closepacked rows. At higher temperatures more adatoms are formed and the diffusion becomes more isotropic as the closepacked rows become progressively more disordered.

The observed diffusion mechanism and the onset of premelting for the (110) surfaces are thus both closely related to the formation of adatom-vacancy pairs. The relatively short lifetime for the adatoms is a severe complication for accurate calculations of their diffusivities in some temperature intervals. A detailed analysis of the diffusive motion of the adatoms seems to indicate that the diffusive motion has no characteristic jump length, this explains why a lattice model for the diffusion is not successful.

Acknowledgments: KWJ would like to express his gratitude to the organizers of the NEC symposium. We appreciate helpful discussions with Lars Hansen and Jens K. Nørskov. Financial support from the Danish Research Councils through the *Center for Surface Reactivity* is gratefully acknowledged.

References

[1] K. W. Jacobsen, J. K. Nørskov, and M. J. Puska, Phys. Rev. B 35, 7423 (1987); K. W. Jacobsen, Comm. on Condensed Matter Physics 14, 129 (1988).

[2] F. R. de Boer, R. Boom, W. C. M Mattens, A. R. Miedema, and A. K. Niessen, *Cohesion in Metals*, North-Holland, Amsterdam (1988).

[3] K. W. Jacobsen and J. K. Nørskov, in *The Structure of Surfaces II*, edited by J. F. van der Veen and M. A. van Hove, pg. 118 (Springer Series in Surface Sciences 11, 1987); K. W. Jacobsen, in *Many-Atom Interactions in Solids* edited by R. M. Nieminen, M. J. Puska, and M. J. Manninen, (Springer 1990).

[4] K. W. Jacobsen and J. K. Nørskov, *Phys. Rev. Lett.* **60**, 2496 (1988), and references therein.

[5] D. J. Coulman, J. Wintterlin, R. J. Behm, and G. Ertl, *Phys. Rev. Lett.* **64**, 1761 (1990); F. Jensen, F. Besenbacher, E. Lægsgaard, and I. Stensgaard, *Phys. Rev. B* **41**, 10233 (1990).

[6] K. W. Jacobsen and J. K. Nørskov, *Phys. Rev. Lett.* **65**, 1788 (1990).

[7] Lars B. Hansen. M. Sc thesis, Laboratory of Allied Physics (1990). Unpublished.

[8] F. Besenbacher, to be published.

[9] J. W. M. Frenken and J. F. van der Veen, *Phys. Rev. Lett.* **54**, 134 (1985); P. H. Fuoss, L. J. Norton and S. Brennan, *Phys. Rev. Lett.* **60**, 2046 (1988); K. C. Prince, U. Breuer and H. P. Bonzel, *Phys. Rev. Lett.* **60**, 1146 (1988).

[10] J. W. M. Frenken, J. P. Toennies and Ch. Wöll, *Phys. Rev. Lett.* **60**, 1727 (1988)

[11] P. Stoltze, J. K. Nørskov and U. Landman, *Phys. Rev. Lett.* **61**, 440 (1988); E. T. Chen, R. N. Barnett, U. Landman, *Phys. Rev. B* **40**, 924 (1989); P. Stoltze, to be published. *J. Chem. Phys.* **92**, 6303 (1990).

Part III

Adsorbed Surfaces

Alkali Metal Ordering on Semiconductor Surfaces and Interfaces

*P. Soukiassian**

Commissariat à l'Energie Atomique, Service de Recherche sur les Surfaces et l'Irradiation de la Matière, Centre d'Etudes de Saclay, F-91191 Gif sur Yvette Cedex, France and Département de Physique, Université de Paris-Sud, F-91405 Orsay Cedex, France

*Also at Department of Physics, Northern Illinois University, DeKalb, Illinois, USA

The formation of alkali metal/semiconductor surfaces and interfaces is investigated by synchrotron radiation using core level and valence band photoemission spectroscopies (XPS - UPS), surface and photoemission extended x-ray absorption fine structure (SEXAFS - PEXAFS) and by scanning tunneling microscopy (STM) experiments. On group IV semiconductors (Si, Ge), the bond between the alkali metal and the surface is found to be a weak and polarized covalent bonding, even at low coverages, with adsorbate metallization at the monolayer. Also, unlike many other metals on silicon surfaces, alkali metals (except Li) do not interdiffuse into the crystal with no silicide formation. On the Si(100)2x1 surface, an ordering transition occurs around 0.5 monolayer, in which the adsorbate-adsorbate interaction appears to be the main driving force, leading to the formation of alkali one-dimensional chains separated by 7.68 Å and parallel to the Si dimer rows at the monolayer coverage. In strong contrast to silicon, alkali metal induces dramatic structural changes of III-V semiconductor surfaces with removal of the relaxation at low coverages. On both GaAs(110) and InP(110) surfaces, the alkali atoms have a double site of adsorption at low coverages. The proposed models and concepts are compared to recent state-of-the-art ab-initio theoretical calculations.

I - INTRODUCTION AND HISTORICAL BACKGROUND

The investigation of alkali metal adsorption on surfaces initiated by Langmuir [1] nearly 70 years ago was a milestone for the beginning of surface science. Since that time, this topic is still being widely investigated for metal as well as semiconductor surfaces. From a fundamental approach, alkali metal/metal or semiconductor surfaces and interfaces represent a prototypical system which allows one to probe basic ideas about interface formation, chemisorption and surface catalysis [2]. Regarding applications, alkali metals are known for their many versatile aspects in surface catalysis, negative electron affinity (NEA) cathodes, thermoelectric and photoelectric emitters, ion propulsion, thermionic conversion and negative ion sources for magnetic fusion devices [3]. On semiconductor surfaces, alkali metals have been shown to be highly efficient promotors of semiconductor oxidation and nitridation with possible removal of the metal catalyst [4,5]. Togother with their special role in the Schottky barrier formation and in monitoring the band line-up at heterojunctions and at insulator/semiconductor interfaces [2,6-8], these unique properties have been a driving force in the recent investigations on the basic understanding of the electronic and structural properties of alkali metal adsorption on elemental and compound semiconductor surfaces. In the historical model of Langmuir and Topping, the alkali metal transfers its valence electron to the surface [1]. The alkali ion forms with its image a dipole, which explains the dramatic decrease of the substrate work function. When the coverage is increased, resulting in the decrease of the alkali-alkali distance, depolarization effects occur leading to a minimum of the work function and to adsorbate metallization near one monolayer [1]. This model was found, until very recently, to be suitable for many kinds of substrate. Since that time, the studies on alkali metal adsorption on surfaces have been considerably developed driven both by fundamental and by technological motivations. More recently, the emergence of new experimental tools as well as more sophisticated theoretical approaches which were not available at Langmuir and Kingdom's time, have brought new insights into the understanding of these systems.

So far, the Langmuir ionic model has been supported by electron energy loss, photoemission and SEXAFS experiments for full d-band and jellium-like metal surfaces such as Cu, Ag and Al [9-11]. However, this picture of an ionic bonding, even in its most sophisticated approaches [12] was unable to explain the electronic properties of alkali metal-transition metal systems such as Mo and W [13,14]. Ab-initio calculations taking into account the true electronic structure of the substrate including localized electronic surface states [13] together with photoemission spectroscopy experiments [14,15] have brought new insights into the understanding of systems which include the presence of localized surface states having a dangling bond character. The nature of the alkali metal-substrate bond was found to be covalent even at low coverages by hybridization between substrate and adsorbate valence electrons [12-16], with no significant charge transfer [14,17,18]. This picture was also found to be in good agreement with other experimental results such as metastable deexcitation spectroscopy - MDS - [19] or nuclear magnetic resonance studies - NMR - [20]. Also, the investigations of the structural properties of alkali metal/metal systems have shown that the adsorbate also induced surface reconstruction, even at low coverages [14,21,22]. Therefore, in view of the above described properties of alkali metal-metal surfaces, it is interesting to investigate alkali metal deposition on semiconductor surfaces which, like transition metals, also have localized surface states related to dangling bonds.

In this article, we will review the latest concepts on the alkali metal/semiconductor interfaces formation on the basis of recent photoemission, SEXAFS, PEXAFS and STM investigations. We will address some important issues such as the nature of the alkali metal/semiconductor bond, the structural changes induced by alkali metals on semiconductor surfaces, the mode of growth during the interface formation, the bond length as well as the adsorbate local ordering on the surface. The results will also be discussed in view of the most recent theoretical calculations.

II - EXPERIMENTAL DETAILS

The angle-integrated photoemission experiments were performed in a ultra-high vacuum chamber at pressures in the low 10^{-11} torr range at the Synchrotron Radiation Center of the University of Wisconsin-Madison. Alkali metal deposition was performed by using very carefully outgassed SAES Getters chromate sources. Other details concerning the experimental set-up and the data analysis can be found elsewhere [23-25].

Alkali metals are known to be highly sensitive to other species such as oxygen, water, atomic hydrogen, carbon monoxide [2,4,26]. It is therefore very important to have a very good vacuum and to outgas very carefully the alkali metal sources which otherwise would release oxygen, carbon monoxide and atomic hydrogen [4]. While the presence of molecules like oxygen or CO could induce charge transfer from the alkali adsorbate and enhancement of their sticking coefficient [2,4], atomic hydrogen has been shown by photon and electron stimulated ion desorption - PSD - [27] & ESD - [28] experiments also to bond to the alkali metal (likely as a hydrate) [27]. This could result in dramatic changes in the electronic as well as structural properties of alkali metal [2,4] and has therefore to be stressed. In the experiments presented here, the pressure during alkali metal deposition never increased by more than 2.10^{-11} torr during the time necessary to grow a monolayer.

III - ALKALI METAL/GROUP IV SEMICONDUCTOR INTERFACES: ELECTRONIC PROPERTIES

a - Silicon surfaces

As mentioned above in the first section, one of the main issues concerning the electronic properties of these systems is the nature of the alkali metal-silicon bond. In addition to the classical model of Langmuir [1], recent pseudopotential calculations [29,30] on the Na/Si(111), K/Si(100)2x1, K/Si(111)2x1 and Na/Si(100)2x1 systems found a very short K-Si bond length at 2.59 Å for the K/Si(100)2x1 system, resulting in a very strong ionic bonding with a complete charge transfer to the surface and silicon surface metallization [30]. However this strong ionic bonding picture, which contrasted with previous models based on theory and experiment [31,32], was questioned almost immediately by other total energy theoretical investigations on the same K/Si(100)2x1 system in which a much larger bond length was found at 3.30 Å by a pseudofunction method [33] and at 3.22 Å by a DMOL calculation using a very large cluster [34]. The K-Si distance was measured experimentally by SEXAFS to be 3.14 Å at one K monolayer which corresponds to the exact sum of K and Si covalent radii [35]. In principle, although it is not easy, from a bond length, to draw a definitive conclusion about the nature of the bonding [36], one can however conclude that the alkali metal/silicon bond is weak, which

Fig.1

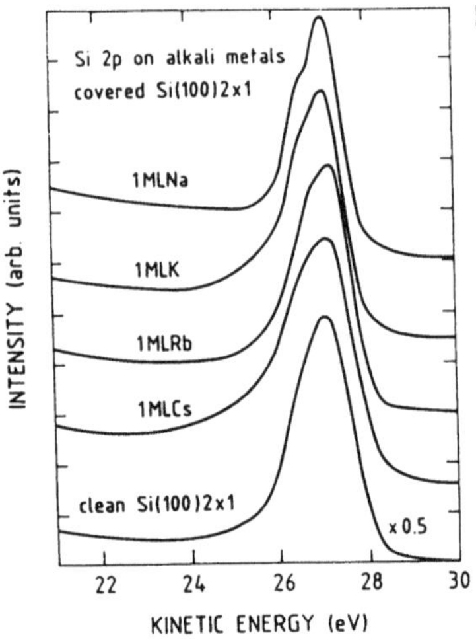

Fig.2

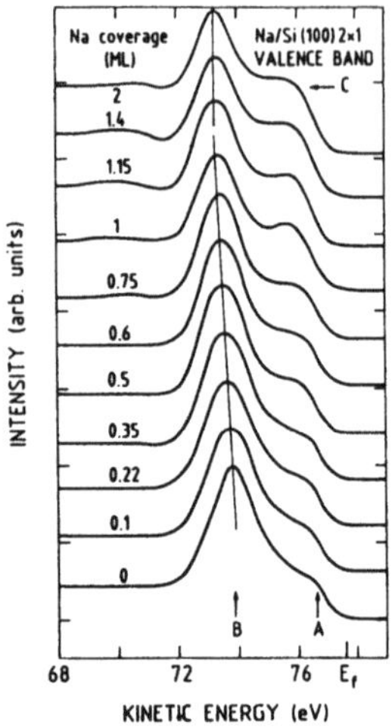

Figure 1: Si 2p core level for a Si(100)2x1 surface covered with 1 monolayer (1ML) of Cs, Rb, K and Na. The photon energy was 130 eV.

Figure 2: Na/Si(100)2x1: valence band spectra for various sodium coverages between 0 and 2 monolayers. The photon energy was 82 eV.

is further confirmed by the rather low temperatures of desorption of alkali metals from silicon surfaces [23,24, 37-39], as compared to e.g. from metal surfaces [40].

One way to probe the importance of a charge transfer in an ionic type of bonding is to look at possible shifts at the Si 2p core level. Very large chemical shifts ($\approx \pm 1$eV) to higher or lower binding energies (for acceptor and donor species respectively) have been previously reported for ionic adsorbate (with very high or very low electronegativity) on silicon surfaces [41-43]. Instead, as can be seen from figure 1 for Cs, Rb, K or Na on Si(100)2x1 surfaces (also for Cs or Na on Si(111)2x1 and Si(111)7x7), there is no significant core level shift at the Si 2p core level, even at low coverages which would indicate an important charge transfer as a result of an ionic type of bonding [2]. This do not favor the strong ionic bond picture based on the pseudopotential calculations [30].

Figure 2 displays the valence band spectra for the representative Na/Si(100)2x1 at various sodium coverages. On the clean Si (undoped) surface, peak A represents the "3p" surface state related to the dangling bond which is occupied for the undoped silicon. In the case of a strong ionic bonding, one should expect destruction of this electronic surface state upon alkali metal deposition. Instead, its intensity increases with the Na coverage up to 1 ML of Na, and then slightly decreases beyond the monolayer coverage (peak C). This behavior is typical of the formation of an electronic interface state (peak C). It is now necessary to identify the electronic levels involved. The only two possible candidates in this energy range are the Si 3p and Na 3s valence electrons. But, due to its very small cross section at the photon energies used here (82 eV), it is not possible to observe in

Table 1: Interatomic distances measured by SEXAFS (a) and PEXAFS (b).

a - SEXAFS [after reference 35]				
System	Auger line	R1 (Å)		
1 ML K/Si(100)2x1	K *KLL*	3.14±0.10		
b - PEXAFS [after reference 46]				
System	photopeak	R1 (Å)	R2 (Å)	R3 (Å)
Clean Si(100)2x1	Si 2p	2.34±0.04	3.87±0.08	4.69±0.10
0.4 ML Na/Si(100)2x1	Si 2p	2.35±0.05	3.86±0.08	4.72±0.10
1 ML Na/Si(100)2x1	Si 2p	2.32±0.06	3.85±0.08	4.75±0.13
0.3 L O_2/1 ML Na/Si(100)2x1	Si 2p	2.50±0.08	3.92±0.08	4.91±0.10
1 ML Cs/Si(100)2x1	Si 2p	2.36±0.06	3.90±0.08	4.78±0.13
0.4 ML Na/Si(100)2x1	Na 2p	2.80±0.10	4.40±0.12	
1 ML Na/Si(100)2x1	Na 2p	2.78±0.10	4.36±0.12	

Note: All measurements were made in p-polarization. R1, R2 and R3 designate the first, second and third nearest neighbour distances.

photoemission the alkali "s" valence electron without a change in its character [44]. The situation would be totally different if the Na 3s electron hybridized with the more localized Si 3p electronic surface state related to the Si dangling bond. In this case, the observed growth of peak C with increasing Na coverages in figure 2 could be easily explained. It clearly indicates that the Na atom is bonded to the Si atom by its dangling bond, through a covalent bonding [44,45]. Further confirmation of this picture could be found by looking at the the behavior of the 3p bulk-like electronic level, peak B (Fig. 2) which is shifted to higher binding energies by 0.6 eV upon sodium deposition, as a result of adsorbate-substrate bonding formation. Similar shifts (≈ 1 eV) have been reported for the d electronic states of W(100), Mo(100), Mo(110) and Ta(100) surfaces upon Cs deposition as a result of hybridization of adsorbate and substrate valence electrons [14,15]. Because of the low temperatures of desorption of alkali metals, 550°C for Na [23,24,37], 600°C for K [35,38,39], 650°C for Rb [37,39] and Cs [23,24,37], the covalent bonding between alkali metal and silicon atoms is weak. This point is further stressed by the large bond length, 3.14 Å for the K-Si distance on Si(100)2x1 measured by SEXAFS [35] and 2.80 Å for the Na-Si one on both Si(100)2x1 and Si(111)2x1 surfaces as measured by PEXAFS [46] at the monolayer coverage. All these values, which are in good agreement with recent ab-initio theoretical calculations [33,34,47] and recent dynamical LEED measurements [48] are summarized in table 1 after phase shift correction of the Fourier transform F(R) of the EXAFS signal at the Na 2p core level for the representative Na/Si(100)2x1 system (Fig. 3). In all cases, the bond length corresponds to the sum of the covalent radii of the adsorbate and substrate atoms which further supports the model of covalent bonding [35,46]. As a result of an important decrease of the work function [49-51], this bond is also polarized.

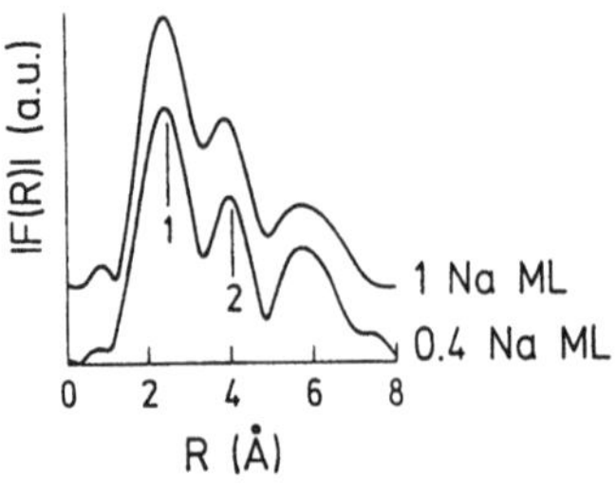

Figure 3: Magnitude of the Fourier transform F(R) measured from PEXAFS oscillations at the Na 2p core level for clean Si(100)2x1, 0.4 ML Na/Si(100)2x1 and 1ML Na/Si(100)2x1.

So far, we have focussed the discussion on the monolayer coverage. It is clear from figure 1 that the 3s-3p hybridized electronic interface state also exists below half a monolayer coverage. This suggests that the nature of the bonding remains the same at low coverages, i.e. covalent. This is further supported by the fact that the Na-Si bond length remains constant at 0.4 and 1 ML Na as measured by PEXAFS at 2.80 Å (see figure 3 and Table 1) [46]. This behavior is in contrast to the one observed in the case of a full d-band metal surface like Ag covered with Cs, where a significant change in the bond length versus the Cs coverage was interpreted as a result of an ionic to non-ionic transition [10]. Further support of a covalent bonding between Si and alkali metal below half a monolayer come from the absence of any significant shift at the Si 2p core level in the low coverage regime. Since the surfaces of Si(100)2x1, Si(111)7x7 and Si(111)2x1 covered with Cs, Rb, K, Na and Li also exhibit formation of similar electronic interface-surface states [45,52,53], the above model of covalent bonding is assumed to be valid for all akali metal/silicon systems. Also, this picture of a covalent bonding [2,4,44,45] is in agreement with recent theoretical calculations on alkali metal covered silicon surfaces [33,34,54]. Finally, recent metastable deexcitation spectroscopy - MDS - experiments (which, in contrast to photoemission spectroscopy, are very sensitive to the alkali "s" valence electron) by Nishigaki et al. [55] have shown for Cs or K covered silicon surfaces that the alkali "s" electron remains occupied even at low coverages, which strongly supports the covalent bonding model described above [45].

Finally, the existence of plasmon within the adsorbate overlayer has been reported for many alkali metal-silicon systems [2,45,56-58]. This indicates metallization of the adsorbate in agreement with the above model of covalent bonding.

b - Germanium surfaces

It is now interesting and challenging to investigate if the above model of covalent bonding between alkali metal and silicon is valid for other group IV semiconductor

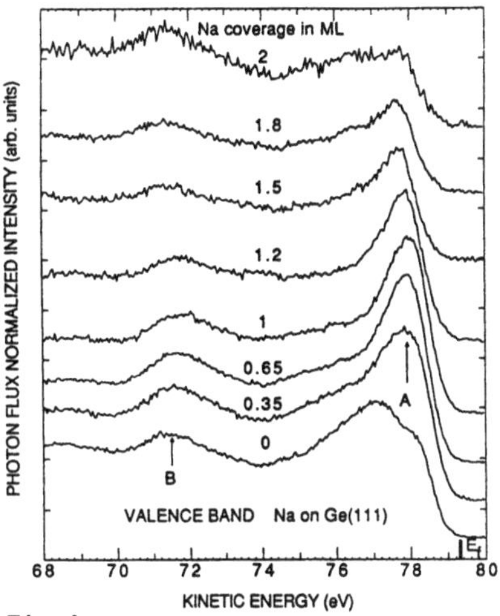

Fig.4

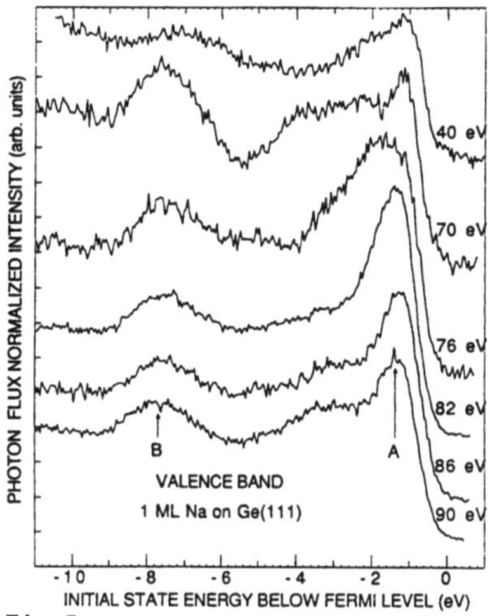

Fig.5

Figure 4: Na/Ge(111)2x1: Valence band spectra for various sodium coverages between 0 and 2 monolayers. The photon energy was 82 eV.

Figure 5: 1 ML Na/Ge(111)2x1: Valence band spectra for various photons energies between 40 eV and 90 eV.

surfaces. A good system to look at is germanium, another important elemental semiconductor, which has also been widely investigated theoretically as well as experimentally [59-61]. Here, we will report core level and valence band photoemission results for the representative Na/Ge(111)2x1 system [62]. Figure 4 displays the valence band results for this system at various Na coverages. A sharp electronic interface state grows up to one Na monolayer and, as demonstrated in the above case of silicon, results from the hybridization of Na 3s and Ge 4p valence electrons.

Figure 5 shows another interesting feature, namely the photon energy dependence of the intensity of this Na 3s - Ge 4p electronic interface state which is found to resonate around 82 eV. This effect does not correspond to any possible resonance within the Ge or the Na atoms. The closest possible transition would be a Na 2s - Na 3s (≈ 63 eV), which is forbidden. However, the situation would be totally different if the Na 3s electron orbital hybridized with the Ge 4p valence electron related to the dangling bond. In this case, the corresponding transition would become allowed. However, our results indicate that this resonance is delayed by nearly 20 eV [62]. Similar delays have been reported in the case of molecules such as N_2 by Dehmer et al. [63]. Therefore, the above resonance appears to be similar to the threshold shape resonance observed in molecules. This explanation is preferred to other potential interpretations in terms of charge density waves - CDW - since this phenomenon generally occurs at the Fermi level [64] instead of 1.5 eV below as observed here [62]. Therefore, this resonance should be very sensitive to the distance between the Na and Ge atoms. Further confirmation of the above interpretation is found in the very high sensitivity to oxygen exposures of this interface resonant state which is destroyed for 0.25 Langmuir of O_2 [62] since, in a recent calculation by Ling Ye, Freeman and Delley for the K/Si(100)2x1 [65], the K atom is found to relax away from the surface by nearly 5% which would explain the destruction of this resonance as well as the Na-Ge covalent bonding. This indicates that the effect of initial oxygen exposure (below the threshold of oxidation) is rather local as opposed to the catalytic oxidation occurring at higher oxygen exposures which has been clearly shown to involve non-local mechanisms [37,66]. Failure to distinguish between these points has recently resulted in some misinterpretation about the nature of the alkali metal-promoted oxidation mechanism [67]. It is interesting to remark here that this resonance would not occur without hybridization between Na and Ge valence electrons which shows a new aspect of resonant photoemission in its ability to probe the nature of a bonding between an adsorbate and a substrate. Finally, the model of covalent bonding between an alkali metal and a germanium surface seems similar to the one described above for silicon surfaces. The other alkali metals like Rb [68], Cs or K [69] have also be shown to be covalently bonded to germanium.

Finally, by looking on the high kinetic energy side of the Ge 3d for the Na/Ge interface, it is found that the recently discovered Ge 3d surface-shifted component [70] has its intensity increased by sodium adsorption. This behavior is likely to indicate that the Na atom interacts with the top Ge surface atom, in agreement with the valence band results.

IV - STRUCTURAL PROPERTIES OF ALKALI METAL-SILICON INTERFACES

a - K/Si(100)2x1: mode of growth and possible sites of adsorption

The structural properties of alkali metal/silicon interfaces have been studied for many years, mainly by the use of rather traditional and well-established techniques such as LEED and/or RHEED [48,51,71-73]. On the basis of LEED measurements on the Cs/Si(100)2x1 system, Levine proposed nearly 2 decades ago that, at 1 Cs monolayer, the Cs atoms form one-dimensional chains with a pedestal site of adsorption [71] (for the various possible sites of adsorption, see figure 6). Tochihara, using LEED experiments also proposed the same structural arrangement for K in the K/Si(100)2x1 system [72]. This linear chain model, with a single pedestal site of adsorption, was found to be consistent with many other experimental and theoretical investigations on these systems [31,35,58]. It is only very recently that this model was questioned on the basis of photoelectron diffraction - PED - results by Abukawa and Kono who proposed that Cs (K) on the Si(100)2x1 surface has two sites of adsorption at the monolayer coverage, the pedestal and the valley bridge sites [74]. Recent thermal desorption experiments - TDS - by Tanaka et al. [75] and RHEED measurements by Makita et al. [76] seems to favor the 2 sites of adsorption model. However, Ling Ye, Freeman and Delley investigated the stability of each

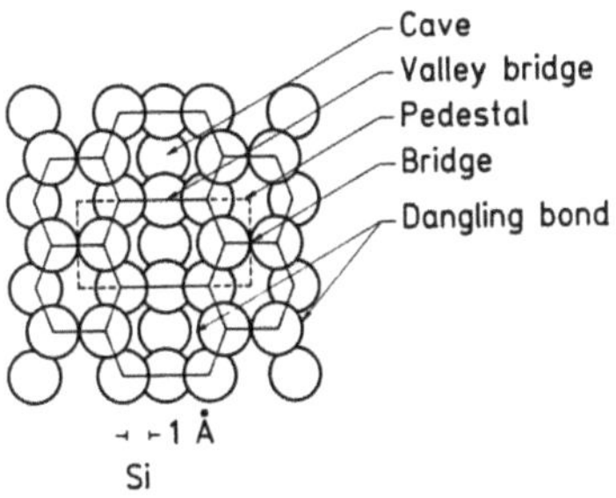

Figure 6: Model of the reconstructed Si(100) surface with all the possible sites of alkali metal adsorption investigated in the literature.

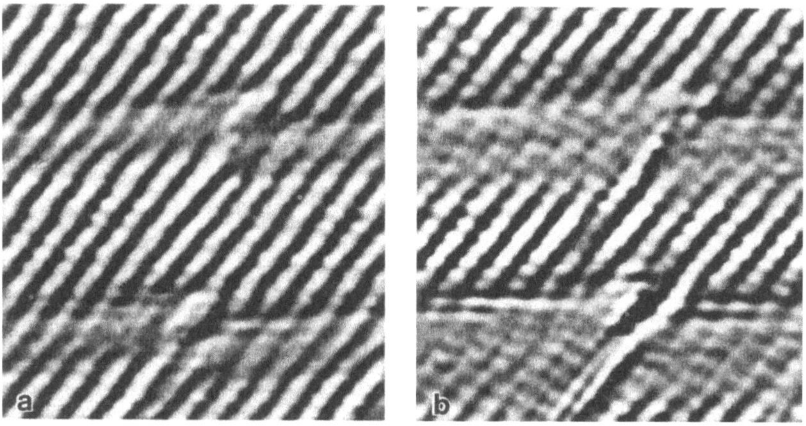

Figure 7a: STM image (filled states) of 1 ML of potassium on the Si(100)2x1 - 4° - stepped surface ($V_t = +1.2$ V). The distance between 2 potassium chains is 7.68 Å.
Figure 7b: STM image (empty states) of 1 ML of potassium on the Si(100)2x1 - 4° - stepped surface ($V_t = -1.2$ V).

possible site of adsorption for K on the Si(100)2x1 surface by total energy ab-initio theoretical calculations [34]. They came to the conclusion that the most stable site for K appeared to be the cave site [34]. Also of interest is the STM investigation of Li adsorption on Si(100)2x1 in the very low coverage regime (≈ 0.1 ML) in which a dangling bond site was claimed [77], in a way somewhat similar to hydrogen atoms known to also bind to the dangling bond. This behavior might look rather normal since Li is the smallest metal with a size and electronic structure close to that of hydrogen. Another interesting feature of this STM study is that the Li mode of growth was found to be perpendicular to the Si dimer rows instead of the expected parallel mode [77].

Since the mode of growth (parallel versus perpendicular to the Si dimer rows) is one of the major issues, STM measurements of K deposition at low and monolayer coverages were performed on a stepped (4°) Si(100)2x1 surface which is now very well known at the atomic scale [78]. This system has (100)2x1 terraces distant by 2 atomic steps and large enough to represent the real surface, with the Si dimer rows oriented in the same direction, i.e. perpendicular to the step edge. Figures 7a and 7b represent the pictures for

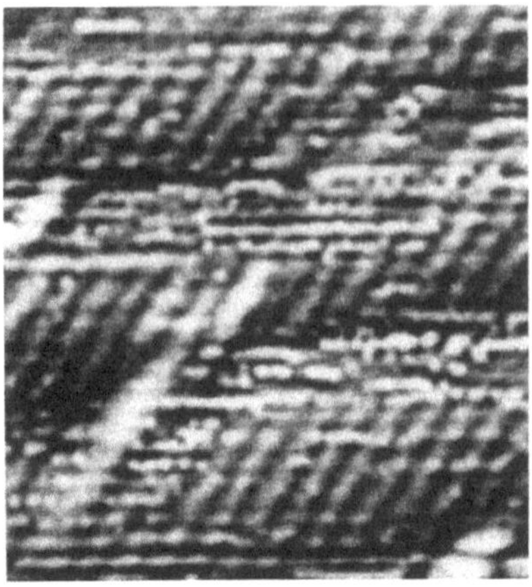

Figure 8: STM image (filled states) of potassium (coverage < 0.4 Ml) on the Si(100)2x1 - 4° - stepped surface (V_t = +2.75 V).

the filled and empty states respectively of 1 ML K/Si(100)2x1 [79]. As can be seen from figure 7, the K atoms are adsorbed as one-dimensional chains distant by 7.68 Å and parallel to the Si dimer rows. The problem is now to know if another layer could be "buried" between 2 chains. In this case, the extreme situation would be for pedestal + cave sites [30] with a height difference of 0.54 Å, calculated on the basis of the K-Si bond length at 3.14 Å measured by SEXAFS [35]. Since the measured corrugations for filled and empty states are 1.5 Å and 0.7 Å respectively, a second layer would be, in principle clearly visible on figures 7a and 7b despite the fact that, one cannot totally rule out possible enhanced corrugation effects which could cloud the existence of a possible second layer. However, we feel confident that these results indicate that K atoms are likely to be adsorbed on a single site at a coverage of 1 ML. In this case, as demonstrated elsewhere [79], this site could only be either the pedestal or the cave site. It is interesting to remark that the K chains from each terrace are connected together over the step edge. This indicates the importance of the K-K interaction versus the K-Si one and further stresses that the latter is weak [24,35,45]. Also of interest on the lower part of figures 7a and 7b is the existence on the step edge of an excess of potassium, likely as a result of a surface defect. This could also be connected to higher sticking probability on step edges as recently suggested [80].

Figure 8 displays the filled states STM image for 0.4 ML of K on the Si(100)2x1 surface [79]. In this case, K atoms seem to have various sites of adsorption (pedestal, dangling bond, cave) with no apparent long range order. This is consistent with the results of theoretical calculations where the possible sites of adsorption were found to be rather close in energy to one another [34]. Also, in agreement with previous STM results, the mode of growth at this lower coverage was found to be perpendicular to the Si dimer rows [77]. However, since the scanning direction is also perpendicular to the <110> direction, possible adsorbate-induced motion by the tip cannot be totally ruled out. Anyway, these results clearly suggest the existence of a phase transition around half a monolayer as recently predicted on the basis of changes in the electronic properties of these systems [45]. As soon as the K-K distance is reduced with increasing coverages, the delicate balance between adsorbate-adsorbate versus the adsorbate-substrate interaction is disrupted, since this latter has been shown above to be a weak bonding [23,24,35,45]. Therefore, the K-K interaction appears to be the main driving force in this ordering transition leading to K arrangement in one dimensional chains at the monolayer coverage as observed above [79].

b - Effects of alkali adsorption and coadsorption ($Na+O_2$) on the structure of the silicon (100) surface

So far, we have mainly discussed the structural properties of the alkali overlayer. Since alkali metals have been shown to induce structural modifications of metal surfaces such as reconstructions [14,21,22], it is interesting to investigate if such changes could also occur in the case of the representative Si(100)2x1 surface. This investigation was performed by PEXAFS experiments using the constant initial state - CIS - mode at the Si 2p core level. As can be seen in figure 9 and Table 1, there is no significant change in the Si-Si first, second and third nearest neighbour distances upon the deposition of 0.4 and 1 ML of Na. This situation is in contrast to the behavior of metal surfaces such as tungsten, molybdenum [14,22] and silver [21] or III-V semiconductors as we will see in the next section. Since the PEXAFS data were taken in p-polarization, the measurements would not be very sensitive to small changes (0.2 Å relaxation) of the Si-Si dimer as recently observed by dynamical LEED experiments and calculations [48]. This relaxation is likely to result from the covalent bonding formation between alkali metal and silicon atoms which induces some changes in charge configuration, leading to the weakening of the Si-Si dimer.

Similarly to Na, a larger alkali atom such as Cs, which also has a lower electronegativity, does not induce significant structural changes in the Si(100)2x1 surface (Table 1). In contrast, as can be seen from figure 12 and also Table 1, a very small oxygen

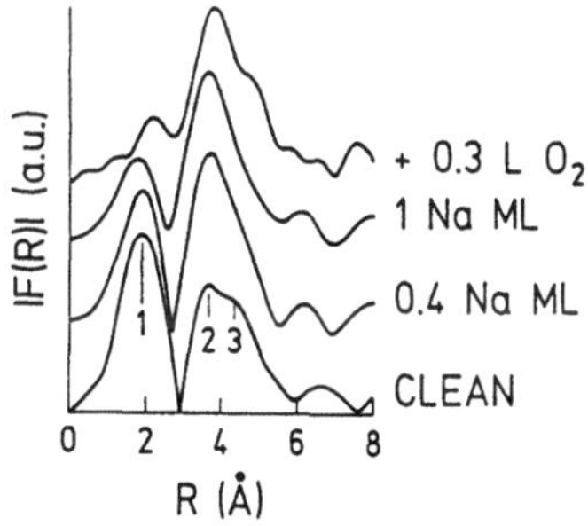

Figure 9: Magnitude of the Fourier transform F(R) measured from the PEXAFS oscillations at Si 2p core level for clean Si(100)2x1, 0.4 ML Na/Si(100)2x1, 1ML Na/Si(100)2x1 and 0.3 L of O_2/1 ML Na/Si(100)2x1.

exposure, (below the threshold of catalytic oxidation [37]) at 0.3 Langmuir of the Si(100)2x1 surface covered by a Na monolayer induces dramatic changes in the Si-Si distances indicating oxygen migration in the sub-surface region [37,46].

Finally, since alkali metals, with the exception of Li, do not give rise to any significant shift at the Si 2p core level, one can easily conclude that, in contrast to many other metal/silicon interfaces [80], the alkali metals do not interdiffuse through the surface with no silicide formation. However, due to its size, the situation is rather different with Li [82].

V - ALKALI METAL/III-V SEMICONDUCTOR INTERFACES: STRUCTURAL PROPERTIES

Studies of the structural properties of alkali metal III-V semiconductor systems were pioneered by LEED measurements more than a decade ago on the GaAs(110) surface covered with Cs by Van Bommel and Crombeen [83] and by Derrien and Arnaud d'Avitaya [84]. It is only recently that the interest in such systems has been renewed because of the model case of alkali metals in the Schottky barrier formation of III-V semiconductors which recently resulted in various experimental [6,7,85,86] and theoretical [87-90] investigations. More recently, the Cs/GaAs(110) system has been successfully investigated in monolayer coverage by STM [91]. Significantly, alkali metals were shown to provoke important band bending (≈ 1 eV) on III-V semiconducting surfaces [85,86,92]. One of the interesting features is provided by investigating alkali metal adsorption on GaAs(110) and InP(110) surfaces with core level photoemission spectroscopy using synchrotron radiation [6,85]. Figures 10 and 11 respectively display the Cs 4d core level for the Cs/GaAs(110) and Cs/InP(110) at various Cs coverages [6,85] and with peak separation treatment. Below 0.4 Cs ML, the Cs 4d clearly exhibits 2 components separated by 0.7 eV in GaAs and 1 eV in InP for both 3/2 and 5/2 spin-orbit splitting. A similar behavior is observed at the Na 2p core level for the Na/InP(110) system where the shifted component is even larger at 1.3 eV

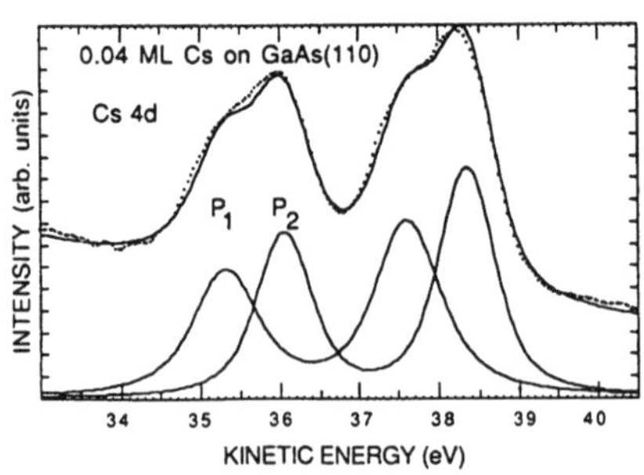

Figure 10: Curve fitting of the Cs 4d core level for the 0.04 ML of Cs/GaAs(110) with 2 doublets corresponding to diffrent adsorption sites. The dot represent the data and the upper solid line is the sum of components P1 and P2. The photon energy was 120 eV.

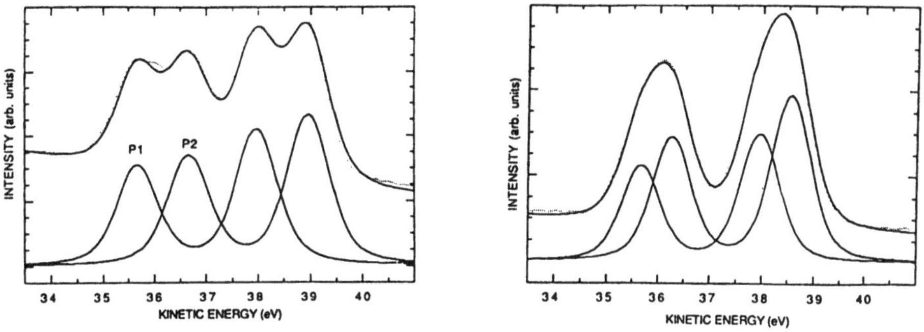

Figure 11: Curve fitting of the Cs 4d core level for 0.1 ML of Cs/InP(110) in the right figure and 0.4 ML of Cs/InP(110) in the left figure with 2 doublets associated with 2 different sites of adsorption. The photon energy was 120 eV.

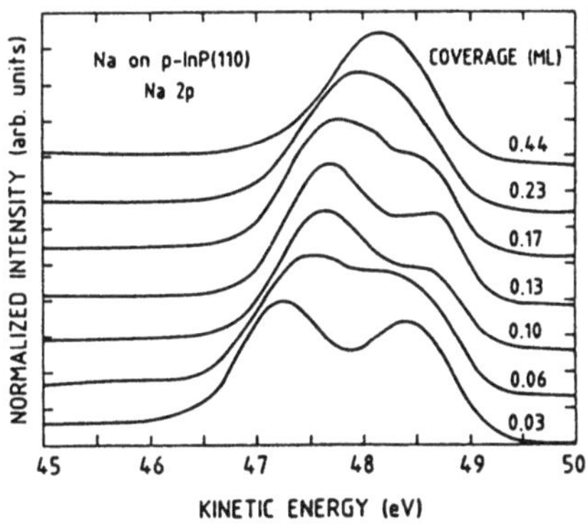

Fig.12

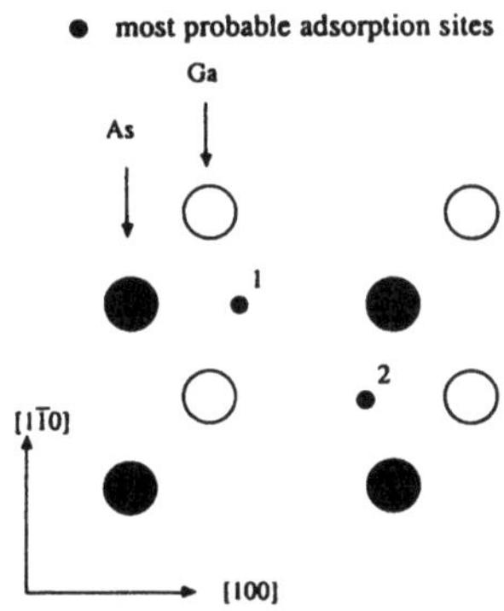

Fig.13

Figure 12: Na 2p intensity normalized spectra of the Na/InP(110) interface at selected Na coverages. The photon energy was 82 eV.
Figure 13: Schematic top view of the GaAs(110) surface with the most likely adsorption sites for Cs marked as 1 and 2. This view is also valid for the Cs/InP(110) system.

(figure 12) [2]. Similar shifted components were also observed at the Cs 5p semi-core level for Cs/GaAs(110) [93]. These features are likely to indicate two inequivalent sites of adsorption as proposed on figure 13 with the alkali atom bonded to one cation and two anions on one site and two cations and one anion on the other one. For each site, the charge transfer would be different since dangling bonds do not have the same degree of occupancy on the anion and on the cation, which explains the observed difference in binding energy. This structural model would be consistent with previous LEED results [83,84] as well as with inverse photoemission experiments where two surfaces states located just above the Fermi level have been observed [94]. It is interesting to note that similar splittings have also been reported for other adsorbates such as Sb on GaAs(110) [95,96].

Another aspect of alkali metal adsorption on III-V semiconductor surfaces would be to look at possible structural changes. The representative Na/InP(110) system has been

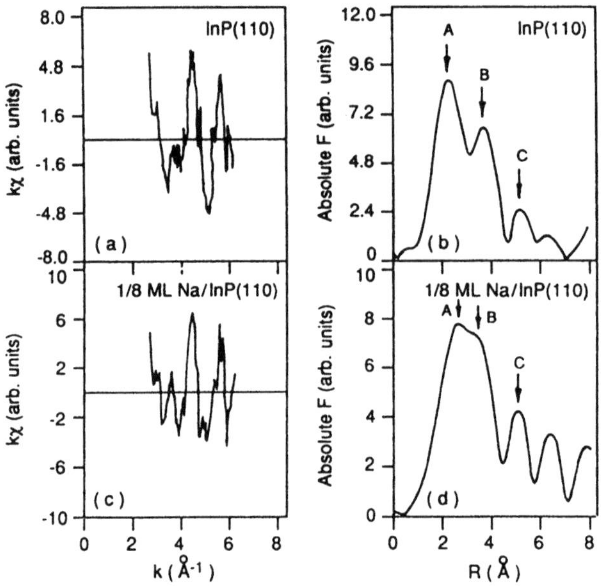

Figure 14: a) PEXAFS signal for the clean InP(110); b) Fourier transform F(R) for clean InP(110); c) PEXAFS signal for 1/8 ML Na/InP(110); d) Fourier transform F(R) for 1/8 ML Na/InP(110).

studied with PEXAFS by using the P 2p core level (Fig. 14). Na deposition at low coverages (1/8 ML) induces dramatic structural changes of the InP(110) surface with removal of the surface relaxation and surface reconstruction [25]. These structural modifications of the InP(110) surface are likely to explain, at least in part, the important band bending provoked by the alkali metal deposition on InP(110) as well as other III-V semiconductor surfaces such as GaAs(110) [6,7,85,86]. For GaAs(110) possible modifications in the surface structure in the presence of Na were also suggested by recent theoretical calculations [97]. Similar changes have been previously reported for metallic surfaces upon alkali metal deposition [21,22]. However, the behavior of InP(110) is in contrast with the one observed above for silicon surfaces. May-be the nature of the alkali metal-surface bond is of some relevance since a bond length shorter than the covalent radii has been measured by preliminary PEXAFS experiments for the Na/InP(110) system which suggests a more ionic type of bonding [98] as opposed to the weak covalent bonding described above for silicon and germanium surfaces [2,4,45,62,68,69]. Also in contrast to silicon, shifted components at the In 4d and P 2p or Ga 3d and As 3d core levels in the presence of Na or K clearly suggest strong reactive interface formation [2,6,7,25,86,99]. This reaction is likely to result from surface defects [2,6,7,86] which have been recently shown to play a key role in alkali metal promoted nitridation of III-V semiconductors like GaAs or InP [100,101].

VI - CONCLUSIONS

We have described the latest concepts in the electronic and structural properties of alkali metal/semiconductor surfaces and interfaces based on XPS, UPS, SEXAFS and PEXAFS experiments using synchrotron radiation and STM investigations. For the group IV semiconductor surfaces, the nature of the bond between the alkali adsorbate and the Si (Ge) surface is found to be a weak and polarized covalent bonding, even at low coverages, by hybridization of adsorbate and substrate valence electrons. The bond length correponds closely to the sum of alkali metal and silicon covalent radii (3.14 Å for K-Si and 2.80 Å for Na-Si, in excellent agreement with the latest ab-initio total-energy self-consistent theoretical calculations) with no change between the low and monolayer coverage regimes. Also, adsorbate metallization is shown to occur near the monolayer coverage. On the Ge(111)2x1 surface, Na was found to induce an electronic interface resonant state which could not resonate without the observed hybridization between the Na 3s orbital and Ge 4d valence electron related to the dangling bond. An ordering transition takes place around half a monolayer coverage for the K/Si(100)2x1 system in which the adsorbate-adsorbate interaction appears to be the principal driving force, leading to a general K atoms rearrangement with formation of one-dimensional linear chains separated by 7.68 Å and parallel to the Si dimer rows. In contrast to silicon, alkali metals are found to induce significant structural changes of III-V semiconductor surfaces such as InP(110) with removal of the surface relaxation. Finally, alkali metals exhibit two inequivalent sites of adsorption at low coverages on the GaAs(110) and InP(110) surfaces.

Acknowledgements

The author is grateful to his numerous collaborators and wishes in particular to acknowledge Krishna Choudhary, Tom Kendelewicz, Sung Kim, Joel Kubby, Pawitterjit Mangat and Hans Starnberg as well as the close collaboration with Art Freeman, Shao Ping Tang and Ling Ye. This work was supported by the U.S. National Science Foundation under contract DMR N° 88-07754 through the Northern Illinois University, by the Commissariat à l' Energie Atomique - CEA, by the North Atlantic Treaty Organization - NATO, by the Dow Corning Corporation and by Xerox.

References

1 K.H. Taylor and I. Langmuir, Phys. Rev. **21**, 380 (1923); I. Langmuir, J. Am. Chem. Soc. **54**, 2788 (1932); Phys. Rev. **44**, 224 (1933); J. Topping, Proc. Roy. Soc. A **114**, 69 (1927).

2 P. Soukiassian and T. Kendelewicz in Metallization and Metal/Semiconductor Interfaces, I.P. Batra editor, Plenum, New York, NATO Advance Studies Institute Series B **195**, 465 (1989) and references therein.

3 N. Hatsopoulos and E.P. Gyftopoulos, in Thermionic Energy Conversion, Cambridge, Massachusets, MIT Press (1977); S.G. Forbes, in Applied Physics in Engineering of Jet Rocket, Nuclear and Electric Propulsion: Theory and Design, edited by W.H.T. Loh, Springer, New York, 7, 442, (1968); H. Sonneberg, Appl. Phys. Lett. **19**, 431 (1971); Chen Chou Wang, J. Appl. Phys. **49**, 147 (1977); M.E. Dry, T. Shingles, L.J. Boshoff and G.J. Ousthuizen, J. Catalysis **15**, 190 (1969); H.J. Drouhin, J.C. Hermann and G. Lampel, Phys. Rev. B **31**, 3859 (1985); ibid, 3872 (1985); D. Paget, H.J. Drouhin and J. Peretti, Appl. Surf. Sci., **41/42**, 38 (1989); C.Y. Su, W.E. Spicer and I. Lindau, J. Appl. Phys. **56**, 1413 (1983); J.R. Hiskes and P.J. Schneider, Phys. Rev. B **23**, 949 (1980); K.N. Leung and K.W. Ehlers, Rev. Sci. Instr. **55**, 803 (1982).

4 P. Soukiassian and H.I. Starnberg in Physics and Chemistry of Alkali Metals Adsorption, Elsevier Science Publishers B.V., Amsterdam, H.P. Bonzel, A.M. Bradshaw and G. Ertl editors, Materials Science Monographs **57**, 449 (1989) and references therein.

5 P. Soukiassian, M.H. Bakshi, H.I. Starnberg, Z. Hurych, T.M. Gentle and K.P. Schuette, Phys. Rev. Lett. **59**, 1488 (1987).

6 T. Kendelewicz, P. Soukiassian, M.H. Bakshi, Z. Hurych, I. Lindau and W.E. Spicer, Phys. Rev. B **38**, 7568 (1988).

7 M. Prietsch, C. Laubschat, M. Domke and G. Kaindl, Europhys. Lett. **6**, 451 (1988); D.A. Evans, A.B. McLean and R.G. Williams, Vacuum **38**, 365 (1988).

8 P. Perfetti, C. Quaresima, C. Coluzza, C. Fortunato and G. Margaritondo, Phys. Rev. Lett. **59**, 2065 (1986); F. Capasso, Science **235**, 172 (1987).

9 S.A. Lindgren and L. Wallden, Sol. Stat. Commun. **25**, 13 (1978); Phys. Rev. B **22**, 5967 (1980).

10 G.M. Lamble, R.S. Brooks, D.A. King and D. Norman, Phys. Rev. Lett. **61**, 1112 (1988).

11 K. Horn, A. Hohlfeld, J. Somers, Th. Lindner, P. Hollins and A.M. Bradshaw, Phys. Rev. Lett. **61**, 2488 (1988).

12 see e.g. N.D. Lang, Phys. Rev. B **4**, 4234 (1971); J.D. Levine and E.P. Gyftopoulos, Surf.Sci. **1**, 171 (1964); A.J. Bennet and L.M. Falicov, Phys. Rev. **151**, 512 (1966); J.W. Gadzuk, Surf. Sci. **6**, 133 (1967); J.P. Muscat and D.M. Newns, ibid, **76**, 355 (1977) and references therein.

13 E. Wimmer, A.J. Freeman, J.R. Hiskes and A.M. Karo, Phys. Rev. B **28**, 3074 (1983); S.R. Chubb, E. Wimmer, A.J. Freeman, J.R. Hiskes and A.M. Karo, Phys. Rev. B **36**, 4112 (1986); Ruqian Wu, Kailai Chen, Dingsheng Wang and Ning Wang, Phys. Rev. B **38**, 3180 (1988).

14 P. Soukiassian, R. Riwan, J. Lecante, E. Wimmer, S.R. Chubb and A.J. Freeman, Phys. Rev. B **31**, 4911 (1985).

15 P. Soukiassian, P. Roubin, J. Cousty, R. Riwan and J. Lecante, J. Phys. C **18**, 4785 (1985).

16 H. Ishida, Phys. Rev. B **38**, 8006 (1988).

17 P. Soukiassian, R. Riwan, J. Cousty, J. Lecante and C. Guillot, Surf. Sci. **152/153**, 290 (1985).

18 D.F. Riffe, G.K. Wertheim and P.H. Citrin, Phys. Rev. Lett. **54**, 571 (1990).

19 B. Woratschek, W. Seeselmann, J. Kuppers, G. Ertl and G. Haberland, Phys. Rev. Lett. **55**, 1231 (1985).

20 U. Memmert and D. Fick, Surf. Sci. **203**, 276 (1988).

21 B.E. Hayden, K.C. Prince, P.J. Davie, G. Paolucci and A.G. Bradshaw, Sol. Stat. Commun. **38**, 325 (1983); C.J. Barnes, M. Lindroos, D.J. Holmes and D.A. King, Surf. Sci. **219**, 143 (1989).

22 R. Riwan, P. Soukiassian, S. Zuber and J. Cousty, Surf. Sci. **146**, 382 (1984).

23 P. Soukiassian, T.M. Gentle, M.H. Bakshi and Z. Hurych, J. Appl. Phys. **60**, 4339 (1986).

24 P. Soukiassian, M.H. Bakshi, Z. Hurych and T.M. Gentle, Phys. Rev. B **35**, 4176 (1987).

25 K.M. Choudhary, P.S. Mangat, H.I. Starnberg, Z. Hurych, D. Kilday and P. Soukiassian, Phys. Rev. B **39**, 759 (1989).

26 P. Soukiassian and R. Riwan, unpublished results (1981).

27 Z. Hurych, H.I. Starnberg and P. Soukiassian, Europhys. Lett. **8**, 567 (1989).

28 T. Yasue, A. Ichimiya and S. Ohtani, Vacuum **41**, 561 (1990).

29 J.R. Northrup, J. Vac. Sci. Technol. A **4**, 1404 (1986).

30 S. Ciraci and I.P. Batra, Phy. Rev. Lett. **58**, 877 (1986); ibid, **60**, 1982 (1987); Phys. Rev. B **38**, 8432 (1988); I.P. Batra, Phys. Rev. B **39**, 3919 (1989).

31 H. Ishida, N. Shima and M. Tsukada, Phys. Rev. B **32**, 6236 (1985); Surf. Sci. **158**, 438 (1985); M. Tsukada, N. Shima, Z. Zhu, H. Ishida and K. Terakura in Ref. 2.

32 H. Tochihara, M. Kubota, M. Miyao and Y. Murata, Surf. Sci. **158**, 497 (1985).

33 R.V. Kasowski and M.H. Tsai, Phys. Rev. Lett. **53**, 372 (1988).

34 Ye Ling, A.J. Freeman and B. Delley, Phys. Rev. B **39**, 10144 (1989).

35 T. Kendelewicz, P. Soukiassian, R.S. List, J.C. Woicik, P. Pianetta, I. Lindau and W.E. Spicer, Phys. Rev. B **37**, 7115 (1988).

36 P.S. Bagus and I.P. Batra, Surf. Sci. Lett. **206**, L 895 (1988).

37 H.I. Starnberg, P. Soukiassian and Z. Hurych, Phys. Rev. B **39**, 12775 (1989).

38 M.C. Ascensio, E.G. Michel, E.M. Oellig and R. Miranda, Appl. Phys. Lett. **51**, 1714 (1987); E.M. Oellig, E.G. Michel, M.C. Ascensio and R. Miranda, Appl. Phys. Lett. **50**, 1660 (1987); E.G. Michel, E.M. Oellig, M.C. Ascensio and R. Miranda, Surf. Sci. **189/190**, 245 (1987).
39 P. Soukiassian, H.I. Starnberg and T. Kendelewicz, Appl. Surf. Sci. **41/42**, 395 (1989).
40 J.L. Desplat and C.A. Papageorgopoulos, Surf. Sci. **92**, 97 (1980).
41 F.R. McFeely, J.F. Morar, N.D. Shinn, G. Landgren and F.J. Himpsel, Phys. Rev. B **30**, 764 (1984).
42 A. Franciosi, J.H. Weaver and D.T. Peterson, Phys. Rev. B **31**, 3606 (1985).
43 Concerning core level spectroscopy on silicon surfaces and interfaces, see the review of F.J. Himpsel, B.S. Meyerson, F.R. McFeely, J.F. Morar, A. Taleb-Ibrahimi and J.A. Yarmoff in Photoemission and Adsorption Spectroscopy of Solids and Interfaces with Synchrotron Radiation, Enrico Fermi School, Varenna, edited by M. Campagna and R. Rosei, North Holland, Amsterdam (1990).
44 P. Soukiassian, Surf. Sci. Lett. **172**, L 507 (1986).
45 P. Soukiassian, M.H. Bakshi, Z. Hurych and T.M. Gentle, Surf. Sci. Lett. **221**, L 759 (1989).
46 S.T. Kim, P. Soukiassian, L. Barbier, S. Kapoor, Z. Hurych, Phys. Rev. B **44**, in press (1991)
47 Recently, the Na/Si(100)2x1 system was investigated by total energy DMOL calculations in which the Na-Si bond length was found to be 2.93 Å for the unrelaxed surface and 3.00 Å for the relaxed one, in excellent agreement with our experimental value of 2.80 Å; S.P. Tang and A.J. Freeman, private communication and to be published.
48 C.M. Wei, H. Huang, S.Y. Tong, G.S. Glander and M.B. Webb, preprint (1990).
49 E.M. Oellig and R. Miranda, Surf. Sci. Lett. **177**, L947 (1986).
50 Y. Enta, T. Kinoshita, S. Suzuki and S. Kono, Phys. Rev. B **36**, 9801 (1987).
51 G.S. Glander and M.B. Webb, Surf. Sci. **222**, 64 (1989); ibid, **224**, 60 (1990).
52 T.M. Gentle, P. Soukiassian, K.P. Schuette, M.H. Bakshi and Z. Hurych, Surf. Sci. Lett. **202**, L 568 (1988).
53 M.H. Bakshi, S.T. Kim, P.S. Mangat, H.I. Starnberg and P. Soukiassian, unpublished results.
54 H. Ishida and K. Terakura, Phys. Rev. B **39**, 11519 (1989).
55 S. Nishigaki, S. Matsuda, T. Sasaki, N. Kawanishi, Y. Ikeda and H. Takeda, Surf. Sci. **231**, 271 (1990); S. Nishigaki, N. Oishi, S. Matsuda, N. Kawanishi and T. Sasaki, Phys. Rev. B **39**, 8048 (1989); S. Nishigaki, S. Matsuda, T. Sasaki, N. Kawanishi, H. Takeda and A. Kawase, Vacuum **41**, 632 (1990).
56 H. Tochihara, K. Kubota and Y. Murata, Sol. Stat. Commun. **57**, 437 (1986).
57 A. Franciosi, P. Soukiassian, P. Philip, S. Chang, A. Wall, A. Raisanen and N. Troullier, Phys. Rev. B **35**, 910 (1987).
58 T. Aruga, H. Tochihara and Y. Murata, Phys. Rev. Lett. **53**, 372 (1984).
59 J.E. Northrup and M.L. Cohen, Phys. Rev. B **27**, 6553 (1983).
60 J.M. Nicholls, G.V. Hansson, U.O. Karlsson, R.I.G. Uhrberg, R. Engelhart, K. Seki, S.A. Flodstrom and E.E. Koch, Phys. Rev. Lett. **52**, 1555 (1984).
61 L.N. Surnev, Surf. Sci. **55**, 625 (1976); L.N. Surnev and M. Tikhov, ibid, **85**, 413 (1979); H.J. Clemens, J. Van Wienskowski and W. Monch, ibid, **78**, 648 (1978); G.R. Riach and W.T. Peria, Surf. Sci. **40**, 479 (1979); L.N. Surnev, J. Elec. Spect. Rel. Phen. **26**, 53 (1982).
62 P. Soukiassian, T. Kendelewicz and Z. Hurych, Phys. Rev. B **40**, 12570 (1989).
63 J.L. Dehmer, A.C. Parr and S.H. Southworth, in Handbook on Synchrotron Radiation, G.V. Marr editor, Elsevier Science Publishers B.V. **2**, 241 (1987).
64 E. Jensen and E.W. Plummer, Phys. Rev. Lett. **55**, 1921 (1985).
65 Ling Ye, A.J. Freeman and B. Delley, private communication and to be published.
66 H.I. Starnberg, P. Soukiassian, M.H. Bakshi and Z.D. Hurych, Surf. Sci. **224**, 13 (1989); Phys. Rev B **37**, 1315 (1988).
67 T.E. Ortega, E.M. Oellig, J. Ferron and R. Miranda, Phys. Rev. B **36**, 6213 (1987); E.G. Michel, J.E. Ortega, E.M. Oellig, M.C. Asencio, J. Ferron and R. Miranda, Phys. Rev. B **38**, 13399 (1988).
68 P. Soukiassian, T. Kendelewicz and Z. Hurych, Physica Scripta **41**, 612 (1990).
69 P. Soukiassian, T. Kendelewicz, S.T. Kim, Aris Papageorgopoulos, S. Kapoor, Y. Baros and Z. Hurych, Vacuum **41**, 571 (1990).
70 K. Hricovini, G. Le Lay, M. Abraham and J.E. Bonnet, Phys. Rev. B **41**, 1258 (1990).
71 J. Levine, Surf. Sci. **34**, 90 (1973).
72 H. Tochihara, Surf. Sci. **126**, 523 (1983).
73 S. Kohmoto, S. Mizuno and A. Ichimiya, Appl. Surf. Sci. **41/42**, 107 (1989); H. Daimon and S. Ino, Surf. Sci. **164**, 320 (1985).
74 T. Abukawa and S. Kono, Phys. Rev. B **37**, 9097 (1988); Surf. Sci. **214**, 141 (1989).
75 S. Tanaka, N. Takagi, N. Minami and M. Nishijima, Phys. Rev. B **42**, 1868 (1990).
76 T. Makita, S. Kohmoto and A. Ichimiya, Surf. Sci. in press.

77 Y. Hasegawa, I. Kamiya, T. Hashizume, T. Sakuraï, H. Tochihara, M. Kubota and Y. Murata, Phys. Rev. B **41**, 9688 (1990); T. Hashizume, Y. Hasegawa, I. Kamiya, T. Ide, I. Sumita, S. Hyodo, T. Sakurai, H. Tochihara, M. Kubota and Y. Murata, J. Vac. Sci. Technol. A **8**, 233 (1990).

78 P.E. Wierenga, J. Kubby and J.E. Griffith, Phys. Rev. Lett. **59**, 2169 (1987).

79 P. Soukiassian and J.A. Kubby, in Structure of Surfaces - III, Springer Series in Surface Science, Vol. **24**, S.Y. Tong, M.A. Van Hove, K. Tagayanaki and X.D. Xide editors, Springer-Verlag, (1991); J.A. Kubby, W.J. Greene and P. Soukiassian, J. Vac. Sci. Technol. B, in press (1991).

80 S. Kennou, M. Karamatos, S. Ladas and C. Papageorgopoulos, Surf. Sci. **216**, 462 (1989).

81 K. Okuno, T. Ito, M. Iwami and A. Hiraki, Sol. Stat. Commun. **34**, 493 (1980); A. Cros, J. Derrien and F. Salvan, Surf. Sci. **110**, 471 (1981); A. D. Katnani, P. Perfetti, T.X. Zhao and G. Margaritondo, Appl. Phys. Lett. **40**, 619 (1982); I. Abbati, G. Rossi, L. Calliari, L. Braicovich, I. Lindau and W.E. Spicer, J. Vac. Sci. Technol. **21**, 409 (1982); J. Derrien and F. Riengeisen, Surf. Sci. Lett. **124**, L 235 (1983); A. Cros, J. Phys. (Paris) **44**, 707 (1983); A. Franciosi, S. Chang, P. Philip, C. Caprile and J. Joyce, J. Vac. Sci. Technol. A **3**, 933 (1985); F.U. Hillebrecht, M. Ronay, D. Rieger and F.J. Himpsel, Phys. Rev. B **34**, 5377 (1986).

82 H. Tochihara and Y. Murata, Surf. Sci. Lett. **215**, L 323 (1989).

83 A.J. Van Bommel and J.E. Crombeen, Surf. Sci. **45**, 308 (1974); **57**, 109 (1976).

84 J. Derrien, F. Arnaud d'Avitaya and M. Bienfait, Sol. Stat. Commun. **20**, 557 (1976); J. Derrien and F. Arnaud d'Avitaya, Surf. Sci. **65**, 668 (1977).

85 T. Kendelewicz, P. Soukiassian, M.H. Bakshi, Z. Hurych, I. Lindau and W.E. Spicer, J. Vac. Sci. Technol. B **6**, 1331 (1988).

86 C. Laubschat, M. Prietsch, M. Domke, E. Weschke, T. Mandel, G. Remmers, J.E. Ortega, C. Xue and G. Kaindl in ref. 2; M. Prietsch, M. Domke, C. Laubschat, T. Mandel, C. Xue and G. Kaindl, Z. Phys. B **74**, 21 (1989).

87 F. Manghi, C. Calandra, C.M. Bertoni and E. Molinari, Surf. Sci. **136**, 894 (1984).

88 J.E. Keipleis and W.A. Harrison, Phys. Rev. B **40**, 5810 (1989).

89 J. Ortega and F. Flores, Phys. Rev. Lett. **63**, 2500 (1989).

90 G. Allan, M. Lannoo and C. Priester, J. Vac. Sci. Technol. B **8**, 980 (1990).

91 P.N. First, R.A. Dragoset, J.A. Stroscio, R.J. Celotta and R.M. Feenstra, J. Vac. Sci. Technol. A **7**, 2868 (1989).

92 P. Soukiassian, M.H. Bakshi and Z. Hurych, J. Appl. Phys. **61**, 2679 (1987); P. Soukiassian, M.H. Bakshi, H.I. Starnberg, A.S. Bommannavar and Z. Hurych, Phys. Rev. B **37**, 6496 (1988).

93 T. Maeda-Wong, N.J. DiNardo, D. Heskett and E.W. Plummer, Surf. Sci. Lett. **208**, L 1 (1989).

94 K.O. Magnusson and B. Reihl, Phys. Rev. B **40**, 5864 (1989); ibid, B **40**,7814 (1989).

95 J.R. Myron, J. Anderson and G.J. Lapeyre, in Proceedings of the 17th International Conference on Physics of Semiconductors, San Francisco (1984), D.J. Chadi and W.A. Harrison editors, Springer, Berlin-New York p. 133 (1985)

96 F. Schaffler, R. Ludeke, A. Taleb-Ibrahimi, G. Hughes and D. Rieger, J. Vac. Sci. Technol. B **5**, 1048 (1987); Phys. Rev. B **36**, 1328 (1987).

97 C.Y. Fong, L.H. Yang and I.P. Batra in Ref. 2.

98 Preliminary PEXAFS measurements for the Na/InP(100) system seems to indicate a rather more ionic type of bonding; K.M. Choudhary, P.S. Mangat, H.I. Starnberg, Z.D. Hurych and P. Soukiassian, recent results.

99 J. A. Schaefer, F. Lodders, Th. Allinger, S. Nannarone, J. Anderson and G. J. Lapeyre, Surf. Sci. **211/212**, 1075 (1989).

100 P. Soukiassian, T. Kendelewicz, H.I. Starnberg, M.H. Bakshi and Z. Hurych, Europhys. Lett. **12**, 87 (1990).

101 P. Soukiassian, H.I. Starnberg, T. Kendelewicz and Z. Hurych, Phys. Rev. B **42**, 3769 (1990).

Surface Electromigration of Au on Si(111) Studied by REM

A. Yamanaka, Y. Tanishiro, and K. Yagi
Physics Department, Tokyo Institute of Technology,
Oh-okayama, Meguro, Tokyo 152, Japan

Abstract. Surface electromigration processes of Au on Si(111) were studied by reflection electron microscopy (REM). Migration processes depended on the adsorbate structure: Au atoms on the $\sqrt{3}\times\sqrt{3}$ adsorbate structure did not show electromigration, while Au atoms on the 5x2 structure showed electromigration in a direction antiparallel to the current. Au atoms migrated not only on the adsorbate structure but also on substrate Si surfaces, resulting in formation of patches of the 5x2 and 7x7 structures. These results were compared with those reported by Yasunaga et al. Preferential spread of one domain of the 5x2 structure were noted by the present REM method.

1. Introduction

It is well known that metal atoms of a wire are driven by a DC current through it. This phenomenon is called electromigration and is physically very interesting. It also contains technically important problems because it causes a failure of integrated circuits[1,2].

In 1982, Zhou et al. first found by RHEED that an In deposit on a Si(111)4x1-In surface moved on the surface in the direction of the current fed through the Si substrate[3]. They reported that the migration direction of the In deposit was opposite to that in the case of the bulk electromigration. They also measured the mobility of the In deposit and found that it was much higher than that of the bulk electromigration. Because of these differences they concluded that the migration of the In deposit on the surfaces was a new type of the electromigration.

Recently, Yasunaga et al. systematically studied this phenomenon for various metal deposits such as In, Ag, Sb, Cu and Au on Si(111) surfaces using scanning Auger microscopy[4,5,6]. They deposited metals in the form of a patch on clean Si(111) surfaces to a thickness of 2-5 monolayers (MLs) and observed expansion of the patch of the deposit induced by the current fed through the Si substrate. They found that the electromigration is generally observed and named this phenomenon surface electromigration. They studied the migration direction and mobility of each metal. They reported that the migration of adatoms on the adlayer is important.

However, the microscopic features of the surface electromigration such as influences of steps, domain structures of the surface and formation of these domains have not been made clear. The present authors applied ultra-high vacuum reflection electron microscopy (UHV-REM), which can characterize surface dynamic processes and surface structures such as steps and surface structure domains[7], to the studies of surface electromigration in the cases of Au, In and Cu deposits on Si(111)7x7 surfaces[8,9]: (1) In the case of Au, migrating atoms form the 5x2 structure and move toward the anode not only on the 5x2 structure but also on the substrate Si 7x7 surface. (2) In the case of In, atoms form the 4x1 structure and move toward the cathode mainly on the 4x1 surface. (3) In the case of Cu,

Springer Series in Materials Science, Vol. 17 **Ordering at Surfaces and Interfaces**
Eds.: A. Yoshimori · T. Shinjo · H. Watanabe

three-dimensional islands of Cu formed on Cu adlayers disappear as the Cu adlayers extend toward the cathode.

Very recently in the case of Au on the Si(111) surface Yasunaga and Sasuga reported an interesting coverage dependence of the surface electromigration as follows: (1) When the coverage of Au deposit was thicker than 1.0ML, judging from the Auger peak height, Au adlayers extended equally toward the anode and the cathode. (2) When the coverage fell to 0.7ML as a result of the extension of the adsorbed region, the Au adlayer spread only toward the anode. (3) When the coverage fell below 0.7ML, an oscillation in the Auger peak height profile of the Au deposit was observed on the cathode side.

In the present paper structural information relating to the above interesting features studied by UHV-REM is presented in addition to the general features of the surface electromigration of Au.

2 Experimental

An UHV electron microscope and its attachment for REM studies and in-situ deposition were what were used in the previous report[9]. Si specimens ($1x10x0.4mm^3$ in size, B doped p-type (a few ohm cm)) with the longer side parallel to the $[1\bar{1}0]$ or $[11\bar{2}]$ direction were cut from well-oriented (111) wafers. The sample surface was cleaned by resistive heating above $1200^{o}C$ at the specimen position of the microscope. After the cleaning a heating current fed through the Si substrate was set to a value at which the electromigration processes were observed. For the deposition of Au the specimen holder was rotated by 90° so that the surface faced a deposition source. The deposition on to the Si surface was made through a slit about 300um in width to get a rectangular patch of the Au deposit. Then the specimen holder was rotated again to the observation position.

3 Results

3.1 General migration features

Figure 1 reproduces a sequence of REM images of a Si(111) surface, showing migration processes of Au. Marks in the lower right of the figure show the observation conditions. A shaded area in a rectangle represents the position of the patch of Au deposit on the surface and a solid circle represents the imaging position. Arrows show the directions of the current (Usually there is one arrow but in this case the direction of current was reversed during observation and two arrows are shown.) A diamond, one half of which is shaded, denotes an orientation of the 7x7 unit cell; a dark triangle means the stacking fault part of the 7x7 unit cell. The current in (a) was in the $[\bar{1}\bar{1}2]$ direction.

The micrograph (a) was taken after the deposition of Au 1ML in thickness on the substrate with a current of 500mA (about $750^{o}C$). The imaging area is about 140um left from the edge of the deposited area. Bright areas are of clean 7x7 surface and curved lines are surface steps of single atom height. Dark areas are 5x2-Au structure regions which spread from the right.

From (a) to (b) (2min) these 5x2 structure regions expanded on the Si 7x7 surface in the direction opposite to the current indicated by arrows. The vertical width of these 5x2 structure regions is seemingly narrow in the images. REM images are however, foreshortened in the direction of the electron beam (see scale marks in the figure). A correction of this effect shows that these 5x2 regions have a width of several microns in the

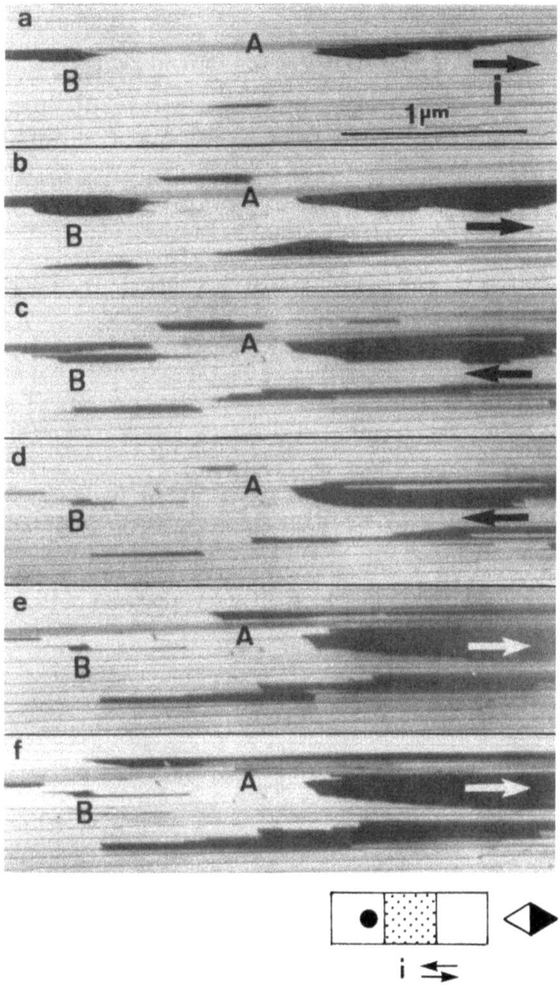

Figure 1. A series of REM images which show spread of 5x2 structure regions (from (a) to (b) and from (e) to (f)) and shrink (from (c) to (d)) depending on the current direction.

vertical direction. The direction of the current was reversed before taking the micrograph (c). From (c) to (d), the 5x2 structure regions shrank and returned to the 7x7 structure (see regions indicated by A and B). This means that Au atoms moved back to the right in the opposite direction to the current. The direction of the current was reversed again before taking the micrograph (e); and from (e) to (f) the 5x2 structure regions expanded again as seen from (a) to (b).

The fact that Au atoms moved to the left (from (a) to (b) and from (e) to (f)) and to the right (from (c) to (d)) means that the step sense has nothing to do with surface electromigration: In the imaging area the right-hand side terrace of each step is higher than the left one. This fact also means that irrespective of the current direction of the $[11\bar{2}]$ and $[\bar{1}\bar{1}2]$ directions, which are different on the three-fold Si(111) surface, the 5x2 domain spreads in the direction opposite to the current.

3.2 Adsorbate structure dependence of migration

Figure 2(a) reproduces a REM image of the edge region of the rectangular patch of Au deposit (3ML) with a current of 100mA (about 450°C). As indicated in the lower right, the current was along the <110> direction. Bright areas are of the 7x7 structure and dark areas are $\sqrt{3}$x $\sqrt{3}$-Au structure regions, which is known from a selected area RHEED shown in the right part of the figure (b). Figure 2(b) reproduces a REM image which was taken 5.5 minutes after (a). The same places in (a) and (b) are marked by arrow heads. From (a) to (b) the $\sqrt{3}$x $\sqrt{3}$ structure regions spread in the current direction. At the same time it was observed that the $\sqrt{3}$x $\sqrt{3}$ region spread in the direction opposite to the current on the other side of the patch of Au deposit. The velocities of the spread of the $\sqrt{3}$x $\sqrt{3}$ structure regions in the two directions were nearly equal. So the spread of the $\sqrt{3}$x $\sqrt{3}$ structure regions are considered not to be due to electromigration but due to thermal diffusion.

Figure 3 reproduces a series of REM images which show changes of a surface under a current of 100mA (about 450°C) in a Au deposited area (3ML). An edge of the deposit is at the right-hand side of the imaging area (65um apart). In (a) three-dimensional islands are seen on the surface of the $\sqrt{3}$x $\sqrt{3}$ structure. They are considered to be Au (or islands containing Au). During the migration experiment some of these islands shrank and disappeared ((b) and (c)). A notable fact is that the islands in the right (closer to the edge) disappeared much faster. This means that Au atoms were emitted from these islands and migrated over the $\sqrt{3}$x $\sqrt{3}$ surface to the edge by thermal diffusion, so that the $\sqrt{3}$x $\sqrt{3}$ structure region spread in both directions.

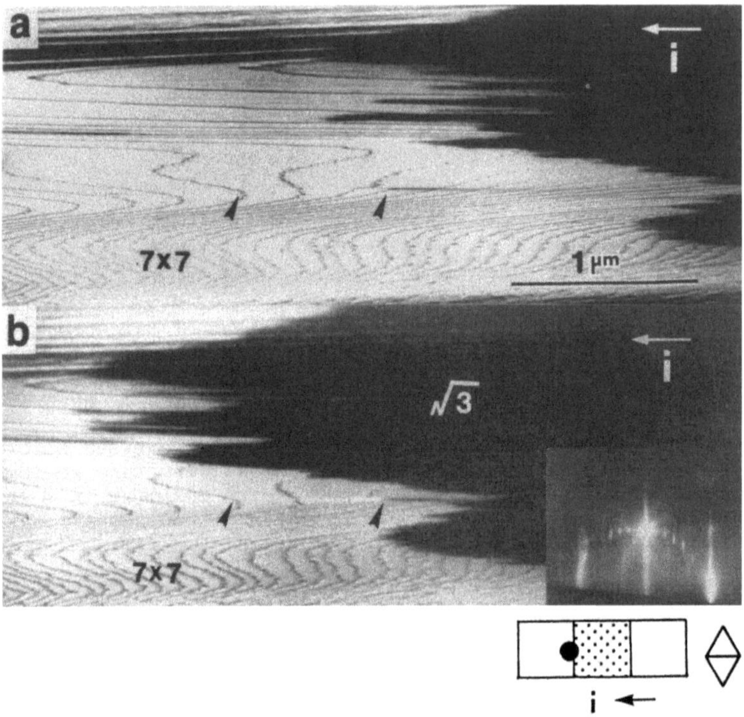

Figure 2. A series of REM images which show a spread of $\sqrt{3}$x $\sqrt{3}$ structure regions (dark regions) as a result of thermal diffusion.

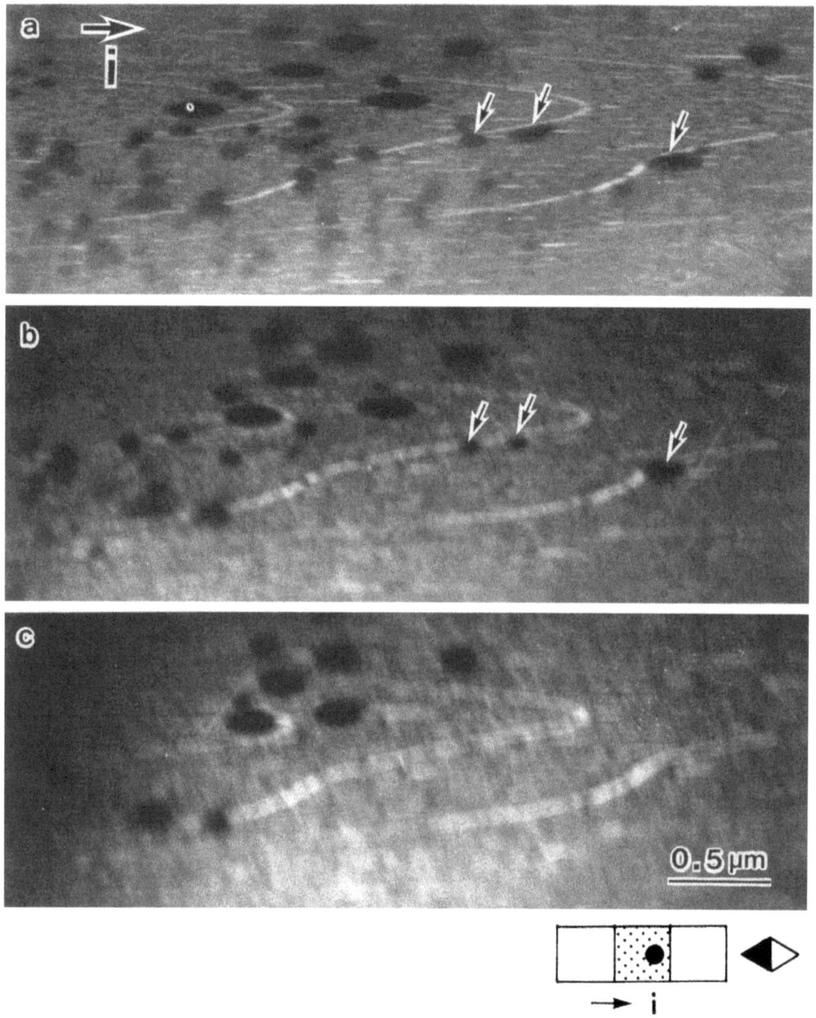

Figure3. REM images of Au deposited region (3ML) taken during a migration experiment. Au islands on the $\sqrt{3}\times\sqrt{3}$ structure regions disappear from lower right part. These islands are the source of Au atoms for the spread of $\sqrt{3}\times\sqrt{3}$ structure regions.

When the current was increased to 500mA (about 750°C) from 100mA, the $\sqrt{3}\times\sqrt{3}$ structure region spread more rapidly in both directions and finally all of the three-dimensional islands in the patch of the Au deposit disappeared. The 5x2 structure was then formed at the boundaries between the spreading $\sqrt{3}\times\sqrt{3}$ structure regions and the 7x7 ones. As seen in Fig. 1 electromigration was observed there: At the anode side, the 5x2-Au structure region spread, and at the cathode side it shrank.

Figure 4 summarizes the migration behavior of Au on the Si(111) surface. A Au deposit of 3ML forms the $\sqrt{3}\times\sqrt{3}$ structure and three-dimensional islands on it (in (a)). The $\sqrt{3}\times\sqrt{3}$ structure regions spread in both directions (thermal diffusion) at the cost of islands under the current of 100mA (about 450°C) (from (a) to (b)). After an increase of the

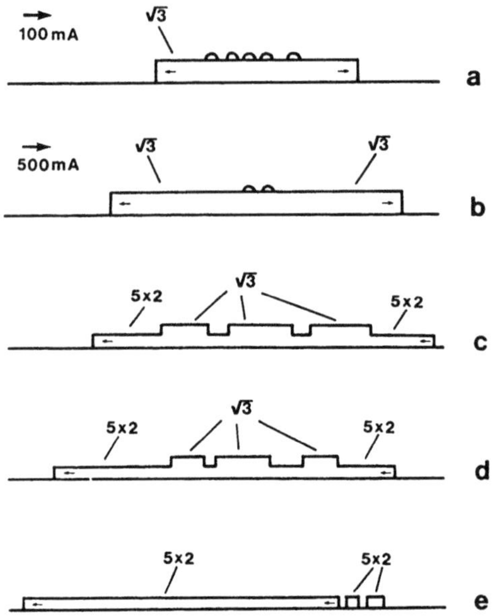

Figure 4. A schematic illustration to show the migration behavior of Au on Si(111) surfaces. On the √3x √3 structure regions thermal diffusion takes place. On the 5x2 structure regions surfaceelectromigration takes place.

current to 500mA (about 750°C), rapid thermal diffusion takes place and the 5x2 structure is formed at the edges (in (c)). There, surface electromigration was seen (from (c) to (d)). In a later stage, clean 7x7 structure regions reappear at the cathode side edge (in (e)).

3.3 Domain structures formed by surface electromigration

Figure 5 reproduces REM images which show an electromigration of Au with a current of 500mA (about 750°C) parallel to the [$\bar{1}\bar{1}$2] direction. The bright areas are of the 7x7 structure and the gray and black areas are domains of the 5x2-Au structure. The black regions correspond to the domain B in the Fig. 6(a) and the gray regions correspond to the domains A and C. It is seen from Fig. 5 (a) and (b) that the gray regions spread toward the anode, by around 1μm. On the other hand, the black regions marked by P and Q spread upward by about 20μm nearly in a direction perpendicular to the current. This direction is parallel to the shorter side of the 5x2 unit cell as shown by a big arrow in Fig. 6(a). It is also seen that the direction is parallel to the steps in those regions.

Figure 7 reproduces REM images and the corresponding RHEED pattern taken during a migration experiment when the current 500mA (about 750°C) flowed parallel to the <1$\bar{1}$0> direction. Micrograph (a) was taken by the specular reflection. The bright areas are of the Si(111) 7x7 structure and the dark areas are of the 5x2 structure which was formed by migrating Au atoms from the left hand side patch of the Au deposit. In the RHEED pattern (c), a reflection marked C from the 5x2 structure is seen. This reflection came from the C domain in Fig. 6(b). Figure 7(b) was taken by this reflection. The dark regions in (a) are almost bright in (b). This means that almost all the 5x2 regions are of the C domain.

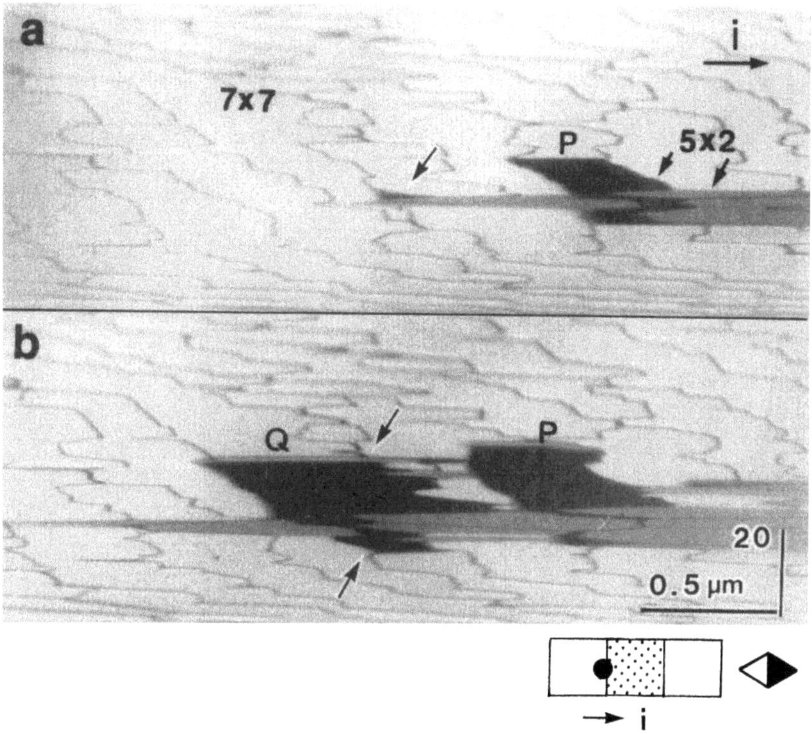

Figure 5. REM images which show preferential growth of dark domains of the 5x2 structure.

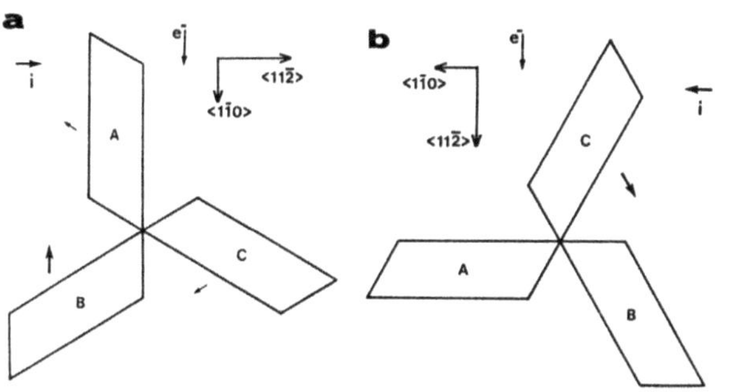

Figure 6. Schematic illustrations which show relations among the current direction, unit cells of the three 5x2 domains and the direction of the preferential growth of one type of domain. In (a) current is along the <112̄> direction. In (b) it is along the <11̄0> direction.

In Fig. 7(b) it is seen that the C domain spreads toward the right but in a direction slightly downward by several degrees. After a correction of the foreshortening of the image, the direction was found to be in a direction 60 degrees from the horizontal direction. It should be noted that the direction is parallel to the shorter side of the 5x2 unit cell of the C domain (see the big arrow in Fig. 6(b)). It is also noted in the C domain regions of Fig. 7(b) that the steps run in the upper-right direction in one region and lower-right direction in the other region.

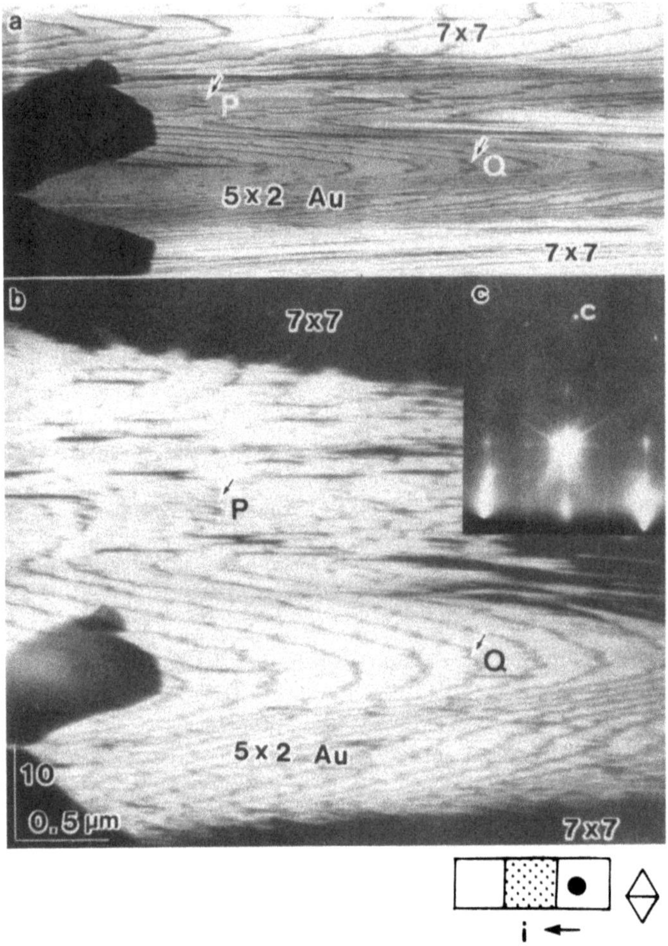

Figure7. REM images ((a)and (b)) of a spreadregion of the 5x2 structure under a current parallel to the $\langle 1\bar{1}0 \rangle$ direction and the corresponding RHEED pattern (c). Note that dark regions in (a) taken by a specular beam are bright in (b) taken by the5x2superlattice reflection marked by C in (c).

Figure 8 shows another example of the single domain formation by surface electromigration. The current was 500mA (about 750°C) parallel to the $\langle 1\bar{1}0 \rangle$ direction. The bright areas in (a) are 7x7 structure regions and dark areas are the 5x2- Au structure regions. Most of the dark areas in (a), are bright in (b) which was taken by the 5x2 superlattice reflection from the C domain, as in Fig. 7. An interesting finding is that all of the isolated Au adsorbed regions indicated by arrows in (a) and (b) are of the C domain. It is also noted that the steps in the regions where the Au adsorbed islands are formed run in the upper-right direction, which is parallel to the direction of the longer side of 5x2 unit cell of the C domain (see Fig. 6(b)). Just after this experiment we carried out another migration experiment using the same sample and under the same conditions except for the direction of the current, where the same C domain spread dominantly again.

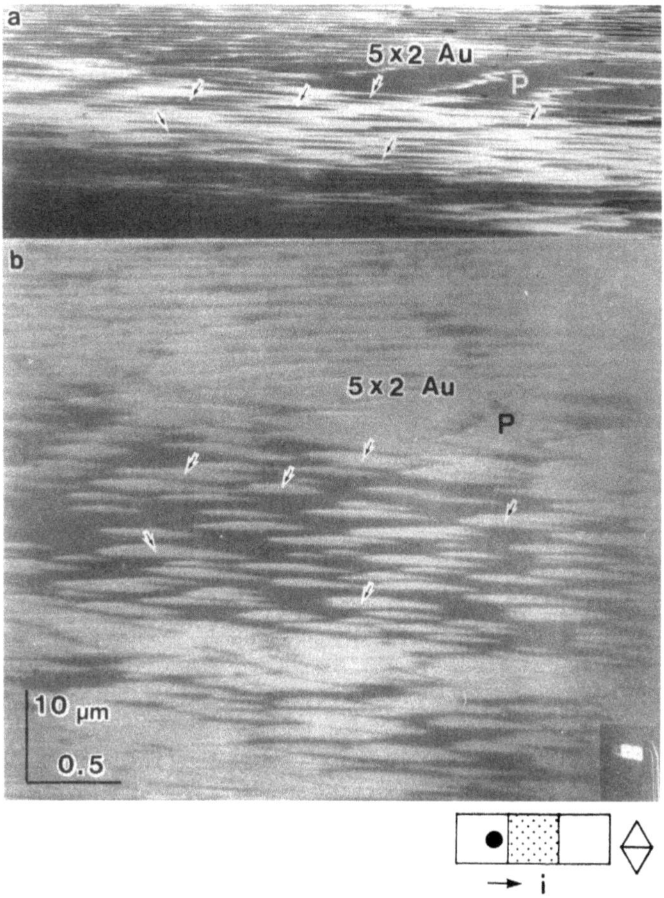

Figure 8. REM images which show preferential growth of the C domain with the current parallel to the <110> direction. Note that all of the isolated 5x2 domains indicated by arrows are C domains.

Figure 9(a) shows a typical RHEED pattern when the C domain spreads dominantly as seen in Fig. 7(c). However, sometimes a RHEED pattern like Fig. 9(b) was obtained. There are reflections marked by A, B and C from three domains, which indicates three domain formation by the surface electromigration. It is not clear at present under what conditions the three domains are formed.

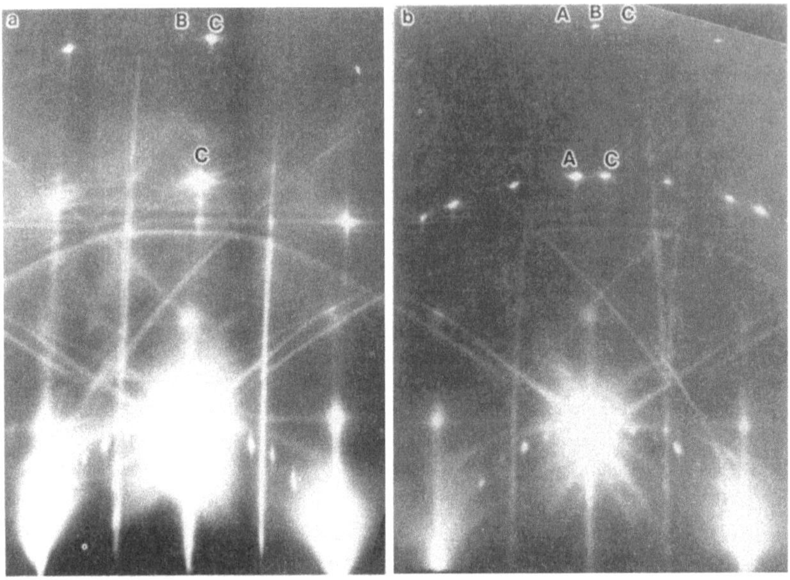

Figure9. RHEED patterns taken during the electromigration of Au with the current parallel to the <110> direction. (a) A typicalpattern when the C domain preferentially spread. (b) A pattern sometimes obtained which shows reflections from three domains.

4 Discussion

4.1 Adsorbate structure dependence of the electromigration

Adsorbate structure dependence of the migration behavior is summarized as follows (see Fig. 4): (1) When the $\sqrt{3}x\sqrt{3}$ structure regions are formed they spread in both directions parallel and antiparallel to the current direction. The velocities of the spreading edge of the Au adsorbate in both directions are nearly equal. (2) The Au islands in the patch are the sources of Au atoms for the spread of the $\sqrt{3}x\sqrt{3}$ structure regions, as in the case of electromigration of Cu[9]. (3) When the 5x2-Au structure regions are formed they spread in the direction opposite to the current. (4) Au atoms move not only on the 5x2 structure surface but also on the substrate Si surface. (5) At the cathode side of the patch of the Au deposit, 5x2 structure regions shrink and the 7x7 structure recovers.

These results are consistent with Yasunaga et al.'s experiment mentioned in section 1; (i) Au adsorbate of 1ML thickness which spread to both sides in their results corresponds to the $\sqrt{3}x\sqrt{3}$-Au structure (present results (1) and (2) mentioned above). (ii) Au adsorbate of 0.7ML thickness which spread in the direction opposite to the current in their experiments corresponds to the 5x2 structure (result (3)). (iii) The oscillation of Auger peak height profiles at the cathode side of the Au adsorbate observed by them corresponds to the formation of coexisting regions of the 5x2-Au structure and the 7x7 structure at the cathode side of the Au adsorbate (results (4) and (5)).

4.2 Single domain formation by surface electromigration

It was found that one of the three domains of the 5x2 structure was preferentially formed during the surface electromigration in both cases where the current was in the $\langle 11\bar{2}\rangle$ and $\langle 1\bar{1}0\rangle$ directions. In both cases such a domain spread mainly in the direction of the shorter side of the 5x2 unit cell. Similar preferential spreading of domains of the 5x2 structure was noted during deposition by Tanishiro et al.[10] and Ozawa et al.[11]. There are two possible explanations for the preferential growth of one domain: (1) The current stabilizes one of the 5x2 structure domains which has a special orientation to the current direction. (2) Steps cause the preferential growth of one of the 5x2 structure domains. However, the two mechanisms cannot explain all of the observations as will be described below.

We observed that for the current in the $\langle 11\bar{2}\rangle$ directions, a domain whose shorter side of the unit cell was perpendicular to the current preferentially spread as shown in Fig. 5. If this had meant that current stabilizes the domain whose shorter side of 5x2 unit cell is perpendicular to the direction of the current, the A and C domains in Fig. 6(b) might have spread equally for the current of the $\langle 1\bar{1}0\rangle$ direction. However, only the C domain preferentially spread, and a reversal of the current also produced the C domain.

We observed that for the current in the $\langle 11\bar{2}\rangle$ direction, the 5x2 structure domain whose shorter side of the 5x2 unit cell was parallel to the direction of the steps spread along the step (Fig. 5). However, for the current of the $\langle 1\bar{1}0\rangle$ direction, our observations were as follows: (1) The C domain preferentially spread irrespective of the direction of the steps (Fig. 7). (2) All of the isolated domains formed on the 7x7 structure region were the C domains but their shorter side of the unit cell was not parallel to the direction of the steps (Fig. 8).

At present we do not have an appropriate explanation for the single domain formation. It should be noted here that in the case of electromigration of Au on Si(001) 2x1 surfaces we observed migration of Au atoms antiparallel to the current, preferential domain growth and anisotropic migration, which will be reported elsewhere[12].

4.3 Mechanism of migration

In the case of the bulk electromigration the atoms were considered to be driven by collisions with the electrons or by the electrostatic force. In the case of the surface electromigration, Yasunaga et al. found that the migration velocity is smaller on the highly doped specimen and considered the importance of the electrostatic force[6].

In this paper it is found that there is little current effect on the $\sqrt{3}x\sqrt{3}$ structure but a strong effect on the 5x2 structure. This means a difference of the electronic states between the $\sqrt{3}x\sqrt{3}$ and 5x2 structures is a reason for the different behavior. At present the two structures have not been solved and we do not know their electronic states. It is important to study electronic states experimentally. At this point it should be noted that the fact that a single domain surface is formed by the electromigration is very useful for the studies of surface electronic sturucture as well as surface atomic structure, as is the case of Si(001)2x1, where nearly single domain surfaces formed by the specimen current effect were used for the atomic structure and electronic structure analysis.[13]

This work was supported by a Grant in Aid from the Ministry of Education, Science and Culture (No. 63609507 and 01609501).

References

1) H. B. Huntington : "Diffusion in Solid (Recent Development)" ed.by A. S. Nowick and J. J. Burton (Academic Press ,New York 1975) Chap. 6.
2) F. M. d'Henrle and P. S. Ho : "Thin Films-Interdiffusion and Reactions", ed. by J. M. Poate, K. N. Tu and J. W. Mayer (John Wiley & Sons. 1978) Chap. 8.
3) J. M. Zhou, S. Baba and A. Kinbara: Thin Solid Films **98** (1982) 109.
4) H. Yasunaga, Y. Kubo and N. Okuyama : Jpn. J. Appl. Phys.1 **25** (1986)L400.
5) H. Yasunaga, S. Sakomura, T. Asaoka, S. Kanayama, N. Okuyama and A. Natori : Jpn. J. Appl. Phys. **27**(1988)L1603.
6) H. Yasunaga and E. Sasuga : Surf. Sci. **231** (1990) 263.
7) N. Osakabe, Y. Tanishiro, K. Yagi and H. Honjo : Surf. Sci. **97**(1980)393.
8) A. Yamanaka and K. Yagi : Ultramicroscopy **29** (1989) 161.
9) A. Yamanaka and K. Yagi : Surf. Sci. **242**(1991) 181.
10) Y. Tanishiro and K. Takayanagi : Ultramicroscopy **31**(1989) 20.
11) S. Ozawa, K. Kobayashi, U. Tanaka, H, Koike and K. Yagi :Proc. 12th Int. Cong. Electron Microscopy (Seattle,1990), vol.1, p340.
12) A. Yamanaka, Y. Tanishiro and K. Yagi : Surf. Sci. submitted.
13) T. Abukawa, S. Kono and T. Sakamoto : Jpn. J. Appl. Phys. **28** (1989) L303.

A Study of Alkali Metal Adsorption on Si Surfaces by Field Ion Scanning Tunneling Microscopy

T. Sakurai[1], *T. Hashizume*[1], *and Y. Hasegawa*[2]

[1]Institute for Materials Research (IMR), Tohoku University, Aoba-ku, Sendai 980, Japan
[2]Institute for Solid State Physics, University of Tokyo, Minato-ku, Tokyo 106, Japan

Abstract. Alkali metal adsorption on the Si(111)7x7 and (100)2x1 surfaces was studied by the field ion-scanning tunneling microscope. Adsorption geometries of Li, K and Cs at the initial coverage were determined.

1. Introduction

Scanning tunneling microscopy (STM)[1] has opened up a new world of microscopy in surface science. By scanning a sharp metallic needle mounted in an electro-mechanical actuator in the lateral direction on a specimen surface, it is possible to map the surface electronic structure on an atomic scale. This novel technique became instantaneously famous when Binnig et al. succeeded in observing the atomic image of the long-sought superstructure of the Si(111)-7x7 surface.[2]

However, a STM with satisfactory reproducibility and stability cannot be realized without precise knowledge of the characteristics of the scanning tip. Very few successful studies have been reported on this topic utilizing the unique power of field ion microscopy[3] to survey the sharp needle specimen surface on an atomic scale. We constructed the field ion-scanning tunneling microscope (FI-STM) in which a tip can be characterized at room temperature, thus, making it possible to operate both the STM and FIM almost simultaneously.[4-5] The instrument has been operating successfully for various surface science investigations.

2. Field Ion Microscopy

Kuk et al. were the first to report a successful construction of the STM combined with a FIM with cryogenic cooling and to study the relationship between the effective tip radius and resolution of the STM.[6] The FIM is reestablishing its unique usefulness in surface science again after latent years.

The FIM is a projection type microscope as is described in Fig. 1. The microscope consists of a specimen shaped to a finely pointed needle tip and a fluorescent screen, which is usually combined with a channel-plate image intensifier, both mounted in a vacuum chamber. By applying a high positive voltage (<20 kV), a field of the order of 5 V/A is produced in the vicinity of the tip whose radius is a few thousand angstrom or less. In the presence of a high field, noble gas atoms, such as He or Ne, can be ionized (field ionization). The ions are formed in the proximity of the specimen surface and projected radially toward

Eds.: A. Yoshimori · T. Shinjo · H. Watanabe

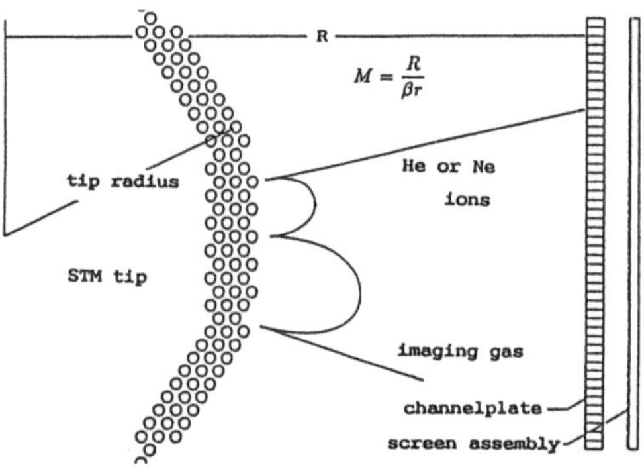

Fig. 1. Principle of field ion microscopy.

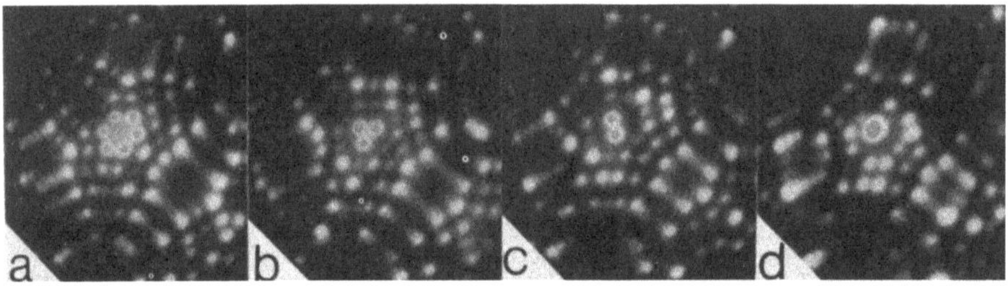

Fig. 2. Sequential FI micrographs of W(111) during field evaporation.

the screen. In the FIM, an atomically resolved image of the needle tip is obtained with a lateral resolution of 2 or 3 A.

Another important phenomenon related to the FIM technique is field evaporation. This is a special mode of desorption of the atoms from the surface under the influence of a strong electrostatic field. Field evaporation takes place when the electric field reaches a few V/A and the electrostatic energy gain in the ionic state becomes comparable to the binding energy of the atom at the surface. By use of field evaporation, the tip can be shaped into a smooth shape since the atoms forming protrusions would be preferentially evaporated as a result of the higher local field. At the same time the tip is cleaned by evaporating the surface contaminants: oxides in many cases.

The most important feature of field evaporation is that it can be employed to modify the tip geometry on an atomic scale. A typical example is shown in Fig. 2. FIM images show sequential changes of the <111> oriented tungsten tip surface during slow field evaporation. In (a), seven atoms form a triangle at the topmost layer of the (111) surface. In (b), the atoms at the three corners of the triangle, where the local tip radius is sharpest, start to field evaporate, leaving the 3 atoms. With further field evaporation, (c) and (d), only a single atom is left at the apex, the ideal situation for a STM tip. Referring to the works by Tersoff and Hamann,[7] or by Tsukada et al.,[8] this single atomic tip with the minimum local radius is capable of providing the highest resolved STM images. The <111> oriented tungsten tip works best as the STM scanning tip for its stability and easy sharpening. Our FI-STM study indeed supports this statement.

3. Field Ion-Scanning Tunneling Microscopy (FI-STM)

There have been many types of the STM design reported, some performing superbly and others working marginally.[5,6] Among the most successful ones, Demuth et al.[9] have employed the differential micrometer with a reduction lever for a stable tip-surface approach. A clever scheme adapted by Demuth's group is the two-step coarse approach made possible by the incorporation of a mechanical stopper. Not only to secure the position of the specimen with respect to the tip mount, thus greatly increasing the resonance frequency of the STM, this contact point between the stopper and the specimen acts as the second hinge for the second-stage fine approach of the sample to the tip. We[4] modified the Demuth-type STM for further simplification and stability and combined it with a FIM operable at room temperature.

The instrument is shown schematically in Fig. 3. The STM device consists of a coarse mechanical adjustment for the specimen and a fine piezo drive adjustment for the STM probe tip. The coarse approach device is a modification of the lever type. The specimen with a dimension of 4x19 mm^2 and thickness of < 1 mm is mounted at the end of the longer lever arm. The final approach is accomplished by a two-step hinge (A, B) system which reduces the sample displacement. The external rotary-linear motion mechanical feedthrough is used to manipulate a micrometer spindle, which pushes the other end of the lever arm by a displacement of 500 μm/turn. This movement is transferred to the sample, which has a coarse approach to the tip with a resolution of 5 mm/turn. As soon as the sample reaches the mechanical stopper, which is positioned near the tip to clamp the specimen in the final position of the coarse approach, the hinge switches from the original position to the second one near the tip, and reduces the displacement rate to 120 μm/turn of the micrometer. The actual operation of the coarse approach is monitored through a 150mm O.D. viewing port using a 100x optical microscope.

The fine drive approach utilizes two types of electro-mechanical drives: electrostriction and piezoelectricity devices. The electrostriction drive is a multilayer type developed for a hysteresis-free, large movement.[10] This drive has a 12 μm vertical z-axis stroke with a resolution of 620 A/V. The sample can reach the tip to within a vacuum tunneling distance without bumping it by using this drive with the above mentioned mechanical approach mechanism in series. The fine

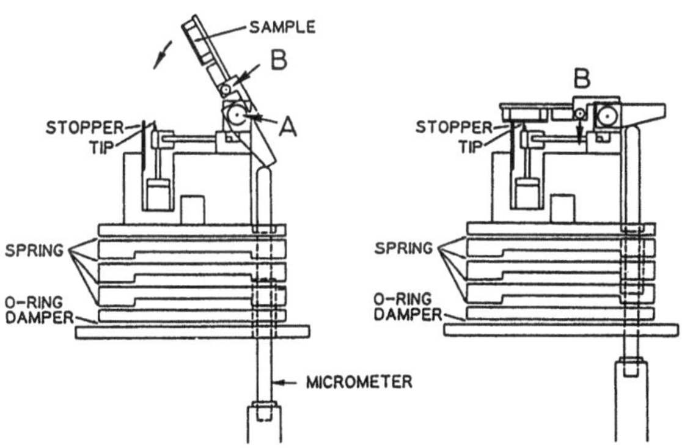

Fig. 3. Schematics of the FI-STM setup showing coarse and fine approach.

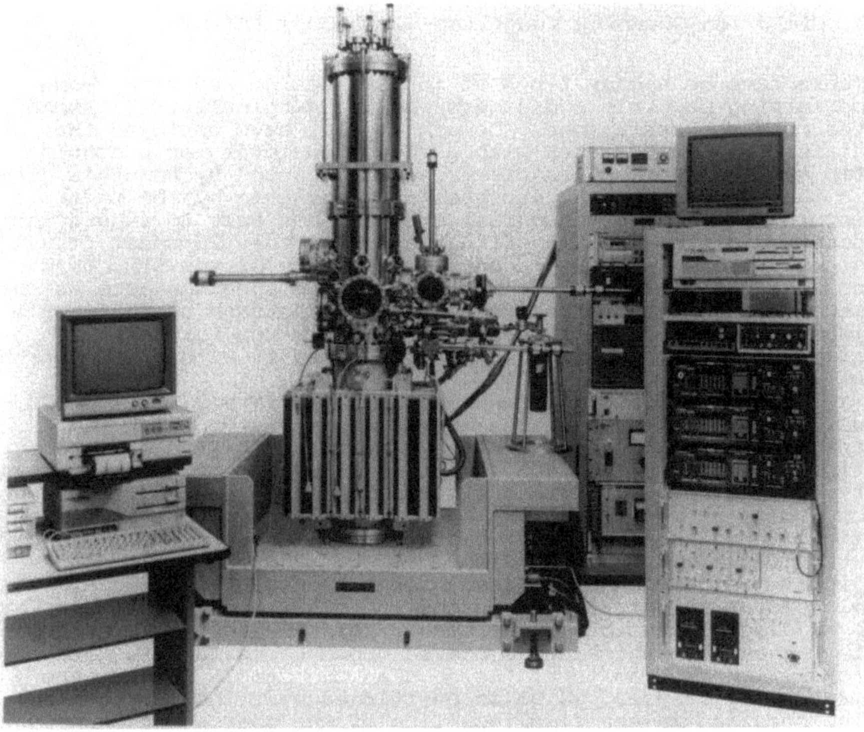

Fig. 4. Photograph of the advanced version FI-STM, available from ULVAC Corp., Japan.

drive is performed by a tripod-type piezoelectric device; 2x4 mm^2 cross section, 40 mm long for the x and y axes and 20 mm long for the z axis. Using these fine motion piezo elements, a 1200-A (accuracy of 120 A/V) linear motion is achieved in the x and y directions and a 500-A stroke is available for the z direction with 50 A/V resolution.

The FIM setup consists of a negative electrode with a 3-mm diameter aperture and 50-mm O.D. chevron channelplate-screen assembly.[11] They are mounted on a single station coupled with an external linear motion mechanical feed-through with a 100-mm stroke. Since the STM is a type of near-field microscope, the FIM mode cannot function simultaneously with the STM. Only after the head of the lever, where the specimen is mounted, is flipped out of the way can the FIM device be brought over the STM tip and the center of the aperture of the negative electrode be positioned exactly above the tip.

The whole system is mounted on the UHV chamber where the base pressure is pumped down to 2 x 10^{-11} Torr by an 1100 l/s ion pump and a 3000 l/s Ti sublimation pump. For evacuation from atmospheric pressure, a 340 l/s magnetic levitated turbo molecular pump backed by another 50 l/s turbo molecular pump and a rotary pump is employed. This pumping system is also used for the evacuation of the imaging gas after observing the FI image. The whole UHV chamber placed on an air-suspended active-damper vibration isolation system, which does not produce any low-frequency resonance characteristic of the air-suspension system (Fig. 4).[12]

4. Experimental

4.1. Sample Preparation

The Si(111) and (100) samples were cut into a 4x19mm^2 piece from P doped (n-type R=1-3 Ωcm), and As doped (n-type, R=0.1 cm) commercial Si wafers, respectively. Prior to introducing the sample into the UHV FI-STM chamber, only washing with methyl-alcohol was administered for sample cleaning. Cleaning of the sample in the UHV chamber (the sample preparation chamber attached to the FI-STM main chamber) was carried out by a series of heatings up to 1200°C following the 5-hour degassing at 700°C. A good quality (111)7x7 and (100)2x1 clean surface was obtained with minimum defect density levels in this manner (Fig. 5).

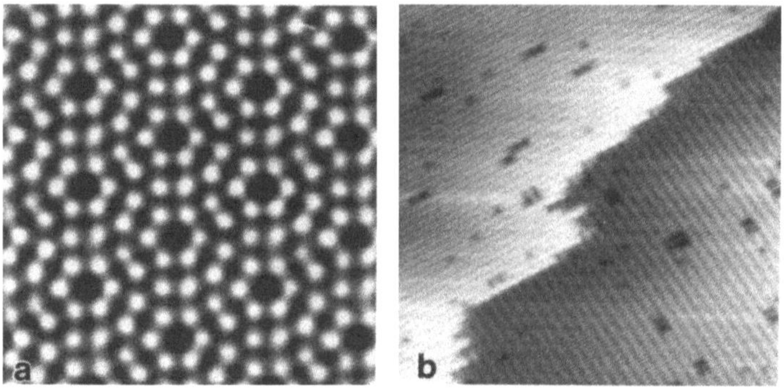

Fig. 5. STM micrographs of the clean Si(111)7x7 and Si(100)2x1 surfaces.

4.2 Alkali Metal Adsorption on the Si(111)7x7 Surface

The clean surface was exposed to alkali metal atoms using the SAES Getters' dispenser. The coverage of K/Cs was estimated by the exposure time with pre-set doser conditions, but the actual coverage was determined by inspecting the STM micrographs.

As for the structure of the clean Si(111)7x7 surface, the DAS model is generally accepted. The 7x7 unit cell contains 12 adatoms at the topmost layer and they are imaged brightly in the STM. Those adatoms can be classified into four groups: six adatoms located around the corner vacancy (corner adatoms), and the other six are away from the corner vacancy (center adatoms). Their properties can be divided into two groups, depending on their location in either the faulted half or unfaulted half. These differences in the electronic properties an be observed in the filled state STM images.

The STM images obtained upon K/Cs exposures were different from those we obtained with the Li adsorption.[13] Figure 6 shows a series of STM micrographs taken during the K adsorption at room temperature. Inspection of the STM micrographs for the Si(111)7x7-K system suggests the following:

1) K atoms adsorbed on the Si(111)7x7 surface are imaged dark initially.

2) K atoms adsorb preferentially at the on-top site of the center adatoms, leaving the corner adatoms free of K.

3) Approximately 60% of adsorbed K atoms reside on the faulted halves.

Fig. 6. STM micrographs of the Si(111)7x7 surface with K.

4) The STM image at the empty electronic states (Fig.6(b)) is almost completely washed out, showing little detail as to the atomic geometries of both K adsorbates and Si substrate.

5) Approximately 10% of K atoms on the surface form brightly imaged clusters, often in the form of trimers.

Almost identical STM images were obtained for the Cs adsorption on the Si(111)7x7 surface. Cs atoms are imaged dark, adsorbed on the center Si adatoms of the faulted half units.

We note that Li atoms adsorbed on the Si(111)7x7 surface also reside only on the faulted halves of the 7x7 unit cells, but they are always imaged bright.[13] This observation was taken as evidence that the Li atoms adsorbed on the Si(111) surface were not ionized completely. However, the present data on Si(111)7x7-K/Cs imply that K and Cs atoms are ionic upon adsorption on the faulted halves. We did not observe this kind of striking difference in the case of alkali metal adsorption on the Si(100)2x1 surface.[14]

When the K covered Si(111)7x7 surface was annealed at 300°C, a new phase with 1x3 superstructure emerged as shown in Fig. 7 in addition to those observed for the room temperature adsorption. The 1x3 K associated phase was previously reported by Daimon and Ino[15] based on their LEED/RHEED study but no structural detail was established. The new phase starts to grow from the atomic steps of the terraces. Close inspection of the 1x3 phase has disclosed it is an incommensurate phase with respect to the 7x7 substrate, with the formation of semi-periodic vacancies (approximately every nine K atoms) along the K linear chains. K atoms essentially occupy the threefold

Fig. 7. STM micrographs of the Si(111)7x7-K system annealed at 300°C.

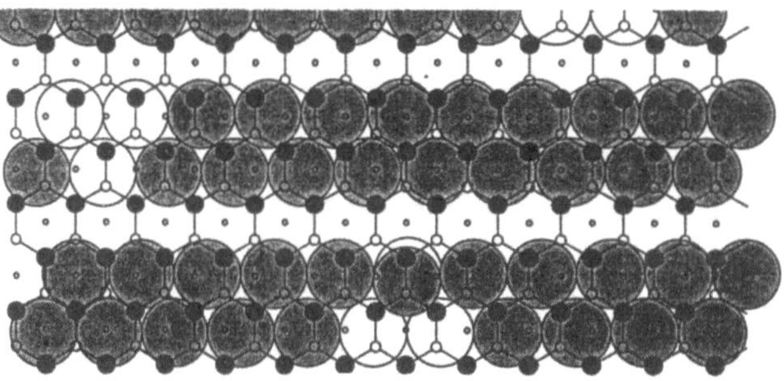

Fig. 8. Schematic of the Si(111)-K 3x1 incommensurate phase.

hollow site (H_3) of the 1x1 substrate (Fig. 8). This atomic geometry of the 1x3 phase can be understood reasonably well based on the atomic size difference (4.53A for K and 3.84A for Si adatom separation).

4.3 Alkali Metal Adsorption on the Si(100)2x1 Surface

Figure 9 shows two typical STM micrographs: (a) the Si(100)2x1 surface with 3 at.% Cs coverage and (b) the magnified micrograph of Fig. 9(a) to show the detail of the adsorption. Cs atoms are imaged brightly in the filled states; Cs atoms adsorb on this surface randomly at this small coverage.

Possible adsorption sites on the Si(100)2x1 dimerization surface have been discussed in detail both experimentally and theoretically. There have been four high-symmetry adsorption sites suggested: the pedestal site, cave site, bridge site and valley-bridge site.[16] Careful study of the STM micrographs suggests that Cs atoms occupy a site different from those four possible high symmetry sites. The micrograph shows that the charge distribution at the Cs adsorption site peaks at a site approximately 1/3-off the adjacent dimer row. By careful study of a series of these micrographs, it is concluded that the Cs atom sits at an off-center position over the two neighbouring dimer atoms.

The suggested position is shown schematically in Fig. 10 together with the adsorption sites for Li and K. We have

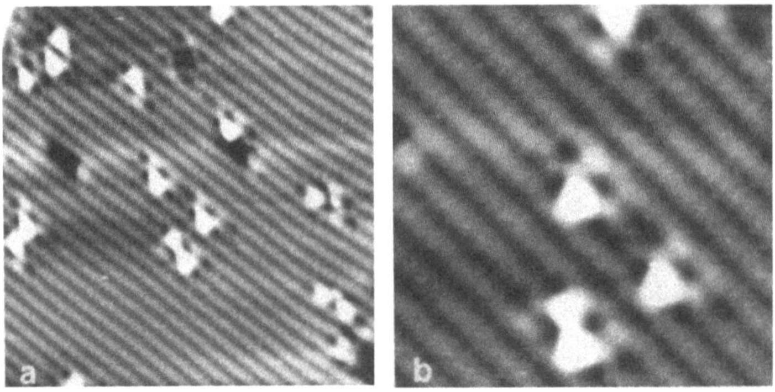

Fig. 9. STM micrographs of the Si(100)2x1 surface with Cs adsorption.

previously reported that Li atoms reside on top of one of the dimer-forming surface atoms and that K atoms occupy a similar site as Li but over the two neighbouring dimer atoms.[14]

The similarities and differences among three alkali metal atoms, Li, K and Cs, can be understood in terms of

1) the strong interaction between the directional dangling bonds of the surface dimer atoms and the s-electron of the alkali metal atoms and

2) size differences of the three alkali metal atoms.

Realizing the existence of active and highly directional dangling bonds at the Si(100)2x1 surface, it is reasonable that alkali metal atoms dominated by the isotropic s-electron are bound via these dangling bonds. The proposed Li adsorption site is consistent with these arguments, even more so when one recalls the well-established experimental results of atomic hydrogen chemisorption on the same Si(100)2x1 surface. The adsorption sites of K and Cs we propose are essentially the same as that of Li in terms of the strong bonding with the dangling bond surface states.

The only difference is the relative position in the direction along the dimer rows. That can be understood by the difference in the atomic size among those three elements. The atomic size in the neutral state in a metal is 3.02A, 4.53A and 5.24A for Li, K, and Cs, respectively. The distance between the adjacent dimer clusters is 3.84A along the dimer rows and 8.7A between the rows. Li atoms are too small to interact with any neighbouring dimer-atom charge distribution in either direction, thus they reside on top of one of the dimer-forming surface Si atoms, consuming the available dangling bond surface states. K atoms (4.53A) are too small to interact effectively with the dangling bond surface states originated at the next dimer rows. However, they are big enough to spill over and interact with the dangling bond surface states originated at the neighbouring dimers along the dimer rows (3.84A). This energy gain may be large enough to compensate the strain energy increase due to the configuration change. Potassium adsorption may be such a case and K atoms shift halfway towards the adjacent dimer atoms along the dimer rows to find the energy mimimum under this configuration. Cs atoms may be large enough even to interact partly with the dangling bond surface states originated at dimer atoms located at the next dimer row. This results in the off-centered valley-bridge site, proposed here and shown in Fig. 10(c). The present experimental results are supported by the recent theoretical investigation by Terakura's group.[17]

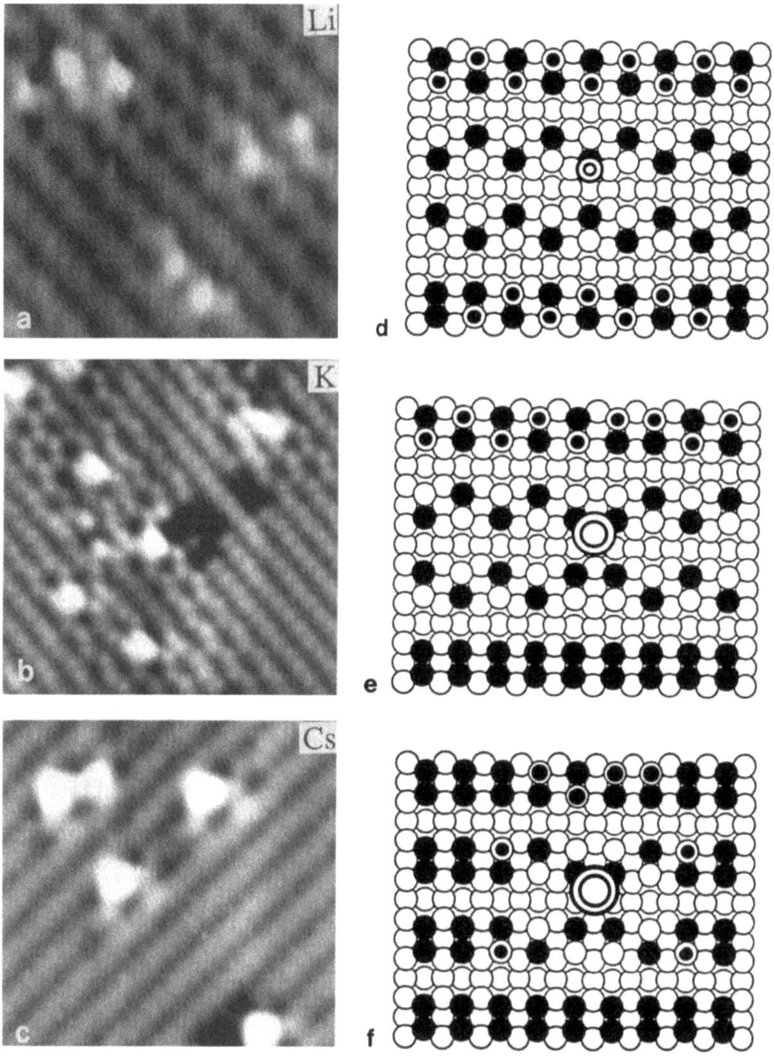

Fig. 10. STM micrographs: a, b, and c of Si(100)2x1 with Li, K and Cs, respectively and their adsorption site models: d, e, and f.

5. Conclusions

Scanning tunneling microscopy is a new powerful analytical technique for surface science. The FI-STM was developed to overcome the critical weakness of poor reliability and unknown resolution of the STM, by incorporating a FIM into the STM for precise characterization of a scanning tip on an atomic scale. The FI-STM was applied for investigation of alkali metal atom adsorption on the Si(111)7x7 and Si(100)2x1 surfaces. Our FI-STM study suggested that the Si(111)7x7 surface and Si(100)2x1 surface behave quite differently initially when alkali metal atoms are deposited: the bonding is more ionic on the (111)7x7 surface while it is more covalent on the (100)2x1 surface.

Acknowledgements

The FI-STM Project was financially supported by the NEC Fundamental Research Laboratry via the University-Industry Cooperation Program initiated by the Department of Education. The alkali metal adsorption study was in collaboration with Professor Murata's group.

References

1. G. Binnig and H. Rohrer, Scientific Am. 253 (1985) 40, Rev. Mod. Phys. 59 (1987) 615.
2. G. Binnig, H. Rohrer, Ch. Gerber and E. Weibel, Phys. Rev. Lett. 50 (1983) 120.
3. T. Sakurai, A. Sakai and H. W. Pickering, "Atom-probe Field Ion Microscopy and Its Applications," Adv. Electronics and Electron Phys., Suppl. 20, pp. 299 (1989) Academic Press, New York, 1989.
4. T. Sakurai, T. Hashizume, I. Kamiya, Y. Hasegawa, T. Ide, I. Sumita, A. Sakai and S. Hyodo, J. Vac. Sci. Technol. A7 (1989) 1684.
5. T. Sakurai, T. Hashizume, Y. Hasegawa, N. Sano, H. W. Pickering, and A. sakai. "Field Ion-Scanning Tunneling Microscopy (FI-STM)," Progress. in Surf. Sci. 33 (1990) 3.
6. Y. Kuk and P. J. Silverman, Rev. Sci. Instrum. 60 (1989) 165.
7. J. Tersoff and D. R. Hamann, Phys. Rev. Lett. 50 (1983) 1998.
8. M. Tsukada and N. Shima, J. Phys. Soc. Jpn. 56 (1987) 2875.
9. J. E. Demuth, R. M. Tromp, and M. E. Welland, J. Vac. Sci. Technol. A4, (1986) 1320.
10. S. Takahashi et al., to be published.
11. T. Sakurai, and E. W. Muller, J. Vac. Sci. Technol. 11 (1974) 878.
12. The details are available by Showa Science Inc.
13. Y. Hasegawa, I. Kamiya, T. Hashizume, T. Sakurai, H. Tochihara, M. Kubota, & Y. Murata, J. Vac. Sci. Technol. A8 (1990) 238.
14. T. Hashizume, Y. Hasegawa, I. Kamiya, I. Sumita, S. Hyodo, T. Sakurai, H. Tochihara, M. Kubota, and Y. Murata, J. Vac. Sci. Technol. A8, (1990) 233
15. H. Daimon and S. Ino, Surf. Sci. 164 (1985) 320.
16. J. D. Levine, Surf. Sci. 34 (1973) 90.
17. K. Terakura, private communications.

Interaction Between Donors and Acceptors on Metal Surfaces

T. Aruga, T. Sasaki, and Y. Iwasawa

Department of Chemistry, Faculty of Science,
The University of Tokyo, Hongo, Bunkyo-ku, Tokyo 113, Japan

We discuss the nature of the interaction between donor and acceptor molecules coadsorbed on metal surfaces. First, the recent experimental result is reviewed for the coadsorption of carbon monoxide (CO) with ammonia (NH_3) and with methylamine (NH_2-CH_3). There seems to exist mutual attraction between CO and ammonia (methylamine) which is typical for the coadsorption of donor and acceptor molecules, where donors (acceptors) denote molecules which donate (accept) electrons through molecule-substrate bonding. We propose a model for the interaction among donors and acceptors based on the concepts of mutual polarization enhancement and depolarization. This model predicts that the donor-acceptor interaction not only causes mutual attraction between a donor and an acceptor but also enhances adsorptive bonding of both molecules. Several experimental results for the CO+ammonia, CO+benzene, and NO+benzene systems are discussed as evidence for this model. We also discuss direct interaction between molecular orbitals of coadsorbed molecules.

1. Introduction

There is a growing interest in the ordering of coadsorbed molecules on transition metal surfaces. Mate and Somorjai [1] found that the coadsorption of benzene (C_6H_6) with CO on Pt(111) results in the formation of new ordered structures. Upon coadsorption, $\begin{pmatrix}4 & 0\\ 2 & 4\end{pmatrix}$ and $\begin{pmatrix}5 & 0\\ 2 & 4\end{pmatrix}$ structures were observed, while no LEED patterns of benzene were observed without coadsorbed CO. Similar observations were also made for CO+benzene on Rh(111) [1], Pd(111) [2], and Ru(001) [3] and NO+benzene on Ru(001) [3]. The structure of CO+benzene overlayers on Pt(111), Rh(111), and Pd(111) was determined by dynamical LEED analysis [2,4-6]. Menzel and coworkers [3,7,8] have conducted several experiments to identify the interaction of benzene with CO and NO on Ru(001). The electrostatic interaction and indirect charge transfer through the substrate have been suggested to play roles in the coadsorption layers.

Recently, we have studied the coadsorption of CO and ammonia on Ru(001) [9,10]. The coadsorption of methylamine with CO was also studied [11] to examine the steric effect of the methyl (CH_3-) group. Both ammonia and methylamine are adsorbed via a lone-pair orbital on the N atom and hence represent donor adsorbates on transition metal surfaces. Carbon monoxide is adsorbed via electron donation from the 5σ molecular orbital to a substrate as well as back-donation from the substrate to the unfilled $2\pi^*$ orbital. The latter interaction is believed to be dominant for Ru, suggesting that CO is adsorbed as an electron acceptor. Therefore attractive donor-acceptor interaction is expected to occur between CO and NH_2-R (R=H, CH_3). The coadsorption of CO with ammonia at 100 K resulted in the formation of two types of intermixed ordered layers of (2x2) periodicity, suggesting that there occurs mutual attraction between CO and ammonia. The results of vibrational spectroscopy and thermal desorption allowed us to construct structure models for the two phases. TPD results indicate that the ammonia dissociation was enhanced in the β-(2x2) phase. HREELS results imply that the ammonia molecular axis is tilted in the coadsorbtion layers, suggesting that the local interaction between ammonia and CO leads to the tilting of ammonia and hence facilitates electron donation from Ru 4d to an antibonding molecular orbital of ammonia.

In the following section, we summarize the experimental results on the ordering of CO and ammonia on Ru(001). Then we propose a model for the interaction between donor and acceptor molecules coadsorbed on metal surfaces. Finally, we discuss the possibility of direct interaction of molecular orbitals between coadsorbed molecules.

2. Ordering of CO and Ammonia on Ru(001)

Ammonia adsorption on Ru(001) at 100 K followed by annealing to successively higher temperatures causes no LEED patterns except for a uniform increase in background intensity [9,10,12]. For CO adsorption, $(\sqrt{3}\times\sqrt{3})R30^0$, $(2\sqrt{3}\times2\sqrt{3})R30^0$, and compressed hexagonal patterns are observed with increasing coverage [13,14]. When CO-precovered surfaces with CO coverage of 0.16 to 0.58 were exposed to ammonia at 100 K, the CO-induced patterns were replaced by (2×2) patterns. For higher CO precoverages, the CO-induced compressed hexagonal pattern coexisted with the (2×2) pattern. The ammonia exposure required for the completion of the (2×2) pattern varied with increasing CO precoverage: 0.6 L (1 L=1×10^{-6} Torr s) at θ_{CO}=0.3 and 0.12 L at θ_{CO}=0.58. For the reverse exposure sequence, CO adsorption on ammonia-precovered Ru(001) at 100 K, no new LEED patterns were observed as already reported [15]. However, annealing this surface to 150 K resulted in the appearance of a (2×2) pattern. HREELS and TPD spectra of this surface were identical to those of NH_3/CO/Ru(001).

We have proposed the models for these phases based mainly on the results of vibrational spectroscopy. The HREEL spectra measured for coadsorption layers were different from those expected by simply adding a spectrum of adsorbed CO to that of ammonia. The spectra also exhibited a remarkable variation as a function of CO precoverage. The spectra at any CO precoverages can, however, be understood by superposing two typical spectra: one at θ_{CO}~0.25 (α-(2×2)) and the other at θ_{CO}~0.5 (β-(2×2)). Figure 1 represents typical HREEL spectra for separate adsorption of NH_3 and CO and for coadsorbed phases of α- and β-(2×2). The assignments of the loss modes were made by referring to the literature [13,16] and confirmed by utilizing the vibrational isotopic effect for NH_3 and ND_3.

Remarkable changes were observed for the C-O stretching (ν_{CO}) mode: ν_{CO} was shifted from ~2000 cm^{-1} for the CO/Ru(001) surface (fig. 1c) to ~1650 cm^{-1} for the α-(2×2) phase (fig. 1d and e) and, for the β-(2×2) phase, ν_{CO} appeared at ~1750 cm^{-1} as well as at ~2000 cm^{-1} (fig. 1f and g). The former result suggests that CO changes its adsorption site from an atop one for the CO/Ru(001) surface to a three-fold hollow one for the α-(2×2) phase. The latter result was interpreted to indicate the coexistence of atop and three-fold-hollow CO species in the β-(2×2) phase. The observed loss peaks for atop and hollow CO for β-(2×2) exhibited almost the same intensity.

The vibrational modes originated from ammonia admolecules also exhibited several differences between the α- and β-(2×2) phases. HREELS measured during anneal-quench cycles indicated that there are two kinds of ammonia species in α-(2×2): One is desorbed at ~200 K and the other persists up to 280 K. The former species exhibits a rocking (ρ_r) mode at 600 cm^{-1} and symmetric deformation (δ_s) at 1150 cm^{-1} while, for the latter species,

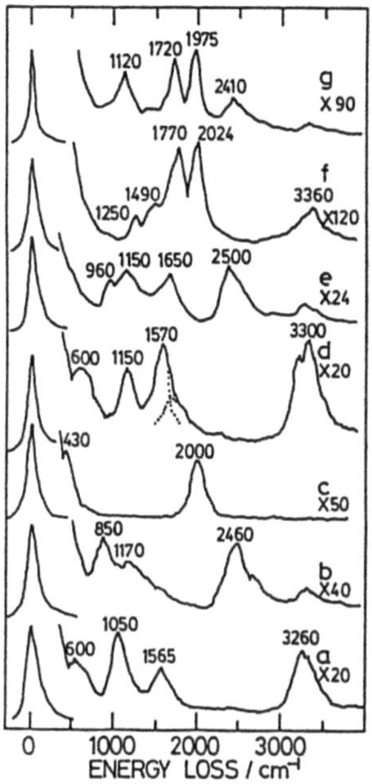

FIG. 1. HREEL spectra recorded at 100 K. (a) NH_3/Ru(001), (b) ND_3/Ru(001), (c) Ru(001)-($\sqrt{3}\times\sqrt{3}$)R30° -CO, (d) Ru(001)-α(2×2)-(CO+2NH_3), (e) Ru(001)-α(2×2)-(CO+2ND_3), (f) Ru(001)-β(2×2)-(2CO+NH_3), and (g) Ru(001)-β(2×2)-(2CO+ND_3).

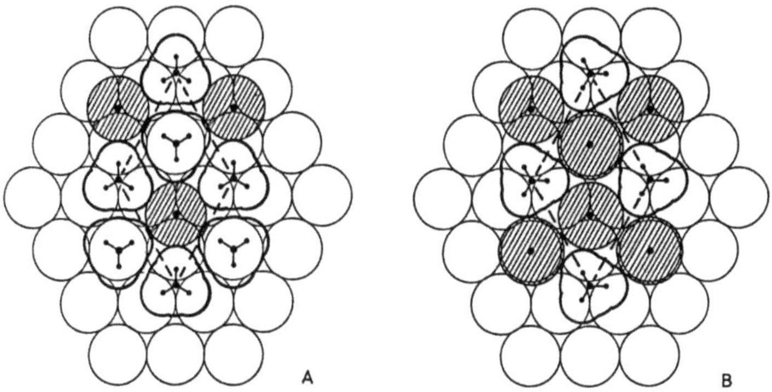

FIG. 2. Models for (A) Ru(001)-α(2×2)-(CO+2NH_3) and (B) Ru(001)-β(2×2)-(2CO+NH_3).

ρ_r is absent and δ_s appears at ~1220 cm^{-1}. Based on the results for the ammonia adsorption on Ru(001) [16,17], the former species was assigned to ammonia adsorbed on an atop site, and the latter on a three-fold hollow site. On the other hand, HREEL spectra for the β-(2×2) phase exhibited only hollow ammonia species.

The HREELS results summarized above and some considerations as to the absolute coverages of ammonia and CO in the α- and β-(2×2) phases allowed us to construct the geometrical models shown in fig. 2. In a unit mesh of the α-(2×2) structure, one CO molecule occupies a three-fold hollow site and two ammonia molecules occupy an atop and a three-fold hollow site. For the β-(2×2) phase, two CO occupy an atop and a three-fold hollow site and one ammonia occupies a three-fold hollow site. Although the CO adsorbed on a three-fold hollow site has not been observed for Ru(001), the large downward shift of ν_{CO} can only be explained in terms of the change of adsorption site. The cause of the site change will be discussed later.

The coadsorption of CO with methylamine was also examined [11]. The results indicate the formation of a similar (2x2) structure at CO precoverages higher than ~0.4 and very incomplete (2×2) ordering at lower CO precoverages. This suggests that the bulky methyl group of methylamine prevents the formation of the α-(2×2) structure, in which two methylamine molecules should be packed together with one CO. On the other hand, the β-(2×2) structure, in which only one methylamine exists, can be formed.

3. Generalized Model for Surface Donor-Acceptor Interaction

Ammonia is adsorbed on Ru(001) via a lone electron pair on the N atom with its C_{3v} axis normal to the surface. Benndorf and Madey [17] measured the work-function change as a function of coverage for the ammonia adsorption on Ru(001). The results indicated that the work function decreases with increasing coverage. The initial dipole moment of an ammonia molecule was estimated to 1.9 D, which is considerably larger than the value for a gas-phase molecule, 1.47 D. This result suggests that the ammonia bonding to the surface is essentially the donation from a lone-pair orbital to unfilled states of the substrate. Thus, ammonia acts as a donor with respect to the bonding with substrates.

As is well known, CO is adsorbed on most transition metals with the C atom down and with its molecular axis normal to the surface. A gas-phase CO molecule has a permanent electric dipole moment (C^-O^+) of 0.11 D. Therefore, if the CO-substrate bonding is accompanied by neither charge transfer nor intramolecular polarization change, the work function should decrease upon CO adsorption. The work-function change upon CO adsorption on Ru(001) [18] exhibited an increase with increasing coverage. The initial dipole moment was estimated to be 0.22 D. Notice that the dipole moment induced upon adsorption orients oppositely to the intramolecular dipole moment. This suggests that the charge transfer from the substrate to CO, or polarization of electrons in the CO-substrate system toward the CO side, occurs upon the adsorption. The widely accepted picture [19] for the CO adsorption on transition metal surfaces suggests that CO interacts with the substrate in two ways: donation from the 5σ molecular orbital to empty states of the substrate and back-donation from filled substrate states to the unfilled CO $2\pi^*$ orbital. The results mentioned above suggest that the back donation is predominant for the CO adsorption on Ru(001). Thus CO is adsorbed as an acceptor.

The interaction between donor and acceptor molecules has been discussed in terms of electrostatic attraction between antiparallel dipoles [20] or indirect charge transfer through the substrate [21]. The former model neglects the mutual enhancement of polarization discussed below and hence underestimates the interaction energy between donors and acceptors. To rationalize the latter model, it was suggested [8] that the increased surface electron density due to the adsorption of donors is responsible for the enhanced charge transfer from the substrate to a neighboring acceptor molecule. This model, however, does not explain why the electron transferred from the donor should be forwarded to the acceptor. The indirect-charge-transfer model would be more clearly understood by taking into account microscopic forces causing the "indirect" charge transfer. We propose that the interaction among donors and acceptors coadsorbed on metal surfaces can be described in terms of enhanced counter-polarization and depolarization. This model is an extension of a model [21,22] for the effect of alkali-metal atoms on coadsorbed molecules. This model suggests that the electrostatic interaction and the indirect charge transfer cannot be separated from each other; these two terms refer to different aspects of the same donor-acceptor interaction modelled below.

Suppose that there is a donor molecule such as ammonia or benzene adsorbed on a surface. This is accompanied by a dipole moment oriented toward the vacuum side. This electric dipole induces an electrostatic potential which shifts downward the electronic states of nearby molecules. If two donor molecules are adsorbed near to each other, the ionization level of each molecule is shifted downward and hence the charge transfer from the molecules to the substrate is reduced, resulting in weakening of the molecule-substrate bonding. This type of phenomenon is most clearly observed for the alkali-metal adsorption and is called "depolarization" [23]. The interaction between acceptor molecules can also be understood in the same way. The electric dipole due to the charge transfer to an acceptor molecule induces electrostatic potential which shifts upward the affinity level of a neighboring acceptor molecule, resulting in the decrease of the charge transfer. This effect is mutual and hence leads to the depolarization of both molecules. Note that electrostatic repulsion between parallel dipoles are reduced to some extent due to the depolarization, while the adsorption energy of each molecule is reduced. On the other hand, if an acceptor molecule comes near a donor, the affinity level of the acceptor is shifted downward and the ionization level of the donor is shifted upward. Both of these effects cause the polarization enhancement of both dipoles, leading to the enhancement of the attraction between antiparallel dipoles. Thus, it is clear that this model describes the microscopic mechanism of the indirect charge transfer in terms of the electrostatic interaction. The important consequence of the mutual enhancement of the polarization, or charge transfer, is the strengthening of the molecule-substrate bonding for both donor and acceptor molecules, which is beyond the scope of the conventional models. All the effects described above stabilize a donor-acceptor pair on surfaces as compared to the case of separate adsorption and, for coadsorption layers, favor intermixing rather than segregation of each component.

The model described above is supported by several experimental facts. First, ultraviolet photoemission measured for (CO+benzene)/Ru(001) [3] indicated uniform binding energy shifts: +0.2 eV for CO peaks and -0.2 eV for benzene peaks. This is in agreement with the expectation of the above model. The shifts were also observed for (NO+benzene)/Ru(001), while shifts due to a change in bonding were superposed [3]. Second, CO coadsorbed with donors such as ammonia [9,10] or benzene [2-6] tends to occupy high-coordination sites: On Ru(001), CO molecules occupy three-fold hollow sites when coadsorbed with ammonia, while CO occupy only atop sites when adsorbed without donors. This tendency has also been observed clearly for CO+benzene monolayers. The CO bonding in high coordination sites suggests enhanced back donation from the substrate. In fact, the C-O stretching frequency is considerably shifted downward when adsorbed on high-coordination sites, indicating C-O bond weakening due to the enhanced occupation of the antibonding $2\pi^*$ orbital. The acceptor character of CO is hence enhanced when adsorbed on high-coordination sites, suggesting stronger donor-acceptor interaction with ammonia or benzene. Third, TPD for the coadsorption of CO and benzene [8] demonstrated that the molecular desorption temperatures of both CO and benzene are higher than for the pure, equally crowded layers, indicating that molecule-surface bonding is strengthened by each other. Fourth, the work-function change upon the formation of a CO+benzene layer was considerably smaller than the sum of those for the individual adsorbates at the corresponding coverages [8], which can be explained by an increased charge transfer between molecules and the surface. Further support for the generalized donor-acceptor model would be obtained by means of various methods. In particular, precise analysis of work-function change and inverse photoemission spectra as a function of composition would provide strong evidence for the model. The calculation using clusters with a varying donor-metal-acceptor angle would also be promising, since this will highlight the role of the electrostatic potential.

4. Direct Interaction

For (CO+ammonia)/Ru(001), the ammonia dissociation probability during temperature-programmed desorption was measured as a function of CO partial coverage [9,10]. The result is reproduced in fig. 3. On clean Ru(001) without coadsorbed CO, ~3% of ammonia molecules are dissociated upon annealing. Careful examination clarified that this is not due to surface defects but is intrinsic to the Ru(001) surface [10]. The rocking motion of ammonia molecules may initiate the dissociation, since this motion enables direct interaction of the lowest unoccupied molecular orbital, $4a_1$, which is of antibonding character with respect to the N-H bonds, with the filled 4d states of the Ru substrate. With increasing CO partial coverage, the dissociation probability decreases slightly and reaches a minimum in the α-(2×2) phase. This decrease may be due to the blocking by CO. As seen in fig. 2a, the rocking motion of ammonia would be reduced by the presence of coadsorbed CO. In fact, the vibrational frequency of the rocking mode is considerably blue shifted in α-(2×2) as compared to that of ammonia adsorbed without CO on Ru(001) [10].

On the other hand, the ammonia dissociation is considerably enhanced in the β-(2×2) phase. What causes the dissociation enhancement? We ascribed this to static tilting of ammonia molecules in β-(2×2). A HREEL spectrum for NH_3 (ND_3) (fig. 1a and b, respectively) exhibits a strong peak due to the δ_s mode at 1050 (850) cm^{-1} and a weak peak due to δ_a at 1565 (1170) cm^{-1}, indicating that ammonia molecules are adsorbed with its C_{3v} axis almost normal to the surface. In α-(2×2) (fig. 1d and e), the δ_s peak at 1150 (960) cm^{-1} and the δ_a peak at 1570 (1150) cm^{-1} exhibit comparable intensity. In β-(2×2) (fig. 1f and g), the intensity of the δ_s peak at 1250 (1000~1050) cm^{-1} is remarkably reduced, while the δ_a peak at 1490 (1120) cm^{-1} is clearly observed. When ammonia molecules are adsorbed with their C_{3v} axis normal to the surface, only the δ_s mode can be dipole active. Hence the observation described above can be explained in terms of tilting of the ammonia C_{3v} axis in the coadsorbed phases, in particular in β-(2×2). This suggests that the tilting of ammonia, which facilitates the direct interaction of the empty $4a_1$ orbital with the Ru 4d states, is responsible for the enhanced ammonia dissociation in β-(2×2), while the donor-acceptor type interaction can also enhance the dissociation of ammonia to a certain extent, however [10].

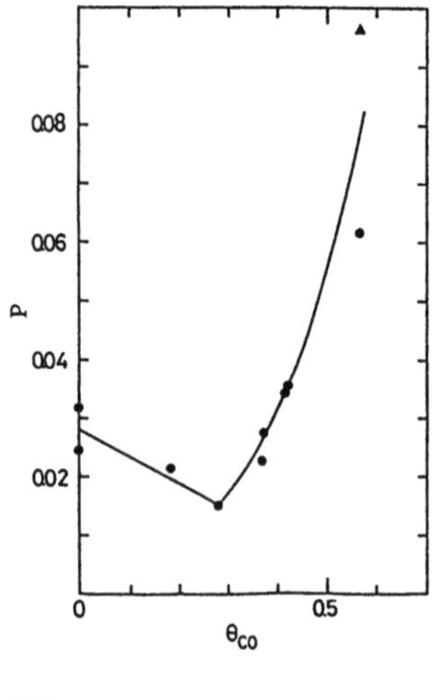

FIG. 3. The ammonia dissociation probability, P, during linear temperature rise (dT/dt=5 K/s) as a function of CO coverage. The dissociation takes place at around 300 K.

Then, the next question arises: What causes the tilting of ammonia? As seen in fig. 2b, in which the van der Waals shape is shown, the tilting seems to be very difficult if there is no direct interaction between ammonia and CO. It should be noted that the sum of the van der Waals radii of hydrogen and oxygen is ~2.7 Å, while the distance between hydrogen-bonded H and O can be reduced to ~2.4 Å [24]. Thus, it is natural to assume that there is direct interaction between molecular orbitals of ammonia and CO. Several contributions to this interaction have been considered [10]. Among these, electron donation from occupied CO 1π to unoccupied ammonia $4a_1$ may lead to attractive interaction between CO and ammonia. This interaction is similar to that proposed [25] for direct K-CO interaction on metals. The difference between α- and β-(2×2) may be ascribed to the fact that there are six CO molecules surrounding an ammonia in β-(2×2) while only three CO exist around an ammonia in α-(2×2).

At any rate, the situation may be a little more complicated than considered above because of bonding change due to the adsorption as well as a possible contribution from substrate states. Therefore, detailed examination by various means such as photoemission experiments and theoretical calculations is required to construct a microscopic model for the direct non-donor-acceptor interaction.

References

[1] C.M. Mate and G.A.Somorjai, Surf. Sci. 160, 542 (1985).
[2] H. Ohtani, M.A. Van Hove, and G.A. Somorjai, J. Phys. Chem. 92, 3974 (1988).
[3] P.A. Heimann, P. Jakob, T. Pache, H.-P. Steinruck, and D. Menzel, Surf. Sci. 210, 282 (1989).
[4] M.A. Van Hove, R.F. Lin, and G.A. Somorjai, J. Am. Chem. Soc. 108, 2532 (1986).
[5] R.F. Lin, G.S. Blackman, M.A. Van Hove, and G.A. Somorjai, Acta Cryst. B43, 368 (1987).
[6] D.F. Ogletree, M.A. Van Hove, and G.A. Somorjai, Surf. Sci. 183, 1 (1987).
[7] H. Rauscher, P. Jakob, and D. Menzel, Surf. Sci. 234, 108 (1990).
[8] P. Jakob and D. Menzel, Surf. Sci. 235, 15 (1990).
[9] T. Sasaki, T. Aruga, H. Kuroda, and Y. Iwasawa, Surf. Sci. 224, L969 (1989).
[10] T. Sasaki, T. Aruga, H. Kuroda, and Y. Iwasawa, Surf. Sci., in press.
[11] T. Sasaki, T. Aruga, H. Kuroda, and Y. Iwasawa, to be published.
[12] L.R. Danielson, M.J. Dresser, E.E. Donaldson, and J.T. Dickinson, Surf. Sci. 71, 599 (1978).
[13] G.E. Thomas and W.H. Weinberg, J. Chem. Phys. 70, 1437 (1979).
[14] H. Pfnur, D. Menzel, F.M. Hoffmann, A. Ortega, and A.M. Bradshaw, Surf. Sci. 93, 431 (1980).
[15] Y. Zhou, S. Akhter, and J.M. White, Surf. Sci. 202, 357 (1988).
[16] J.E. Parmeter, Y. Wang, C.B. Mullins, and W.H. Weinberg, J. Chem. Phys. 88, 5225 (1988).
[17] C. Benndorf and T.E. Madey, Surf. Sci. 135, 164 (1983).
[18] H. Pfnur and D. Menzel, J. Chem. Phys. 79, 2400 (1983).
[19] T.N. Rhodin and J.W. Gadzuk, in "The Nature of the Surface Chemical Bond" ed. T.N. Rhodin and G. Ertl (North-Holland Publ., 1974) p. 207.
[20] C.T. Kao, C.M. Mate, G.S. Blackman, B.E. Bent, M.A. Van Hove, and G.A. Somorjai, J. Vacuum Sci. Technol. A6, 786 (1988).
[21] J.K. Norskov, S. Holloway, and N.D. Lang, Surf. Sci. 137, 65 (1984).
[22] T. Aruga, H. Tochihara, and Y. Murata, Phys. Rev. B34, 8237 (1986).
[23] T. Aruga and Y. Murata, Prog. Surf. Sci. 31, 61 (1989).
[24] L.N. Kuleshova and P.M. Zorkii, Acta Cryst. B37, 1363 (1981).
[25] D. Heskett, I. Strathy, E.W. Plummer, and R.A. de Paola, Phys. Rev. B32, 6222 (1985).

Reactivity of Surface Hydroxyl Groups with Metal Complex Compounds

T. Yamada, Y. Kuroda, A. Fukuoka, M. Ichikawa, and K. Tanaka

The Institute for Solid State Physics, The University of Tokyo, 7-22-1 Roppongi, Minato-ku, Tokyo 106, Japan

The surface reaction of the adsorbed hydroxyl group OH(a) with $Rh(CO)_2(\pi\text{-}C_3H_5)$ delivered from the gas phase:

$OH(a) + Rh(CO)_2(\pi\text{-}C_3H_5)(g) \rightarrow O\text{-}Rh(CO)_2(a) + C_3H_6(a/g)$

has been studied on Ni(110), Ni(100) and ZnO(10$\bar{1}$0) mainly by HREELS and TPRS. On Ni(110), p(2 × 1)-OH(a) was completely diminished by adsorption of $Rh(CO)_2(\pi\text{-}C_3H_5)$ below 300 K. C_3H_6 stayed on the surface after this reaction. On ZnO(10$\bar{1}$0) OH(a) was deposited with surface defects introduced by Ar ion bombardment. The same reaction occurred with OH(a) on ZnO(10$\bar{1}$0), but not completely on Ni(100) which was covered with oxide layers.

1. INTRODUCTION

The hydroxyl group on solid surfaces (which we denote by OH(a); (a) means adsorbed species, and (g) gas-phase species) is one of the best studied molecular fragments in surface science [1]. Formation of OH(a) has been reported upon adsorption of H_2O on the surface of transition metals when they are pre-covered by oxygen adatoms. The chemical reaction between H_2 and O_2 on some well-defined metal surfaces also produces OH(a).

Hydroxyl groups have also been extensively studied in the field of practical catalysis. Metal catalysts are usually deposited on "supports" of porous inorganic oxides such as SiO_2, Al_2O_3, TiO_2, etc. The surfaces of these oxides are in most cases covered by a monolayer of hydoxyl groups and multilayer-physisorbed water.

An important feature is that each OH(a) can attach to a molecule of a metal complex compound by forming a chemical bond with it. Yermakov et al. [2] first noticed that the amount of OH(a) on SiO_2 support was reduced by adsorption of π-allyl ($\pi\text{-}C_3H_5$) complex compounds of Zr, Cr, Mo and W. Formation of Si-O-metal bonds and elimination of $\pi\text{-}C_3H_5$ were concluded by analogy to organometallic allyl compounds in the solution phase. It should be emphasized that each π-allyl complex molecule reacts with one OH(a) and forms one admolecule. This is an advantage in preparing supported catalysts because the metal is atomically dispersed on the support in this situation. Special catalytic reactivity and selectivity are expected on this kind of ultimately dispersed metal. Iwasawa [3] reviewed the detailed structural analysis and reactivity on the catalysts prepared in this manner, the so-called "fixed catalysts". There are various fixed catalysts derived by changing the ligands around the metal atom or taking multinuclear complex compounds.

In this work we studied the deposition reaction of dicarbonyl π-allyl rhodium $Rh(CO)_2(\pi\text{-}C_3H_5)$ delivered from the gas phase onto three well-defined surfaces:

(1) Ni(110) p(2 × 1)-OH(a) prepared by $O(a) + H_2O(g) \rightarrow 2OH(a)$

(2) Ni(100) Hex-OH(a) prepared by $H_2(g) + O_2(g) \rightarrow 2OH(a)$
(3) Zn(10$\bar{1}$0)-OH(a) prepared by H_2O adsorption.

We adopted this complex compound in the anticipation that the π-C_3H_5 ligand is substituted by surface OH and it retains only CO ligands. The expected deposition reaction is designated as

$$OH(a) + Rh(CO)_2(\pi\text{-}C_3H_5)(g) \rightarrow O\text{-}Rh(CO)_2(a) + C_3H_6(a/g) \quad \{1\}$$

We applied high-resolution electron energy loss spectroscopy (HREELS) to detect and identify the admolecules and temperature-programmed reaction spectroscopy (TPRS) with a mass spectrometer to pursue the elementary reaction steps. A similar deposition system of Fe(100)-OH(a) and molybdenum π-allyl complex has been examined by Bernasek [4].

2.EXPERIMENTAL

An ultrahigh vacuum system equipped with AES, LEED, HREELS and a mass spectrometer was used in this study. This apparatus contains a part to treat the specimen in pressurized gases without disturbing ultrahigh vacuum experimentation. Its details are described elsewhere [5]. The Ni(110) and (100) crystals were shaped in disks (8 mm in diameter, 1mm thickness). They are suspended with Ta wires for resistive heating by directly applying dc voltage on the ends. The Zn(10$\bar{1}$0) crystal piece (5mm × 5mm × 2mm, polished and treated in H_3PO_4) was wrapped around with a Ta foil, and this foil was heated by a dc current through it. It could reach 800 K at the maximum. The temperature was measured by a chromel-alumel thermocouple in all cases. These surfaces were cleaned by argon ion bombardment and annealing cycles.

$Rh(CO)_2(\pi\text{-}C_3H_5)$ was synthesized from $RhCl_3$ by (1) substitution with CO to derive $Rh(CO)_4Cl_2$ [6] and then by (2) Grignard reaction with C_3H_5MgCl [7]. The crude crystalline powder of $Rh(CO)_2(\pi\text{-}C_3H_5)$ was purified by sublimation in an argon atmosphere. The pure form of this material is a volatile dark yellow powder. This vapor pressure at room temperature is estimated to be 10^{-4} to 10^{-5} Torr. It can be introduced into UHV from an air-tight container through a conventional variable leak valve. The pressure of $Rh(CO)_2(\pi\text{-}C_3H_5)$ can be controlled from 10^{-9} Torr to 10^{-5} Torr measured by a B-A gauge.

3. RESULTS AND DISCUSSION

3.1. Ni(110) p(2 × 1)-OH(a)

The p(2 × 1)-OH(a) overlayer on Ni(110) was reported by Benndorf et al.[8] A stoichiometric surface reaction of pre-adsorbed O(a) and $H_2O(g)$ was found in TPRS, which yields a full coverage (0.5 per one surface Ni atom) of OH(a) when the precoverage of O(a) is 0.25. Ultraviolet photoemission and electron-stimulated desorption were utilized to identify OH(a) and to determine the structure.

We took Benndorf et al.'s procedure with H_2O and D_2O to deposit OH(a) and OD(a) respectively. First 0.5 L of $O_2(g)$ were introduced at room temperature, and then dosed with 2.0 L of H_2O or D_2O. Figs. 1 and 2 contain HREELS of freshly prepared p(2 × 1)-OH/OD(a). O-H stretching at 3400/3520 cm^{-1}, O-D stretching at 2460/2580 cm^{-1}, and O-Ni stretching at 620/600 cm^{-1} are seen. The scissors vibration mode of H(D)-O-H(D) was not detected, namely there was no significant amount of $H_2O(D_2O)(a)$.

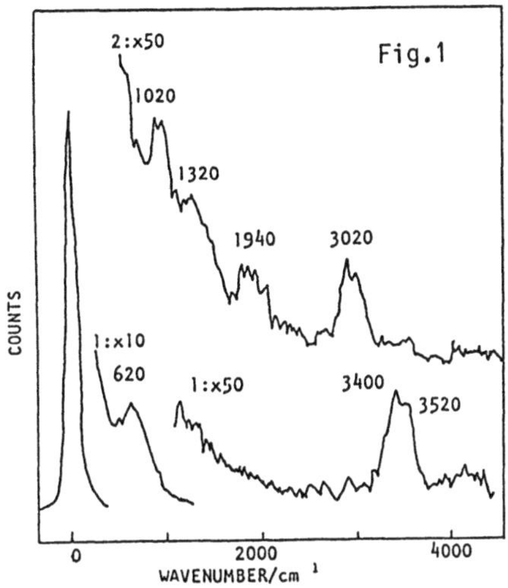

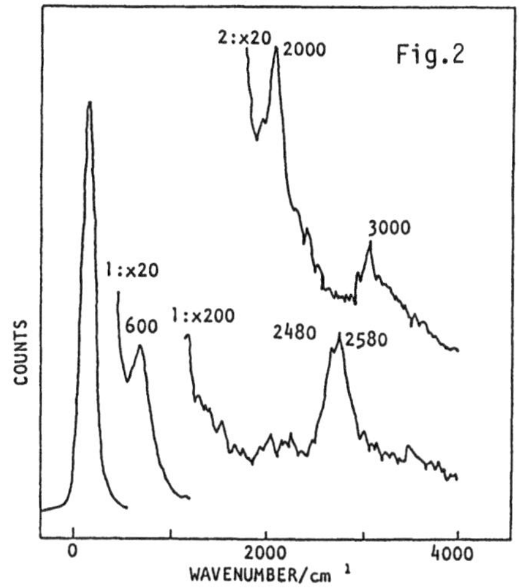

Fig. 1 HREELS of Ni(110). **1:** OH(a) prepared by adsorption of O_2 (0.5 L, 300 K) and then H_2O (2 L, 300 K). The LEED pattern was sharp p(2 × 1). **2:** After adsorption of $Rh(CO)_2(\pi\text{-}C_3H_5)$ at 1×10^{-7} Torr (by BA gauge) for 10 sec at 300 K. The LEED pattern was diffuse p(1 × 1) with a strong background. Rh(309eV)/Ni(852eV) = 0.39. (Hereafter the AES signal intensity is indicated by the ratio to that of the strongest peak of the substrate element. In parentheses is indicated the AES peak position of the element.)

Fig. 2 HREELS of Ni(110). **1:** OD(a) prepared by adsorption of O_2 (0.5 L, 300 K) and then D_2O (2 L, 300 K). **2:** After adsorption of $Rh(CO)_2(\pi\text{-}C_3H_5)$ at 1×10^{-7} Torr (by BA gauge) for 10 sec at 80 K.

Then $Rh(CO)_2(\pi\text{-}C_3H_5)$ was introduced onto freshly prepared p(2 × 1)-OH/OD(a). The intensity of the elastic peak of HREELS, shown in Figs. 1 and 2 was reduced to ca. 1/100. In the spectra the peaks of O-H and O-D stretching was completely diminished, and stretching of C-O (1940, 2000 cm^{-1}), stretching of C-H (3020, 3020 cm^{-1}), bending of C-H (1320 cm^{-1}), and stretching of C-C (1020 cm^{-1}) emerged. The LEED pattern at this stage was diffused p(1 × 1) with a strong background. The overlayer was now disordered. Rh(MVV) and C(KVV) signals were prominently seen in AES. The incident electron of AES (2 keV) destroyed the overlayer. HREELS after recording AES did not show strong C-O stretching.

TPRS of $Rh(CO)_2(\pi\text{-}C_3H_5)$/Ni(110)-OD(a) from 80 K to 300 K is shown in Fig. 3, in which desorption of H_2(2 AMU), HD(3), D_2(4), CO(28) was detected. The desorption pressure of propylene (44, 45 AMU) was below the background level. Desorption of hydrogen at 220 K is produced from adsorbed C_3H_5 or C_3H_6 by thermal cracking.

These results are summarized as follows by interpreting the spectra with other authentic spectra:

(1) Both at 80 K and 300 K reaction {1} proceeded and O-H/O-D bonds were broken. This is proved by diminishment of the O-H/O-D signal in HREELS and no comparable evolution of D_2O in TPRS.

(2) C_3H_6 produced by reaction {1} stayed on the surface, and was decomposed into H_2(g) and C(a) upon TPRS.

(3) HD and D_2 desorbed simultaneously with H_2 when the $Rh(CO)_2(\pi\text{-}C_3H_5)$-deposited overlayer was decomposed in TPRS. This indicates that D atoms from O-D are assembled in C_3H_6 by reaction {1}, i.e. the product was C_3H_5D(a).

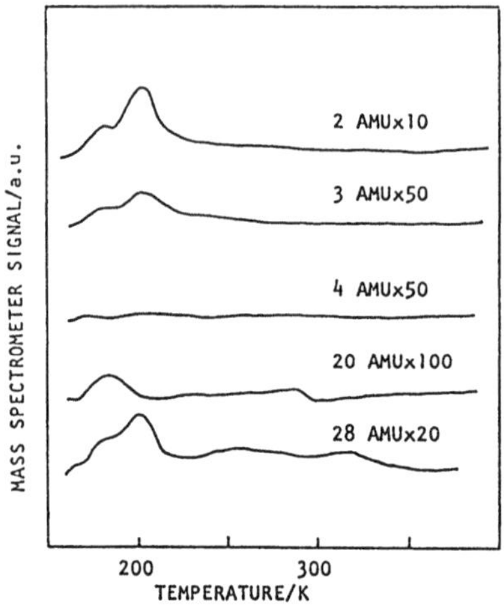

Fig. 3 TPRS of Ni(110) recorded after taking HREELS of Fig. 2-2. Heating rate: 5 Ks^{-1}. 2 AMU = H_2, 3 AMU = HD, 4 AMU = D_2, 20 AMU = D_2O, 28 AMU = CO. After this, AES gave the peak ratios: Rh/Ni = 0.63, C(280eV)/Ni(852eV) = 0.32.

(4) By HREELS CO on a Rh atom and CO on Ni were not distinguished. The stretching frequencies of CO in $Rh(CO)_2(\pi\text{-}C_3H_5)$ /Ni(110) p(2 × 1)-OH(a) and CO/Ni(110) were similar to each other.

3.2. Ni(100) Hex-OH(a)

The 2-domain hexagonal LEED pattern on Ni(100) ($(2\sqrt{3},0 \mid -1/\sqrt{3},1)$ in matrix notation; we term it "Hex".) has been frequently observed after a large exposure to O_2 [9-13], after an exposure to a mixture of O_2 and H_2 [12-15], or after the $H_2 + O_2$ reaction [12,13]. On this surface a few NiO layers cover the metal and OH(a) sits on the outermost surface. In HREEL spectra the O-H stretching is seen at ca. 3700 cm^{-1} as well as a strong phonon peak of NiO at ca. 500 cm^{-1} [11,16]. We may say that OH(a) is dispersed on the oxide in this case. An interesting fact about this OH(a) is that H in OH(a) originates from hydrogen dissolved in the Ni bulk and not from gas-phase H_2 directly [16]. Even when 99.9% H_2 was delivered with O_2, OD(a) was dominantly formed on the crystal pretreated in D_2 at > 1 Torr at 800 K to dissolve D in Ni bulk, and vice versa [16]. This OH(a) is stable up to 650 K, at which it decomposes into H_2O(g) and surface oxygen.

In this study Ni(100) Hex-OH(a) was prepared by (1) pretreatment of clean Ni(100) with H_2(g) at 10 Torr, 800 K for 30 min., and then (2) adsorption of 500 L of O_2(g) at 400 K. Fig. 4 shows HREEL spectra of Ni(100) Hex-OH(a) at several exposures of $Rh(CO)_2(\pi\text{-}C_3H_5)$, and spectra after successive heating treatment in vacuum, together with AES data. The facts seen in these spectra are:

(1) The sticking probability of $Rh(CO)_2(\pi\text{-}C_3H_5)$ was ca. 100 times smaller than on clean or OH(a)-covered Ni(110). It needs hundreds of Langmuirs to deposit a similar amount of Rh to that on Ni(110).

(2) The peak of O-H stretching at about 3700 cm^{-1} did not disappear at any stage of the experiment. The signal intensity was reduced to a half of the initial value at the furthest during adsorption or during heating below 600 K.

(3) The peak of C-O stretch is very small compared to that of C-H stretch. The intensity is negligible in contrast with the intact molecules on Ni(110).

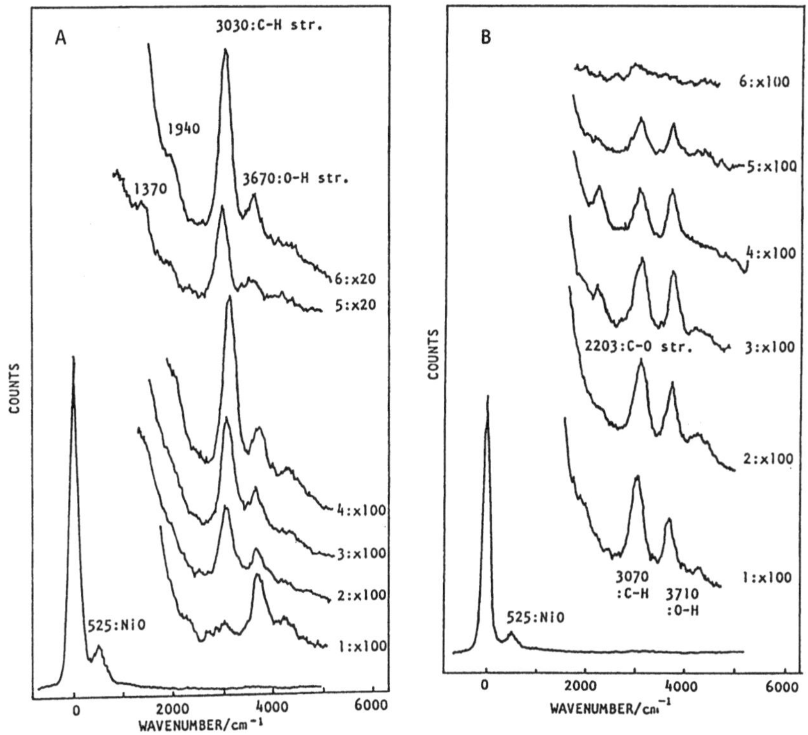

Fig. 4 HREELS of Ni(100). **A:** 1: OH(a) prepared by O_2 treatment at 1×10^{-6} Torr at 400 K for 20 min. (hydrogen dissolved in the bulk). Then exposed to $Rh(CO)_2(\pi\text{-}C_3H_5)$ at ca. 1×10^{-5} Torr for 2: 40 sec (AES Rh/Ni = 0.42), 3: 80 sec (0.47), 4: 120 sec (0.64), 5: 200 sec (0.90), 6: 520 sec (0.92). **B:** After heating at the given temperature in vacuum. HREELS was recorded at 300 K. 1: 465 K (AES Rh/Ni = 0.66), 2: 519 K (0.62), 3: 568 K (0.63), 4: 602 K (0.62), 5: 640 K (0.61), 6: 687 K(0.15).

Fact (2) indicates that Ni(100) Hex-OH(a) is much less reactive than Ni(110) p(2 × 1)-OH(a) or hardly reactive at all. This is related to the thermal stability of OH(a). Fact (3) is due to desorption of CO upon adsorption of $Rh(CO)_2(\pi\text{-}C_3H_5)$. C-O stretching at ca. 2200 cm^{-1} at heating stage comes from CO(a) formed by oxidation of -C_3H_5 ligands. TPRS of this surface from 300 K showed evolution of CO_2 during heating. The ligands were completely oxidized in this case.

Consquently, reaction {1} was not dominant on Ni(100) and the $Rh(CO)_2$ species could not remain undecomposed.

3.3. ZnO(10$\bar{1}$0)-OH(a)

Adsorption of H_2O on ZnO powders has been well studied via infrared measurement and other techniques. In brief, H_2O is dissociated into OH(a) in the first layer, which gives infrared bands at 3620 cm^{-1} and 3670 cm^{-1}. This OH(a) is stable up to 900 K [17], whereas some photoemission showed that clean single crystal surfaces of ZnO can hardly dissociate H_2O [18,19] Effects of the defects on crystal surfaces were pointed out [1]. Formation of OH(a) via adsorption of CH_3OH was reported on ZnO(10$\bar{1}$0) and ZnO(0001) [20].

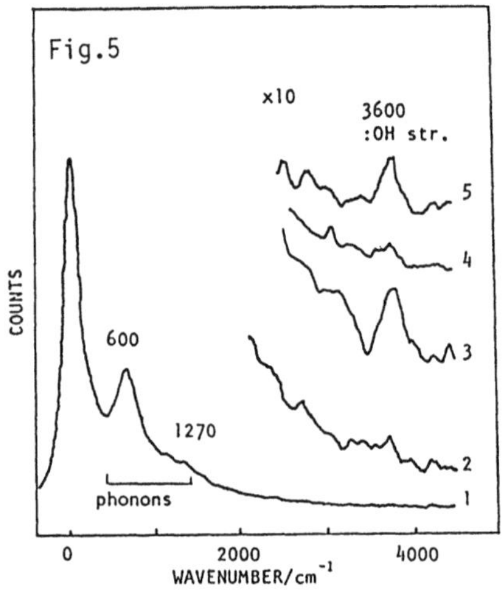

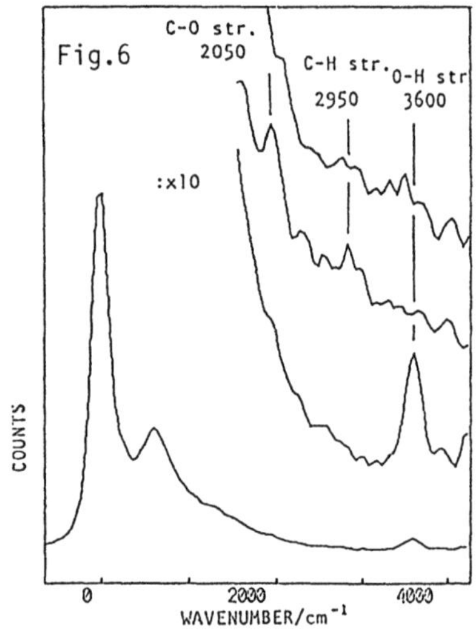

Fig. 5 HREELS of ZnO(10 $\bar{1}$0).
1: A clean surface prepared by Ar ion bombardment and heating at 750 K.
2: = 1 × 10
3: OH deposited during argon ion bombardment.
4: A clean surface prepared by the same procedure as in 1.
5: After H_2O adsorption at 2×10^{-7} Torr for 30 min. at 300 K.

Fig. 6 HREELS of ZnO(10 $\bar{1}$0).
1: After OH formation by Ar ion bombardment.
2: After exposure to $Rh(CO)_2(\pi\text{-}C_3H_5)$ at 2.5×10^{-6} Torr (by a B-A Gauge) for 150 sec.
3: Then after electron irradiation (20μA for 500 sec) for AES. C(280eV)/Zn(998eV) = 1.50, Rh(280eV)/Zn(998eV) = 0.90.

We first prepared clean ZnO(10$\bar{1}$0) by Ar ion bombardment (700 eV incident ion energy, estimated sample current -1μA, the treatment time 120 min. total) and flashing at 900 K. HREELS of the ZnO(10$\bar{1}$0) clean surface (Fig. 5) includes at least two intense phonon-loss peaks at 640 cm^{-1} and 1270 cm^{-1}.(Our incident and exit angles are both 63.5°.) The peak count of the elastic peak was ca. 4000 cps, which is 10 to 50 times smaller than in the cases of metal surfaces. Ibach [21] reported four equally spaced losses with 555 cm^{-1} interval at 45° of both angles.

The freshly prepared clean ZnO(10$\bar{1}$0) was subjected to Ar ion bombardment again. Then we found an intense O-H stretch loss right after Ar ion bombardment at room temperature (Fig. 5-3). The residual H_2O seemed to adsorb during bombardment. This peak was diminished by heating at 700 K. After the total Ar bombardment time reached ca. 500 min. and annealing below 800 K. the surface now readily adsorbed H_2O (2×10^{-7} Torr, 30 min.) to form OH(a) at room temperature (Fig. 5-5). The LEED pattern did not change all the way from diffused p(1 × 1). OH(a) seems to be distributed randomly on the surface lattice points. This result is explained by introduction of surface lattice defects by Ar ion bombardment. Annealing below 800 K did not work to remove this kind of defect.

The reaction of OH/ZnO(10$\bar{1}$0) with $Rh(CO)_2(\pi\text{-}C_3H_5)$ was observed by HREELS. ZnO(10$\bar{1}$0)-OH(a) was prepared by H_2O adsorption and exposed to $Rh(CO)_2(\pi\text{-}C_3H_5)$ vapor at 2.5×10^{-6} Torr for 150 sec. The obtained spectrum (Fig. 6-2) shows that OH was completely diminished, and that CO and C_xH_y were deposited. The intensity of C-O stretch (2000 cm^{-1}) was comparable to that of C-H (2900 cm^{-1}) in this case. It is concluded that π-C_3H_5 species from the intact molecules stay on the surface as C_3H_6(a) or something else. AES assured adsorption of C and Rh. Since ZnO cannot adsorb CO at room temperature, the CO detected by HREELS was fixed on Rh. The formation of the fragment Zn-O-Rh$(CO)_2$ is now postulated. After electron irradiation for AES, CO stretch was greatly weakened (Fig. 6-3), indicating that this fragment on the surface is not very stable at room temperature. The same thing occurred just by heating the crystal up to 350 K.

4. CONCLUSION

(1) Adsorption of Rh$(CO)_2$(π-C_3H_5) on Ni(110) p(2 × 1)-OH(a) breaks O-H bonds, and forms C_3H_6 adsorbed on the surface, which is decomposed into H_2(g) and C(a) by annealing. The adsorption site of CO (Rh or Ni) could not be determined.
(2) Adsorption of Rh$(CO)_2$(π-C_3H_5) on Ni(100) Hex-OH(a) does not significantly react with O-H bonds. Oxidation of C_3H_5 ligands proceeds via annealing up to 600 K.
(3) OH(a) on ZnO($10\bar{1}0$) is formed by adsorption of H_2O on the surface defects induced by argon ion bombardment.
(4) Adsorption of Rh$(CO)_2$(π-C_3H_5) on ZnO($10\bar{1}0$)-OH(a) breaks O-H bonds. Formation of the fragment Zn-O-Rh$(CO)_2$ was evidenced.

ACKNOWLEDGEMENT

Taro Yamada acknowledges generous financial support by a Corning Research Grant in 1986.

REFERENCES

1. P.A.Thiel and T.E.Madey; *Surf.Sci.Reports* **7** (1987) 211.
2. Yu.I.Yermakov; *Catal.Rev.-Sci.Eng.* **13** (1976) 77.
3. Y.Iwasawa; *Advances in Catalysis* **35** (1987) 187.
4. S.L.Bernasek, IUVSTA Workshop Meeting on Structure and Reactivity of Small Molecules on Surfaces; (Ofir, Portugal, 1989) .
5. T.Yamada, T.Misono, K.Tanaka and Y.Murata; *J.Vac.Sci.Technol.* **A7** (1989) 2808.
6. J.A.McBleverty and G.Wilkinson; *Inorganic Synthesis* **8** (1966) 211.
7. S.O'Brien; *Chem.Commn.* **1968** (1968) 757.
8. C.Benndorf, C.Nobl and T.E.Madey; *Surf.Sci.* **138** (1984) 292.
9. P.H.Holloway and J.B.Hudson; *Surf.Sci.* **43** (1974) 123.
10. D.F.Mitchell, P.B.Sewell and M.Cohen; *Surf.Sci.* **61** (1976) 355.
11. S.Andersson and J.W.Davenport; *Solid State Commn.* **42** (1978) 677.
12. J.Nakamura, T.Yamada and K.Tanaka; *Surf.Sci.* **185** (1987) L515.
13. T.Yamada, J.Nakamura, I.Matsuo, I.Toyoshima and K.Tanaka; *Surf.Sci.* **297** (1989) 323.
14. H.E.Farnsworth and J.Tuul; *Phys.Chem.Solids* **9** (1958) 48.
15. H.E.Farnsworth, R.E.Schlier and J.Tuul; *Phys.Chem.Solids* **9** (1958) 57.
16. T.Yamada, J.Nakamura, S.Kawamura and K.Tanaka; *J.Elec.Spec.Relat.Phenom.* **44** (1987) 79.
17. W.Hirshwald, Zinc Oxide in: Current topics in materials science, Vol.7, Ed. E.Kaladis; (North-Holland, Amsterdam, 1981) ch. 3, p.148.
18. G.Zwicker and K.Jacobi; *Surf.Sci.* **131** (1983) 179.
19. C.T.Au, M.W.Roberts and A.R.Zhu; *Surf.Sci.* **115** (1982) L117.
20. C.T.Au, W.Hirsch and W.Hirshwald; *Surf.Sci.* **221** (1989) 113.
21. H.Ibach; *Phys.Rev.Lett.* **24** (1970) 1416.

Dynamics of Adsorbed Molecules at Surfaces

A. Okiji and H. Kasai
Department of Applied Physics, Osaka University,
Suita, Osaka 565, Japan

1 Introduction

This paper briefly reviews a few selected theoretical investigations on dynamical features of molecule-surface interactions. In the first topic, the results of numerical calculations are shown for the thermal desorption spectra on the basis of the absolute reaction rate theory within the framework of the lattice gas model [1,2]. The experimental results for NO, CO and their coadsorbed system on Pt(111) are reproduced satisfactorily [3,4]. The dependence of the dynamics of thermal desorption on the coverage of adsorbed molecules is clarified. In the second topic, we propose coupled differential equations in order to describe the kinetic oscillation of CO oxidation on Pd(110) [5]. The numerical results agree with the experimental observation qualitatively [6]. It becomes clear that the quantity variation of subsurface oxygen influences strongly the dynamics of the kinetic oscillation at surfaces. In the last topic, we show the results of numerical calculations for the vibrational excitation of desorbing molecules on the basis of the concept of a reaction path with the use of the kinetic theory of gas molecule-surface scattering [7-10]. A good account of the experimental results is given for vibrational cooling of NO in trapping-desorption from Pt(111) and for vibrational heating of H_2 in associative desorption from Cu.

2 Dynamics of Thermal Desorption

2.1 Lattice gas model for adsorption system

In order to describe adsorbed states of NO and CO, we use the following lattice gas model:

$$H = E_b \sum_{i_b} n_{i_b} + E_t \sum_{i_t} n_{i_t} + V_{tb} \sum_{<i_b, j_t>} n_{i_b} n_{j_t} + V_{tt} \sum_{<i_t, j_t>} n_{i_t} n_{j_t}, \tag{2.1}$$

where E_b (E_t) is the binding energy for an adsorbate at the bridge bonded (on-top bonded) site (see refs. 1,2), V_{tb} (V_{tt}) the lateral interaction between adjacent molecules occupying on-top bonded and bridge bonded sites (on-top bonded and on-top bonded sites) and $n_i = 1(0)$ if the i-th site is

Eds.: A. Yoshimori · T. Shinjo · H. Watanabe

occupied (unoccupied) by the adsorbate. The summation $\sum_{<i_b,j_t>}(\sum_{<i_t,j_t>})$ extends over the nearest neighbor pairs of bridge bonded and on-top bonded sites (on-top bonded and on-top bonded sites). Since there are two kinds of adsorbed molecules, E_b has two values, *i.e.*, the binding energy for CO and NO molecules at a bridge bonded site, and V_{tb} has three values depending on adjacent molecules and adsorption sites, NO or CO and bridge bonded or on-top bonded sites, and so on. Firstly, through the calculation of the free energy for the adsorption layer described by the above model, the coverage dependence of the adsorbate position observed experimentally at low temperatures is reproduced for NO/Pt(111) and CO/Pt(111) and their coadsorption system. In the calculation of free energy, the Monte Carlo method is used [11]. The size of the two-dimensional system and the Monte Carlo steps are increased until the numerical results attain stability. Thus obtained typical sizes of the two-dimensional system and the number of the Monte Carlo steps are 675 - 900 sites and 1000 steps, respectively. The pattern behavior of NO is rather involved (see refs.1,2): At low initial coverages, the molecules occupy bridge bonded sites on the surface. Because of the repulsive interaction between adsorbed molecules, they stand apart from one another and form a (2×2) structure at low surface temperatures. Since this ordered structure of the adsorption layer, which produces the additional spots in the LEED pattern, is completed at the surface coverage $\theta = 1/3$, NO molecules are forced to stand close to one another for higher coverages. It is noted that the conversion of adsorption positions starts at this stage from bridge bonded to on-top (near top) bonded sites in order to avoid the strong repulsive interaction between bridge bonded NO molecules. The occurrence of such a conversion is confirmed experimentally from the coverage dependence of the NO stretching vibration [12]. The pattern behavior of adsorption for CO on Pt(111) is simpler as compared to NO. CO molecules adsorb at on-top bonded sites for low coverages. Subsequently, as the coverage increases, bridge bonded sites are occupied. The conversion of adsorption sites does not occur in this case at an intermediate coverage.

2.2 Thermal desorption spectra

On the basis of the absolute reaction rate theory, the thermal desorption spectra (TDS) are calculated numerically. For that purpose we introduce a transition-state layer, which is located above the adsorption layer and constructs a triangular lattice from where adsorbed molecules desorb. In order to describe the molecules in the transition-state layer we can use a model similar to the model (2.1). For simplicity, we assume that $E_b = E_t = \tilde{E}$ and $V_{tb} = V_{tt} = 0$ in the transition-state layer, and that the molecules in the transition-state layer and in the adsorption layer are in thermal equilibrium at the surface temperature T_S. Then one can see that the occupation probability of the triangular lattice site in the transition-state layer $\tilde{n}$ is given by

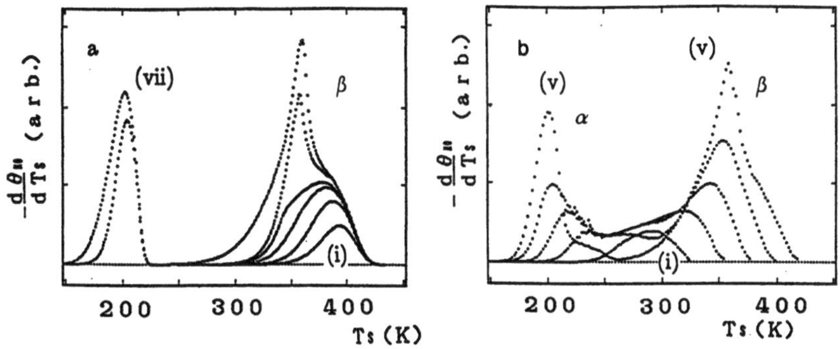

Fig.1a: Thermal desorption spectra of NO molecules from a Pt(111) surface. Initial coverages are 0.08, 0.16, 0.29, 0.32, 0.46, 0.87 and 0.99 for (i) through (vii).

Fig.1b: Thermal desorption spectra for the coadsorption system, NO+CO / Pt (111). Initial coverages of NO molecules are 0.10, 0.30, 0.50, 0.70 and 0.90 for (i) through (v).

$$\tilde{n} \simeq \exp[-(\tilde{E} - \mu)/k_B T_S]. \tag{2.2}$$

Here μ is the chemical potential for the molecules in the adsorption layer which can be determined by the Monte Carlo method as a function of T_S and θ. It is taken into account in the calculation that the density of the molecules is very small in the transition-state layer, which is the basic assumption of the absolute reaction rate theory. Then the desorption rate R can be calculated by

$$R = -\frac{d\theta}{dt} = \nu \tilde{n}, \tag{2.3}$$

where t is the time and ν is the so-called prefactor. In Fig.1a the numerical results are shown for the thermal desorption spectra for NO/Pt(111). Here we take $\tilde{E} = 20.5$kcal/mol and $\nu = 1.0 \times 10^{12}$(1/sec) . Following the experimental procedure, the surface temperature is increased linearly as a function of time, *i.e.*, $T_S = T_0 + at$, from the initial surface temperature T_0 ($a = 10$K/sec and $T_0 = 140$K). The experimental results [3,4] are reproduced well by the present calculation. Through the investigation of the pattern behavior of adsorption sites during the thermal desorption, the complicated structure of the main peak β for high coverages can be considered to originate from the occurrence of the conversion of bridge bonded to on-top bonded NO molecules during the thermal desorption. Similarly the experimental results for CO/Pt(111) [3,4] can be reproduced well by the numerical calculation.

Furthermore, the present calculation can be extended to the coadsorption system of NO and CO. As the additional parameter, the magnitude of the interaction between the adsorbed NO and CO is necessary to describe the coadsorption system. We take it tentatively as the arithmetic mean value of the pair interactions between adjacent NO molecules and between adjacent CO molecules. The thermal desorption spectra of NO molecules on a Pt(111) surface covered with preadsorbed CO molecules are shown in Fig.1.b for several CO coverages. Here, the total coverage $\theta_{NO} + \theta_{CO}$ is fixed to be 0.95. For low coverage θ_{NO} , the spectrum contains a single peak β, of which the peak position is at about 290K. Because of the lateral interaction between CO and NO, the peak position is located at temperatures lower than that in the case of Fig.1a. As the initial coverage of NO increases, the peak position shifts toward high temperatures and a second peak appears at low temperatures. The height of the two peaks α and β increases at the same time as the initial coverage of NO molecules increases. The desorption rate existing between the two peaks gets larger in the coadsorption system as compared to that in single adsorption system. In this way, the preadsorbed CO molecules strongly influence the desorption dynamics of NO molecules from a Pt(111) surface. The numerical results show a good account of the experimental results of TDS for the coadsorption system where CO molecules are preadsorbed and then NO molecules are adsorbed [4]. The shoulder of the main peak β appearing for high coverages θ_{NO} originates from the occurrence of the conversion of adsorption sites during the thermal desorption in this case as well.

3 Kinetic Oscillation of CO Oxidation on Pd(110)

Under properly chosen conditions, it is observed that the rate of the catalytic CO oxidation and the work function can exhibit temporal oscillations on well-defined Pt(100), Pt(110) and Pd(110) [6,13,14]. In the case of Pt(100) and Pt(110), these oscillations are associated with periodic changes of surface structure. The theoretical modeling of the oscillatory phenomenon on Pt(100) has been achieved by an adequate set of coupled nonlinear differential equations including detailed knowledge about the elementary steps (adsorption, desorption, surface reaction, surface phase transition) [15]. In addition, the effects of low-amplitude periodic modulation of oxygen pressure on the kinetic oscillation of the catalytic oxidation on Pt(110) are investigated by experiments [14] and theories [16]. In contrast to Pt single crystal, the clean Pd(110) surface does not need to change the surface structure drastically. In the following, on the basis of a mechanism for kinetic oscillations on Pd (110) [6], in which the formation/removal of subsurface oxygen may cause periodic changes in the catalytic properties and the reaction rate, a set of coupled phenomonological differential equations is proposed and the numerical results are obtained which reproduce the oscillatory phenomenon.

3.1 Model

It is considered that the catalytic CO oxidation on Pd(110) proceeds through the following steps [6]:

$$CO + * \rightleftharpoons CO_{ad} \tag{3.1}$$

$$O_2 + 2* \rightarrow 2O_{ad} \tag{3.2}$$

$$CO_{ad} + O_{ad} \rightarrow CO_2 + 2* \tag{3.3}$$

$$O_{ad} \rightleftharpoons O_{sb}. \tag{3.4}$$

Here * symbolizes a free adsorption site, CO_{ad} and O_{ad} denote adsorbed CO and adsorbed O, respectively and O_{sb} denotes subsurface oxygen. Step (3.1) represents kinetics of adsorption and desorption of CO. Step (3.2) represents adsorption of oxygen and neglects oxygen desorption. The surface reaction occurs via the Langmuir-Hinshelwood mechanism whereby CO_2 desorbs immediately (step (3.3)). The kinetics of migration of adsorbed oxygen to the subsurface and of subsurface oxygen to the surface are described by step (3.4). A suggested model by Ladas *et al.* [6] for kinetic oscillations on Pd(110) which includes the steps mentioned above is as follows: a. Starting with a bare Pd(110) surface under oscillatory conditions this surface will immediately be covered with oxygen due to the large excess of oxygen in the gas phase. b. Because of the large oxygen coverage, oxygen adatoms will diffuse into bulk Pd and the filling of the subsurface oxygen reservoir begins. c. The formation of subsurface oxygen has the consequence that the oxygen sticking coefficient becomes smaller and, at some point, CO adsorption will dominate over oxygen adsorption, leading to the formation of a CO adlayer on the surface. d. The oxygen adlayer has been replaced by a CO adlayer, but subsurface oxygen is now slowly diffusing to the surface and reacting off with adsorbed CO. e. The depletion of the subsurface oxygen will in turn restore the high oxygen sticking coefficient and, at some point, oxygen adsorption will again dominate over CO adsorption. The CO adlayer will be replaced by an oxygen-covered surface and the oscillatory cycle may go back to a.

Based on this model, we end up with a set of coupled differential equations for kinetic oscillations. The equations are as follows:

$$\frac{dX}{dt} = k_1 p_{co} - k_2 X - k_4 XY, \tag{3.5}$$

$$\frac{dY}{dt} = k_3 p_{o_2} - k_5 Y + k_6 Z - k_4 XY \tag{3.6}$$

and

$$\frac{dZ}{dt} = k_5 Y - k_6 Z. \tag{3.7}$$

Here, the coverages of adsorbed CO and adsorbed O are denoted by X and Y, respectively. The amount of subsurface oxygen is denoted by Z . Equation (3.5) denotes the variations in the amount of CO_{ad} . The first term describes CO adsorption, the second term CO desorption and the third term the reaction between CO_{ad} and O_{ad} . Equation (3.6) describes the variations in the amount of O_{ad}. The first term describes O adsorption, the second term the diffusion of O_{ad} to the subsurface and the third term the segregation of O_{sb} to the surface. The last term describes the surface reaction. Equation (3.7) describes the variations in the amount of subsurface oxygen and the two terms describe the same as those in eq.(3.6). This model includes six parameters, namely $k_1 - k_6$, and reliable values of $k_1 - k_4$ can be derived from individual experimental data [17-21]. Unlike $k_1 - k_4$, k_5 and k_6 cannot be obtained from the experimental data now, so we use appropriate values for k_5 and k_6 . In addition, the coverage dependence of k_1 is given experimentally. So we describe it approximately by means of Kisliuk's model [22]. Besides, it is known that the formation of subsurface oxygen and of the CO adlayer make the oxygen sticking coefficient k_3 smaller. This property can be considered to be one of the main reasons leading to the kinetic oscillation in this system and may be described by using a Kisliuk type model. That is, we express it as

$$k_3 = k_3^0 \frac{a_3 - S}{a_3 - S + K_3 S}. \tag{3.8}$$

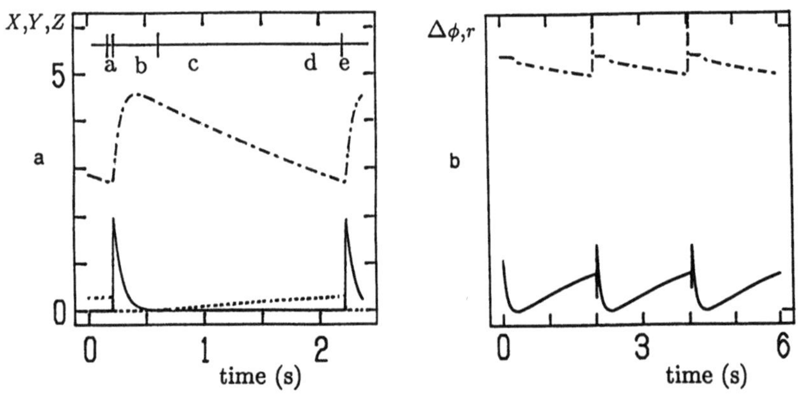

Fig.2a: Calculated variations of the adsorbate coverage X (dashed line) and Y (solid line) and the amount of the subsurface oxygen Z (dashed-dotted line) during kinetic oscillations on Pd (110).

Fig.2b: Calculated variations of the reaction rate r (dashed-dotted line) and the work function change $\Delta\phi$ (solid line) during kinetic oscillations.

Here, $S = X + Z$, a_3 and K_3 can be determined by fitting the experimental results.

Numerical results calculated with the use of the equations (3.5)-(3.8) are presented in Fig.2a. The regions a-e of the figure correspond to a-e in the model for kinetic oscillations. For comparison with the experimental results, the work function change and the reaction rate are calculated (see Fig.2b). The numerical results give a good account of the experimental results [6].

4 Vibrational Excitation of Desorbing Molecules : Heating vs Cooling

With the aid of the kinetic theory of gas molecule-surface scattering, vibrational excitation of desorbing molecules is investigated on the basis of the concept of a reaction path. As an example of a non-dissociative adsorption system, NO molecules interacting with a Pt surface are considered. In this system, vibrational cooling is observed in desorption, *i.e.*, the vibrational excitation probability of desorbing molecules becomes smaller than expected from the Boltzmann distribution at T_S. On the other hand, as an example of a dissociative adsorption system, H_2 molecules interacting with a Cu surface are considered. In this system, vibrational heating is observed in associative desorption, *i.e.*, the vibrational excitation probability of desorbing molecules becomes larger than expected from the Boltzmann distribution at T_S. As one can see in the following, it is suggested [10] that the occurrence of vibrational heating (vibrational cooling) in desorption may be coordinated generally with the occurrence of vibrationally assisted (vibrationally hindered) sticking.

4.1 Vibrational excitation of NO desorbing from Pt(111)

From the kinetic theory of gas molecule-surface scattering, one can write down the desorption probability $D(\epsilon, \nu)$ as follows [7,10]:

$$D(\epsilon, \nu) = \frac{S(\epsilon, \nu)}{< S >_S} \frac{\exp[-(\epsilon + \epsilon_\nu)/k_B T_S]}{k_B T_S Z(T_S)}. \qquad (4.1)$$

Here, ϵ and ν are the translational energy and quantum number of the vibrational state for the molecule (ϵ_ν is its vibrational energy), and $Z(T_S)$ is the partition function of vibrational states at T_S. $S(\epsilon, \nu)$ is the sticking probability of the molecule with translational energy ϵ and in the νth vibrational state. $< S >_S$ is the thermal-averaged sticking probability at T_S. From the above expression for the desorption probability, one can see that the vibrational excitation probability of desorbing molecules shows deviation from the value expected by the Boltzmann distribution at T_S if $S(\epsilon, \nu)$ depends on the

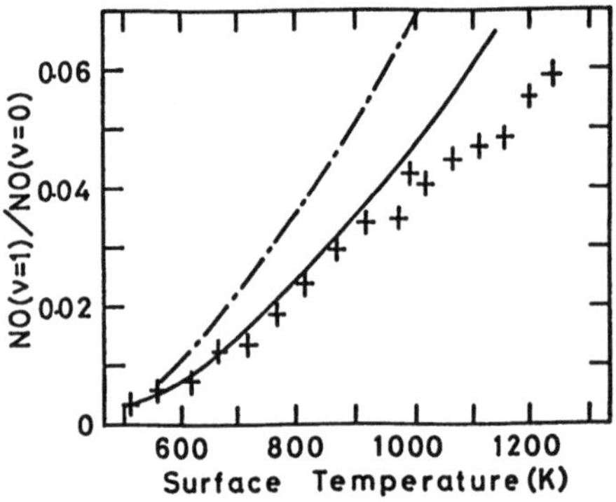

Fig.3: Vibrational excitation probability, $NO(\nu = 1)/NO(\nu = 0)$, of NO molecules as a function of T_S in the trapping-desorption process from a Pt(111) surface. The solid line is the numerical result and the cross points are the experimental results. The dashed-dotted line corresponds to the case where the sticking probability is independent of the vibrational state of the molecule and the desorbing molecules have Boltzmann distribution at T_S.

vibrational energy of the molecule. Vibrational cooling (heating) can be seen in desorption if the sticking probability is a decreasing (increasing) function of the vibrational energy. In order to obtain the vibrational excitation probability of desorbing molecules, the following integral equation for $S(\epsilon, \nu)$ is solved firstly under the boundary condition $\lim_{\epsilon \to -\infty} S(\epsilon, \nu) = 1$:

$$S(\epsilon, \nu) = \int_{-\infty}^{0} \sum_{\nu'} S(\epsilon', \nu') q(\epsilon', \nu'; \epsilon, \nu) d\epsilon', \qquad (4.2)$$

where q is the transition probability in a single scattering process of the molecule from the surface. Then, the obtained results are inserted into eq.(4.1). In Fig.3 the numerical results for vibrational excitation probability of NO in trapping-desorption from Pt(111) are shown as a function of T_S . The numerical results give a good account of vibrational cooling observed experimentally [23].

4.2 Vibrational excitation of desorbing hydrogen

In the recent experimental results for H_2/Cu [24,25], vibrationally assisted sticking seems to have been observed. The sticking probability of particles which are initially in the vibrationally excited states shows considerable enhancement at low translational energies as compared to particles in the ground state. As for the vibrational excitation of associatively desorbing molecules, the occurrence of vibrational heating in desorption [26,27] may be considered to originate from vibrationally assisted sticking if one uses the argument of the previous subsection. Since the sticking probability is an increasing function of initial vibrational energy in the present case, the desorption probability deviates from the value expected from the Boltzmann distribution at T_S and strong vibrational excitations can be seen (see eq.(4.1)). Such vibrational heating in desorption can be reproduced by coupled channel calculations based on the concept of a reaction path. A few percent of vibrational excitation probability can be obtained for desorbing H_2 from Cu at $T_S \sim 850K$ [8-10]. This value agrees with the experimental results [26] within given experimental error bars. This means that the vibrational excitation probability of associatively desorbing hydrogen is almost 2 orders of magnitude larger than expected from the Boltzmann distribution at T_S . Such vibrational heating in associative desorption can be seen in the case of D_2/Pd as well [27].

References

[1] H.Kasai, S.Enomoto and A.Okiji, J.Phys.Soc.Japan **57** (1988) 2249.

[2] M.Mizuno, H.Kasai and A.Okiji, Surf.Sci. **242** (1991) 464.

[3] J.L.Gland and B.A.Sexton, Surf.Sci. **94** (1980) 355.

[4] D.A.Mantell, R.R.Cavanagh and D.S.King, J.Chem.Phys. **84** (1986) 513.

[5] T.Yamamoto, H.Kasai and A.Okiji, J.Phys.Soc.Japan **60** (1991) 982.

[6] S.Ladas, R.Imbihl and G.Ertl, Surf.Sci. **198** (1988) 42.

[7] H.Kasai, S.Enomoto and A.Okiji, J.Phys.Soc.Japan **55** (1986) 2508.

[8] W.Brenig and H.Kasai, Surf.Sci. **225** (1989) 170.

[9] W.Brenig, S.Küchenhoff and H.Kasai, Appl.Phys. **A51** (1990) 115.

[10] H.Kasai, A.Okiji and W.Brenig, J.Electron Spectrosc. Relat.Phenom. **54 / 55** (1990) 153.

[11] K.Binder and D.Stauffer, In Topics in Current Physics **36**, 1984, Applications of the Monte Carlo Method in Statistical Physics (Springer, Berlin), ed.K.Binder.

[12] B.E.Hayden, Surf.Sci. **131** (1983) 419.

[13] R.Imbihl, M.P.Cox and G.Ertl, J.Chem.Phys. **84** (1986) 3519.

[14] M.Eiswirth and G.Ertl, Surf.Sci. **177** (1986) 90.

[15] R.Imbihl, M.P.Cox, G.Ertl, H.Müller and W.Brenig, J.Chem.Phys. **83** (1985) 1578.

[16] H.Kasai, T.Yamamoto and A.Okiji, Surf.Sci. **220** (1989) L709.

[17] M.Ehsasi, C.Seidel, H.Ruppender, W.Drachsel, J.H.Block and K.Christmann, Surf.Sci. **210** (1989) L198.

[18] M.Eiswirth, P.Möller and G.Ertl, Surf.Sci. **208** (1989) 13.

[19] J.W.He and P.R.Norton, J.Chem.Phys. **89** (1988) 1170.

[20] J.W.He, U.Memmert, K.Griffiths and P.R.Norton, J.Chem.Phys. **90** (1989) 5082; 5088.

[21] J.Goschnick, M.Grunze, J.Loboda-Cackovic and J.H.Block, Surf. Sci. **189 / 190** (1987) 137.

[22] P.Kisliuk, J.Phys.Chem.Solids **3** (1957) 95.

[23] M.Asscher, W.L.Guthrie, T.H.Lin and G.A.Somorjai, Phys.Rev.Lett. **49** (1982) 76.

[24] G.Anger, A.Winkler and K.D.Rendulic, Surf.Sci. **220** (1981) 1.

[25] B.E.Hayden and C.L.A.Lamont, Phys.Rev.Lett. **63** (1989) 1823.

[26] G.D.Kubiak, G.O.Sitz and R.N.Zare, J.Chem.Phys. **83** (1985) 2538.

[27] H.Zacharias, Appl. Phys. **A47** (1988) 37.

Hetero-electromigration on Stepped Si(111)

H. Yasunaga and Nan-Jian Wu

Department of Electronics Engineering,
The University of Electro-Communications,
Chofu-shi, Tokyo 182, Japan

Abstract. A hetero-electromigration on semiconductor surfaces, which has been investigated by the authors' group since its discovery, is described, paying particular attention to morphology and driving force in the case of Ag on Si(111). The electromigration gives a unique contribution to 'ordering at surfaces and interfaces'. The driving force is dominated by the direct force. In the presence of steps on a surface the migration is greatly enhanced for Ag but reduced for In over vicinal Si(111) inclined towards $[1\bar{1}0]$. The step structure of the vicinal surface is revealed with LEED.

1 Introduction

Surface migration of adatoms is an essential process for the formation of "ordering at surfaces and interfaces". Recently it has been recognized that we are able to apply some electric and elastic forces externally on adatoms and to drive them in an arbitrary direction over semiconductor surfaces. Now we hopefully expect that the research into the externally driven migration will provide a breakthrough in the physics and technology of surface and interface processes.

The electrical drive of In ultrathin film on Si(111)7×7 was first observed with RHEED by Zhou et al.[1] and with scanning Auger microscopy(SAM) by Yasunaga et al.[2] independently. Thenceforth, we have made a systematic investigation of the drive for a variety of elemental overlayers on several semiconductors [3-6]. As a result, the mass transport driven by some electrical force has been widely observed. Now we have good reason to take it as a universal phenomenon and it is referred to as the electromigration on semiconductor surfaces. Some other electrical effects on surface migration have also been noticed such as a predominance of either domain 2×1 or 1×2 on Si(100) [7] and rearrangement between a step bunching and a regular step array [8] depending on the direction of DC current. These effects seem to manifest an electrical drive of the homo-adatom on the semiconductor surface. Webb also observed that an externally applied strain produces a reversible asymmetry in the relative population of the 2×1 and 1×2 domains on Si(100) [9].

This paper gives a general introduction to the hetero-electromigration on semiconductor surfaces in sec. 2 with suggestions of its applicability to some fabrication techniques of semiconductor devices. The driving force which distinguishes the present electromigration from the conventional one is clarified. The electromigration is greatly enhanced or reduced in the presence of surface steps. In sec. 3 we describe SAM observation of the contrast in the influence of the step structure on the electromigration of Ag and In on a vicinal Si(111). The step structure of the vicinal surface revealed with LEED is also described briefly.

2 Electromigration on semiconductor surfaces

Electromigration is an electrically-driven mass transport. The conventional electromigration is caused by an extremely high DC current density and causes deformation in bulk conductors [10] and hazardous damage in thin films [11]. The present electromigration is fundamentally different in several respects from the conventional one as described below and should be taken as a new type of surface mass transport. It is schematically demonstrated with comparison to the surface diffusion in fig. 1. Surface diffusion causes an isotropic lateral spread of deposited material on heated Si(111)7×7 as indicated in (a). On the other hand the electromigration exhibits a preferential spread of deposited material towards either the cathode or the anode when the semiconductor substrate is heated by a DC current and field as shown in (b). The amount of the spread is proportional to the time of the DC power supply rather than the square root of the time. Another interesting feature of the electromigration is that the spreading-out adlayer exhibits a homogeneous coverage and structure which are proper to the electromigration. The characteristic coverage and structure are not sensitive to conditions of the formation so that the reproducibility of the characteristic surface phase can be ensured.

A time evolution of the Ag overlayer profile originally deposited in the area of 100x200 μm on Si(111)7×7 is indicated by an Auger line analysis along the DC current direction in fig. 2 [3]. The surface phase spreading towards the cathode proved to be exactly 1.0 ML in coverage

Fig. 1

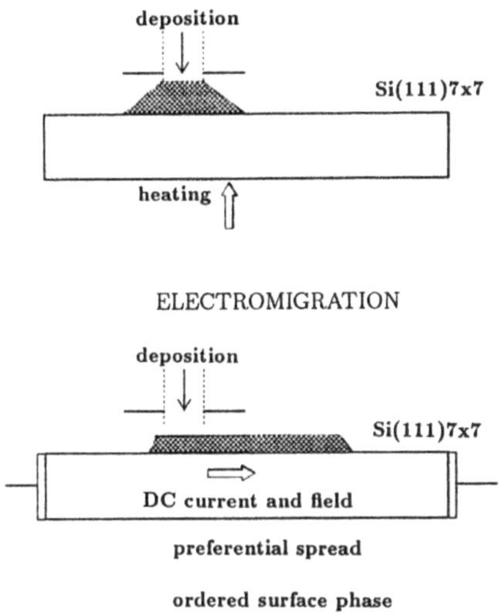

Fig. 2

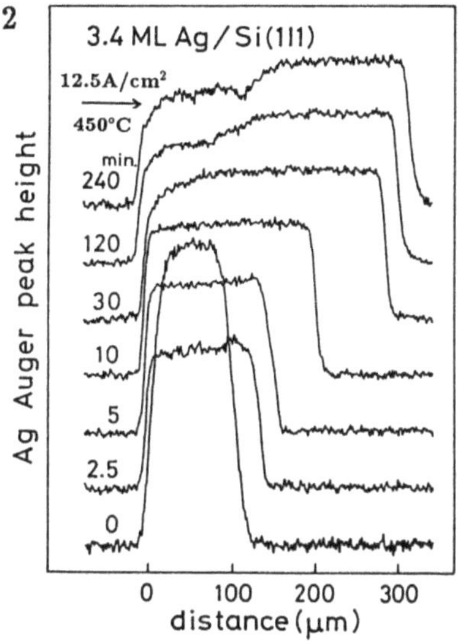

Fig. 1 Electromigration on semiconductor surfaces compared to surface diffusion.

Fig. 2 Time evolution of an Auger line profile of a Ag patch on Si(111)7×7 heated to 450 °C by a DC current density of 12.5 A/cm^2.

of Ag and $\sqrt{3} \times \sqrt{3}$ in structure. The surface structure was observed with the conventional LEED. Towards 30 min mobile Ag atoms released from the islands on the $\sqrt{3} \times \sqrt{3}$ intermediate layer were exhausted for the fast spread of the leading edge and a gradual reduction in coverage started on the anode side and extended towards the cathode. The loss of the coverage here is correlated with a slow spread of the leading edge. A mixed phase of the $\sqrt{3} \times \sqrt{3}$ and 3×1 structures was observed in the region of the reduction in coverage. It is noted that among three possible domains with the 3×1 structure the one with the shorter period of the surface unit cell parallel to the current in the direction $[1\bar{1}0]$ was preferentially observed. This seems to be just an example demonstrating that the electromigration is able to give a unique contribution to the ordering at surfaces and interfaces.

The electromigrative surface flux of a variety of elements on Si(111) 7×7 is summarized in fig. 3 as a function of the temperature of the substrate heated by DC current [6]. The flux is the product of the number density and the electromigrative velocity of the mobile atoms. The open circles denote the mass transport towards the cathode and the solid circles towards the anode. The electromigration has also been observed on

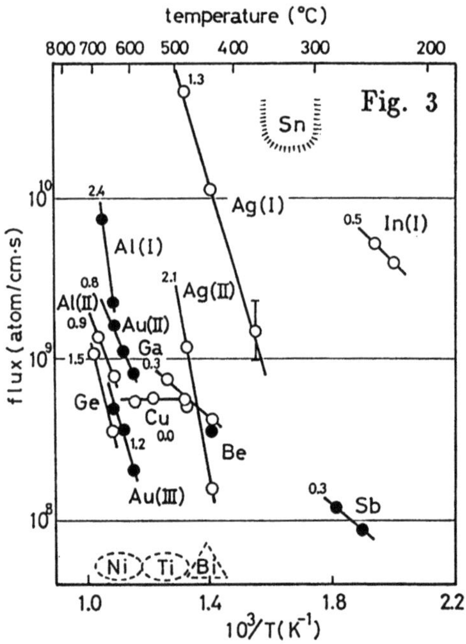

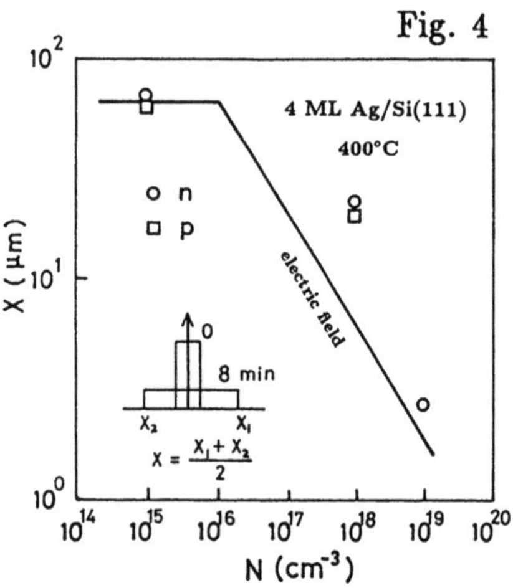

Fig. 3 Electromigrative surface flux of various adatoms on Si(111)7×7 as a function of the temperature of the substrate heated by DC current. The numbers denote the activation energy in eV.

Fig. 4 The displacement of the center of gravity of the Ag overlayer over 8 min. X is plotted as a function of the doping level N of the Si substrate. All the substrates used were heated to the same temperature 400 °C by DC power supply with the electric field indicated by the solid line.

semiconductor surfaces other than Si(111)7×7 such as Si(100)2×1 and Ge(111)8×2.

A possible driving force of the electromigration can be inferred from the relation of the atomic velocity with the DC current and field applied in the substrate. Since so far we have not been able to measure the velocity, the displacement of the center of gravity of the overlayer during a given time is employed instead of the velocity. The DC current and field can be varied with the doping level of the substrate N. The current and field required to keep the substrate at a given temperature must be proportional to $N^{1/2}$ and $N^{-1/2}$, respectively, if the substrate lies in the extrinsic range, while they are independent of N in the intrinsic range. In fig. 4, the displacement X is plotted as a function of N. The electric field is also given here in an arbitrary unit. Evidently the electromigrative velocity is proportional to the electric field rather than the current. Therefore the most plausible driving force is the direct force, namely, the electric field acting upon supposedly-charged mobile atoms. It is interesting to note here that the sign of the charge of the adatom, judged from a change in the work function of Si(111) on adsorption, is consistent with the direction of the mass transport for several elements such as Ag, In, Au and so on, provided that the direct force is acting upon a charged single adatom. The direct force should dominate the wind force on semiconductor surfaces because the electric field available here is greater than that of the conventional electromigration in metals where current density can be made overwhelmingly higher and therefore the wind force, namely, momentum transfer from current carriers to an adatom on their scattering, can dominate the direct force.

3 Electromigration on stepped Si(111)

It has been noticed that a surface step can enhance the mobility of adatoms on metal surfaces [12] and semiconductor surfaces [13]. We have investigated an effect of steps on the electromigration and recently found that the surface migration of Ag adatoms was enhanced by more than a hundred times on stepped Si(111) in the direction parallel to a step edge [14] while the presence of the step reduced the mobility of In adatoms on Si(111). The influence of steps on the electromigration is not only an interesting phenomenon in itself but also promising for the technical application of the electromigration.

Vicinal Si(111) employed as a stepped surface was misoriented towards $[1\bar{1}0]$ with a resistivity of 0.02 Ωcm (n-type). The step structure of the surface cleaned at 1200 °C proved to be sensitive to the following heat treatment. An approximately regular array of steps with a single-atom height (3.1 Å), which was consistent with the vicinal off-angle, was obtained when the specimen was quenched fast enough from 1000 to below 500 °C or so. When it was cooled slowly, a step band structure, like the electron energy band structure of solids, with an alternating array of a step-allowed band and a step-forbidden band of 7×7 terrace was obtained. The density of the step in the allowed band seems to be independent of the vicinal angle between 3 and 6°. LEED patterns enlarged around the (01) spot are indicated in fig. 5. A voltage-dependent splitting of the integral spot and the existence of the fractional 7×7 spot in (a) suggest a step band structure indicated in (b). The pattern is reproducibly observed on heat treatment of Joule heating at 1200 °C for

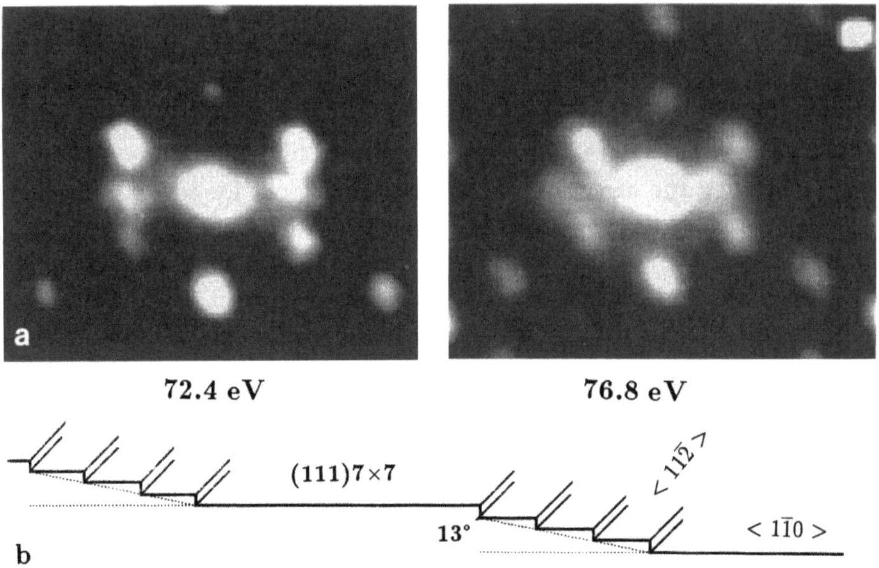

Fig. 5 LEED patterns enlarged around (01) spot (a) and the step band structure (b) of slowly cooled 6°-off Si(111).

30 s followed by a quenching to 900 °C and a subsequent slow cooling with a rate of 10 °C/s to room temperature. The step band structure is consistent with Phaneuf's result [15].

Surface migration of Ag and In adatoms over the surfaces of Si(111) with this step band structure was observed with SAM. Since the substrate was heavily doped, the electromigrative force was very weak and thus surface diffusion was predominantly observed. It is noted that the Einstein relation turned out to hold well between the surface diffusion constant and the electromigrative mobility [6]. The observed Auger maps of Ag on 3°-off Si(111) are indicated in fig. 6 with the configuration of the DC current direction and the step orientation. The mapped areas were always arranged so that the deposited patch was located approximately in the center of each area as shown in (a). It is evident that spreading is anisotropic and predominantly preferential along the edge of the step regardless of the configuration. Similar anisotropy was also observed for vicinal surfaces with different off-angles. The result is summarized in fig. 7 as a relation of a spreading rate with the inverse temperature. The spreading rate denotes the slope of the linear relation between the diffusion spread and the square root of the DC power supply time and should be proportional to the diffusion constant as well as the number density of mobile adatoms. The 3°-off surface exhibited the

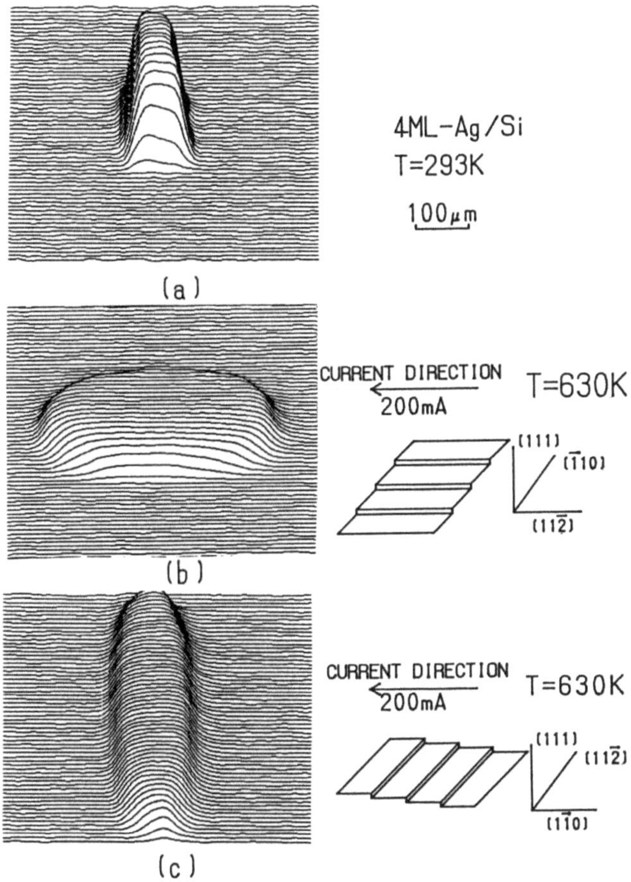

Fig. 6 Surface migration of Ag on 3°-off Si(111). Deposited 4 ML Ag patch (a) exhibits a preferential movement along the step edge irrespective of the current direction either parallel (longitudinal) (b) or perpendicular (transverse) (c) to the step edge orientation.

greatest anisotropy, reaching a hundred or so. The activation energy of the spreading rate amounts to 1.3 to 1.6 eV.

The anisotropy of the surface migration between parallel and perpendicular to the step edge was also observed in the case of In adatoms on the vicinal Si(111) as shown in fig. 8. Time evolution of the In overlayer profile is also indicated here by the Auger line analysis. As far as the anisotropy is concerned, both Ag and In are evidently quite similar. However, the spreading velocity along the step edge is greatly enhanced by the surface step for Ag while it is reduced for In as shown in fig. 9. This contrast in the influence of the step is not understood so far.

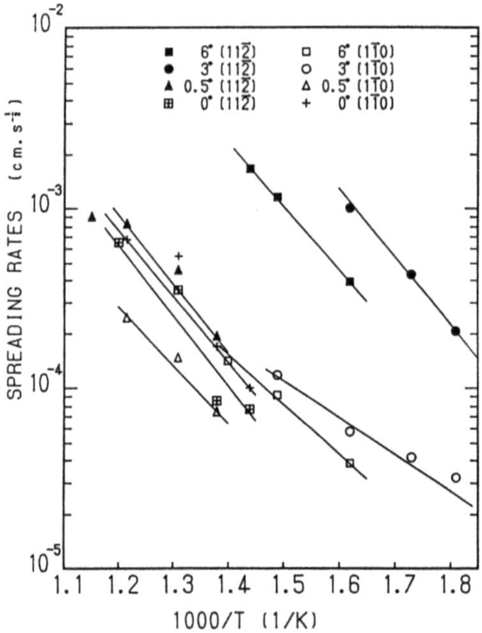

Fig. 7 The spreading rate of surface diffusion vs the inverse temperature for the longitudinal (solid marks) and the transverse (open marks) configurations.

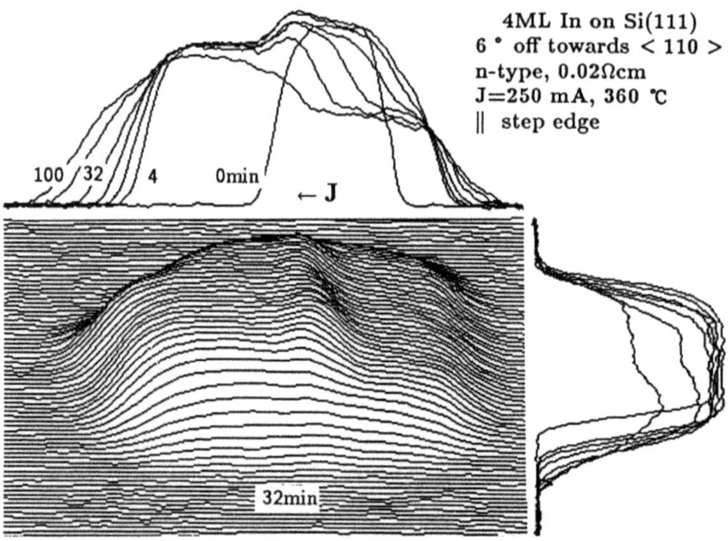

Fig. 8 (a) Auger profiles of In on 6°-off Si(111) for DC current parallel to the step edge (the longitudinal configuration).

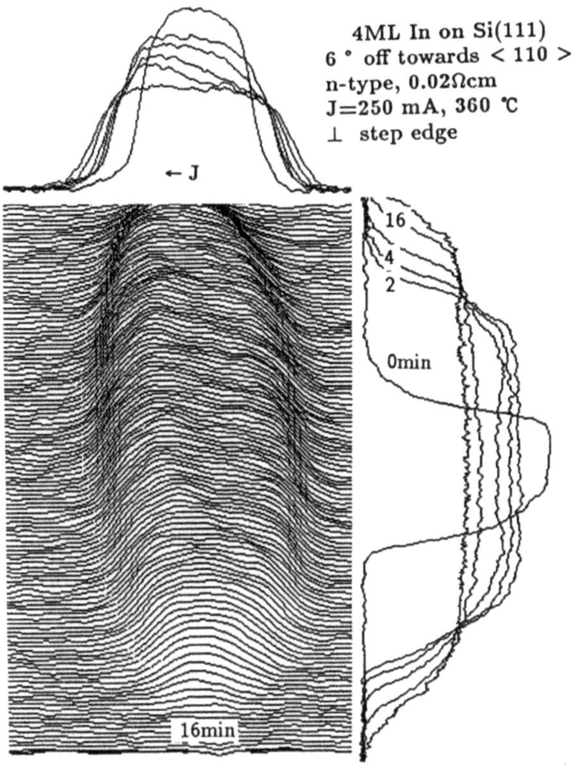

Fig. 8 (b) Auger profiles of In on 6°-off Si(111) for DC current perpendicular to the step edge (the transverse configuration).

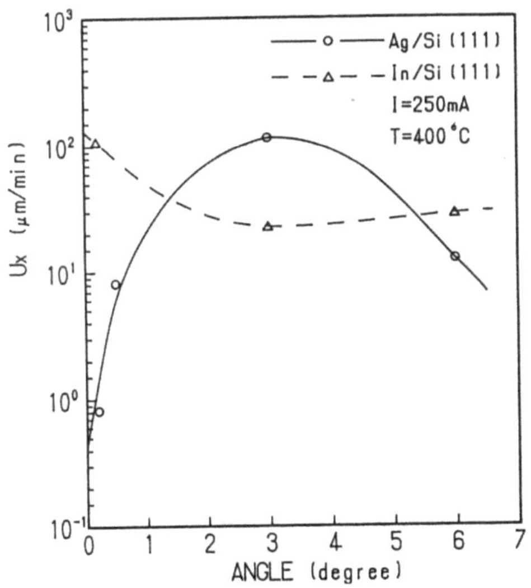

Fig. 9 The spreading velocity of the spread Ux vs the vicinal angle of Si(111).

References

[1] Z. M. Zhou, S. Baba and K. Kinbara, Thin Solid Films **98**(1982)109.

[2] H. Yasunaga, Y. Kubo and N. Okuyama, Jpn. J. Appl. Phys. **25**(1986)L400.

[3] H. Yasunaga, S. Sakomura, T. Asaoka, S. Kanayama, N. Okuyama and A. Natori, Jpn. J. Appl. Phys. **27**(1988)L1603.

[4] A. Yamanaka, K. Yagi and H. Yasunaga, Ultramicroscopy **29**(1989)161.

[5] H. Yasunaga and E. Sasuga, Surf. Sci. **231**(1990)263.

[6] H. Yasunaga, Surf. Sci. **242**(1991)171.

[7] H. Kahata and K. Yagi, Jpn. J. Appl. Phys. **28**(1989)L858.

[8] A. V. Latyshev, A. L. Aseev, A. B. Krasilnikov and S. I. Stenin, Surf. Sci. **227**(1990)24.

[9] M. B. Webb, F. K. Men, B. S. Swartzentruber, R. Kariotis and M. G. Lagally, Surf. Sci. **242**(1991)23.

[10] H. B. Huntington:*Diffusion in Solids(Recent Developments)*, ed. by A. S. Nowick and J. J. Burton (Academic Press, New York, 1975) Chap. 6.

[11] F. M. d'Heurle and P. S. Ho:*Thin Films–Interdiffusion and Reactions*, ed. by J. M. Poate, K. N. Tu and J. W. Mayer (John Wiley, New York, 1978) Chap. 8.

[12] R. Butz and H. Wagner, Surf. Sci. **87**(1979)85.

[13] F. Jentzch and M. Henzler, Appl. Phys. **A46**(1988)119.

[14] Nan-Jian Wu, A. Natori and H. Yasunaga, Surf. Sci. **242**(1991)191.

[15] R. J. Phaneuf, E. D. Williams and N. C. Bartlet, Phys. Rev. **B38**(1988)1984.

Symmetry and Ordering of Metal Deposits on the Si(111) 7×7 Surface

K. Takayanagi and Y. Tanishiro

Tokyo Institute of Technology, Department of Materials Science and Engineering, 4259 Nagatsuda, Midori-ku, Yokohama 227, Japan

Abstract. Surface structures formed by deposition of metals, Ag, Pd, Cu, Al on the Si(111)7×7 reconstructed surface analyzed by ultra-high-vacuum electron microscopy and transmission electron diffraction are reviewed. The Si(111)7×7 surface is reordered by the depositions, and the number density of Si atoms in the restructured layer changes from 2.08 monolayer (ML) of the 7×7 to the specific value for each metal deposit. In the $\sqrt{3}\times\sqrt{3}$ structures of Pd, Au and Ag, trimer configurations of metal atoms compose the frame-work. In Pd (and Cu), rotation of each trimer occurs about the normal axis of the surface, resulting in lowering of the surface symmetry.

1 Introduction

It is well known that the Si(111) surface is restructured by adsorption of the metal deposits to ordered structures. The commensurate $\sqrt{3}\times\sqrt{3}$ structure is most often formed by many kinds of metal deposits, while incommensurate structures are formed for Cu and Al. In them, metal and Si atoms form compounds which are specific for the surfaces and are not found in the bulk. Despite much detailed work, the structure of these surface phases had not been solved satisfactorily until recently, except in cases where only one metal atom exists per unit cell such as $\sqrt{3}\times\sqrt{3}$- Al [1]. Particularly, the $\sqrt{3}\times\sqrt{3}$-Ag structure has been studied intensively [2-5].

In this paper we report metal-Si surface structures analyzed by UHV electron microscopy, and discuss orderings of metal atoms and restructured Si atoms.

2 Experimental

UHV electron microscopy was used for "in-situ" observation of the condensations of metal deposits on a Si(111)7×7 surface [6]. Depositions were done from a tungsten filament at rates of about a few ML per minute. The evaporated amounts were monitored by a quartz oscillator to find the coverage of metals. The growth processes were recorded on photographic films or on video recording tapes. We observed the movements of steps during the depositions of metals, since excess or deficient Si atoms on the structural change from the 7×7 to a new metal-Si phase should migrate to step edges. The density of Si atoms in the restructured metal-Si layer, N, can be known from the areas of the new phase growing over the higher-side terrace, S, and over the lower-side terrace, s, provided that the excess Si atoms from a domain of the former area migrate to be incorporated in the domain of the latter area [7];

$$N = N_7 - 2s/(s + S),$$

where N_7 is the density of the 7×7 surface of the DAS structure [8].

Atomic positions were determined from transmission electron diffraction (TED) experiments, where the TED intensities from the restructured layer were used to find a model structure which gives the smallest reliability factor.

3 Ordering of metal and Si atoms in the restructured layer

3.1 Restructuring of Si surface layer

Fig.1 shows reflection electron microscope images (REM) of a growth of Al on the Si(111)7×7 surface at 760 °C [9]. As indicated by vertical lines, movements of the steps occur from one phase to the other. As the deposition proceeds, structures of a 7×7, $\sqrt{3}\times\sqrt{3}$, 1×1, and an incommensurate "9.6 × 9.6" structure appear in this order [9]. During the phase change from the 7×7 to the $\sqrt{3}\times\sqrt{3}$, steps move slightly towards the higher-side terrace. We consider that the Si layer of the $\sqrt{3}\times\sqrt{3}$ structure has almost the same atomic density as the 7×7 surface. Since the 7×7 surface of DAS configuration has a density of 2.08 ML, the density of the $\sqrt{3}\times\sqrt{3}$ surface is 2 ML, although the movement of the step suggests a slightly higher density than 2.08.

In a similar manner, atomic densities of Si layers were determined for the cases listed in Table 1, where 1 ML (monolayer) unit is the density of Si-1×1 surface, 7.84x10^{-14} atoms/cm^2. For the $\sqrt{3}\times\sqrt{3}$-Ag, the number density of Si atoms, which was reported to be 1ML in ref.[4], was deduced to be 1 - 1.67ML [5], while reinvestigation of REM images has suggested 0.7 - 1ML [10]. The

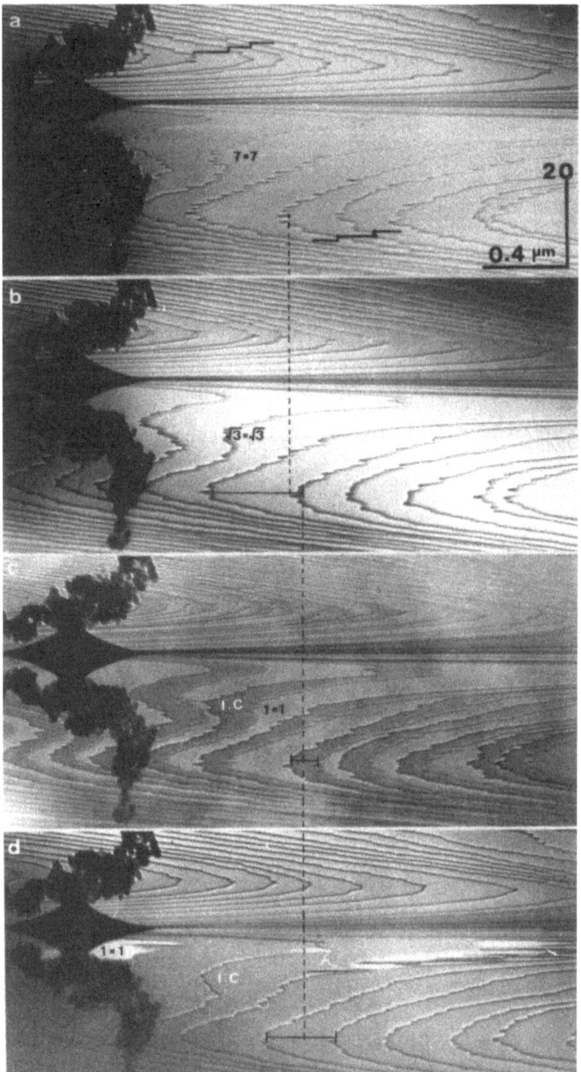

Fig.1 Growth of Al on Si(111)7×7. The surface structure changes to a $\sqrt{3}\times\sqrt{3}$, a 1×1 and to an incommensurate "9.6 × 9.6". Notice the movement of steps from (b) to (d).

number density is 2ML (six Si atoms within the $\sqrt{3}\times\sqrt{3}$ unit cell) for Pd [11] and for Al [9]. The number density reduces even to 1 ML (three Si atoms) in the case of the incommensurate "5.3 × 5.3" structure formed by Cu [9], and of "9.3x9.3" structure formed by Al [9]. Restructurings of the Si surface occur, as well as ordering of the metal atoms. It is noteworthy here that the

Table 1. Structure and atomic density of metal-Si(111)

metal	structure	s/S	N(Si)	N(metal)
Ag	$\sqrt{3}\times\sqrt{3}$	0.25	1.67 - 0.7	1
Pd	$\sqrt{3}\times\sqrt{3}$	0	2	1
Al	$\sqrt{3}\times\sqrt{3}$	0	2	-
Cu	"5.3 × 5.3"	1.2	1	1
Al	"9.6 × 9.6"	1	1	-

missing-top configuration of Si has a density of 1ML, the honeycomb on the missing-top layer, 1.67 ML, and the double layer, 2ML.

3.2 Structure models

Structures of the $\sqrt{3}\times\sqrt{3}$ phases of Ag, Pd and the incommensurate "5.3 × 5.3" phase of Cu are shown in figs. 2-4, respectively [5,11,9]. The coverage of metal atoms for each structure is also given in Table 1.

An interesting feature is that metal atoms form networks slightly distorted from the hexagonal lattice. Particularly in the case of the "5.3 × 5.3" Cu, the hexagonal lattice is rotated from the Si-1×1 lattice by 30+3.5° about the normal axis of the surface. Such rotation of the metal layer also occurs in the case of Pd, where three Pd atoms forming a trimer located at each corner of the $\sqrt{3}\times\sqrt{3}$ unit cell rotate about +6°. However, in these cases the symmetry of the surface is lowered to p3m1 from the highest p31m symmetry of the Si-1×1 surface (the symmetry is with respect to the $\sqrt{3}\times\sqrt{3}$ unit cell), by the rotation in the metal layers. In the $\sqrt{3}\times\sqrt{3}$ structure of Ag there is no rotation of the metal lattice, but a honeycomb-chained-trimer configuration proposed first by Takahashi et al. [2] occurs. In the model the three Ag atoms forming a trimer configuration rotate, as indicated in fig.2, but the symmetry of the surface remains p31m.

Thus, there are two kinds of ordering of metal atoms, with and without the rotation, or, the symmetry lowering.

Fig.2

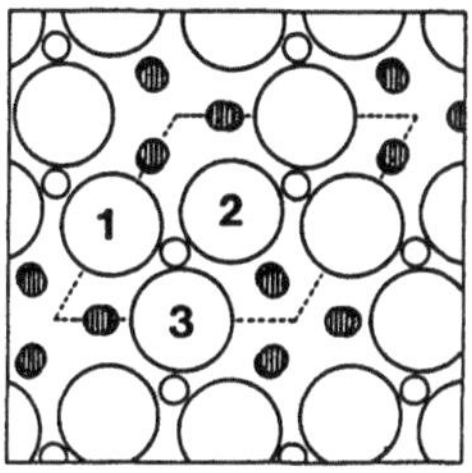

Fig.3

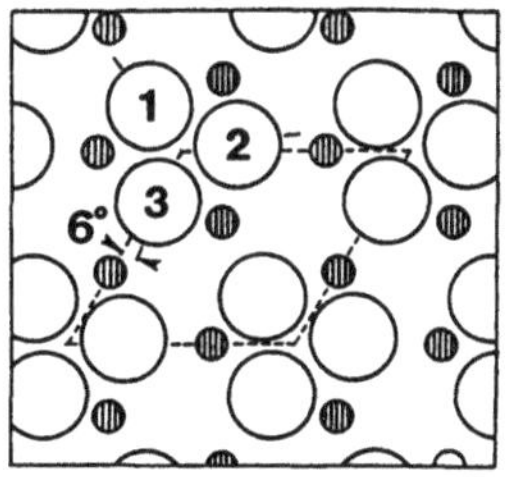

Fig.4

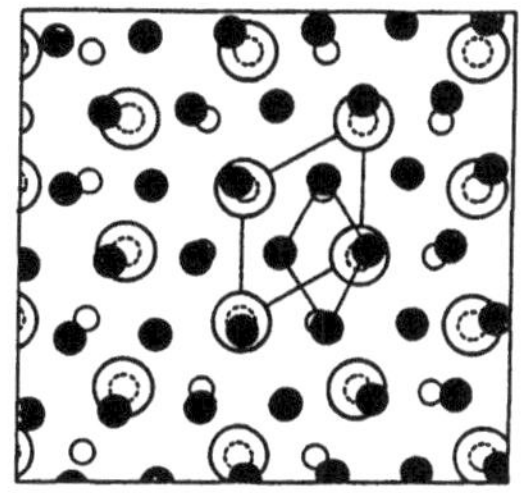

Fig.2 Structure model of $\sqrt{3}\times\sqrt{3}$-Ag. Open circles are Ag atoms forming a honeycomb-chained-trimer [2]. Small hatched and cross-hatched circles are Si atoms, respectively forming a missing-top layer. Small open circles are Si atoms forming a honeycomb. Re-analysis of REM experiments [10] suggested less Si atoms than 1 ML, that is, no honeycomb Si atoms. Note the trimer of Ag atoms marked 1-3 rotates slightly, but the symmetry of the surface is p31m.

Fig.3 Structure model of $\sqrt{3}\times\sqrt{3}$-Pd. Open circles are Pd atoms, and a trimer marked 1-3 rotates about 6° to have a surface symmetry of p3m1.

Fig.4 Structure of the incommensurate "5.3 × 5.3"-Cu. Solid circles are Cu atoms forming an almost hexagonal network of a lattice parameter, 0.245 nm. The hexagonal lattice is rotated by 30 + 3.5°. The symmetry of the surface is lowered to p3m1(with respect to the $\sqrt{3}\times\sqrt{3}$-Si unit cell) by the rotation from the Si-1×1 lattice forming the missing-top configuration.

4 Summary and Discussion

In some of the metal-Si(111) surfaces, symmetry is lowered from the Si-1×1 (Si-7×7) surface. This is caused by the rotation of the metal trimers about the surface normal. As well as the ordering of the metals, the Si layer is restructured and has a different density specific for each structure.

The atomic structures in three dimensions will be investigated further to understand the physical properties of these metal-Si phases which are only stabilized as the surface phase.

note added in proof: The number density of Si atoms in the $\sqrt{3}\times\sqrt{3}$ Ag structure was re-investigated by REM, resulting in values of 0.7 - 1.0 ML[10].

Reference

1. J.E.Northrup, Phys. Rev. Lett. **53**, 683(1984).
2. T.Takahashi, S.Nakatani, N.Okamota, T.Ishikawa and S.Kikuta, Jpn. J. Appl. Phys. **27**, L753(1988).
3. E.Vlieg, A.W.Denier van der Gon, J.F.van der Veen, J.E.Macdonald and C.Norris, Surface Sci. **209**,100(1989).
4. M.Aono, in Proceedings of the 3rd Intl. Conf. on Structure of Surfaces (ICSOS-3), Milwaukee, 1990: to be published in The Structure of Surfaces III, ed. by S.Y.Tong, M.A.Van Hove, K.Takayanagi, X.D.Xie, Springer Ser. Surf. Sci., Vol.24 (Springer, Berlin, Heidelberg 1991)
5. T.Yoshida, Master Thesis at Tokyo Institute of Technology, 1990; K.Takayanagi, in Proceedings of the ICSOS-3, Milwaukee, 1990.
6. K.Takayanagi, Y.Tanishiro, K.Kobayashi, K.Yagi and K.Akiyama, Jpn. J.Appl. Phys. **26**, L957(1987).
7. Y.Tanishiro and K.Takayanagi, Ultramicroscopy **31**, 20(1988).
8. K.Takayanagi, Y.Tanishiro, S.Takahashi and M.Takahashi, Surface Sci. **164**, 367(1985), and references there in.
9. T.Ishitsuka, Master Thesis at Tokyo Institute of Technology; T.Ishitsuka, Y.Tanishiro and K.Takayanagi, Surface Sci. submitted.
10. Y.Tanishiro, Surface Sci. submitted.
11. K.Akiyama, K.Takayanagi and Y.Taniashiro, Surface Sci. **205**, 177(1988).

$(\sqrt{3}\times\sqrt{3})$B Structure Covered with Molecular Beam Epitaxial Si

T. Tatsumi[1], *I. Hirosawa*[2], *T. Niino*[1], *H. Hirayama*[1], and *J. Mizuki*[2]

[1]Microelectronics Research Laboratories, NEC Corporation, 4-4-1 Miyazaki, Miyamae-ku, Kawasaki 213, Japan

[2]Fundamental Research Laboratories, NEC Corporation, 34 Miyukigaoka, Tsukuba, Ibaraki 305, Japan

Abstract. The electric activation efficiencies for the a-Si/B$\sqrt{3}$x$\sqrt{3}$/Si(111) system was measured by Hall measurement. The result strongly suggests that almost all boron atoms are activated at the a-Si/Si(111) interface because of the $\sqrt{3}$x$\sqrt{3}$ structure formation. A ($\sqrt{3}$x$\sqrt{3}$)B structure was also found to be formed on a (5x5) Ge_xSi_{1-x}/Si(111) surface on which Ga or Sn atoms did not form any superstructures. A Si epitaxial over-layer can be grown on the ($\sqrt{3}$x$\sqrt{3}$) B/50A $Ge_{0.4}Si_{0.6}$/Si(111) structure at a growth temperature of 300 °C while preserving a large fraction of ($\sqrt{3}$x$\sqrt{3}$)B structure due to strain compensation.

1. Introduction

When group-III metal atoms such as Al, Ga, or In are deposited onto a clean Si(111)7x7 surface, various superstructures appear.[1)-3)] It has been known that they show quite similar behavior on a Si(111) surface. However, few studies have been reported for B/Si(111) systems because the low vapor pressure of B makes it difficult to deposit B with good controllability. We have previously reported that HBO_2 has a high vapor pressure and is applicable for controlled B doping in SiMBE.[4)] Using this B source, we have found that it also forms a $\sqrt{3}$x$\sqrt{3}$ structure on a Si(111) surface.[5)] Recent studies have revealed that this B$\sqrt{3}$x$\sqrt{3}$ structure was different from the $\sqrt{3}$x$\sqrt{3}$ structures formed by other group-III elements. Akimoto et al. have reported that B$\sqrt{3}$x$\sqrt{3}$ structures can be preserved on Si(111) after subsequent Si deposition.[6)] Headrick et al. have reported that B occupies the S_5 subsurface site rather than the T_4 adatom site.[7)] The possibility of preserving a two-dimensional structure at a buried interface raises possibilities for new materials which have, for example, high mobility at very low temperatures, where ion scattering is dominant.

In this letter, we report the electric activation efficiencies for a-Si/B$\sqrt{3}$x$\sqrt{3}$/Si(111) and Si epitaxial growth on B$\sqrt{3}$x$\sqrt{3}$/Ge_xSi_{1-x} structure.

2. Experimental

Boron-containing interface structures were formed on 100mm-diameter n-type (111) and (100) Czochralski silicon wafers with 10-20 Ωcm resistivity. Wafers were precleaned using a chemical solution (NH_4OH:H_2O_2:H_2O=1:6:20) to form a thin protective oxide layer. The wafers were then loaded into MBE equipment (ANELVA MBE-430) in which the base pressure was $1.2x10^{-10}$ Torr. A thin amorphous silicon layer (10Å) was then formed on the oxide layer. The wafer was preheated at 780 °C to evaporate the oxide layer. This cleaning technique has previously been reported elsewhere.[8] Boron was deposited in the form of HBO_2, an accurately controllable doping source for Si-MBE,[4] using a Knudsen cell. The boron deposition rate was $2.9x10^{12}$ atoms$\cdot$cm^{-2}s^{-1}, which value was determined from the doping concentration of Si-MBE growth layers by Hall measurement and secondary ion mass spectroscopy(SIMS). An a-Si layer was then deposited on this boron-deposited surface at room temperature. The a-Si layer thickness was 100Å. Al electrodes were evaporated onto the a-Si layer after the sample was taken out from the MBE chamber. The sheet carrier concentration was determined by Hall measurment using the van der Pauw technique.

3. Results and Discussion

Figure 1 shows, for a-Si/B/c-Si structures, the sheet carrier concentration dependence on the substrate temperature at which HBO_2 was deposited. Also indicated here are AES peak height for oxygen and the RHEED pattern, both of which were observed just after HBO_2 deposition. Oxygen was detected when the substrate temperature was below 700 °C, as previously reported.[4] The surface structure did not change and the RHEED patterns of Si(111) and Si(100) remained 7x7 and 2x1, respectively. When HBO_2 was deposited at above 700 °C, oxygen was not detected, and the RHEED pattern of Si(111) changed from 7x7 to $\sqrt{3}x\sqrt{3}$. On the other hand, on Si(100), the RHEED pattern remained 2x1 even at temperatures above 700 °C. While the sheet carrier density for (100) stayed essentially the same when temperatures were increased to over 700 °C, that for (111) increased drastically, from the level equivalent to the density for (100) to the level approximately six times larger. Figure 2 shows the sheet carrier concentration

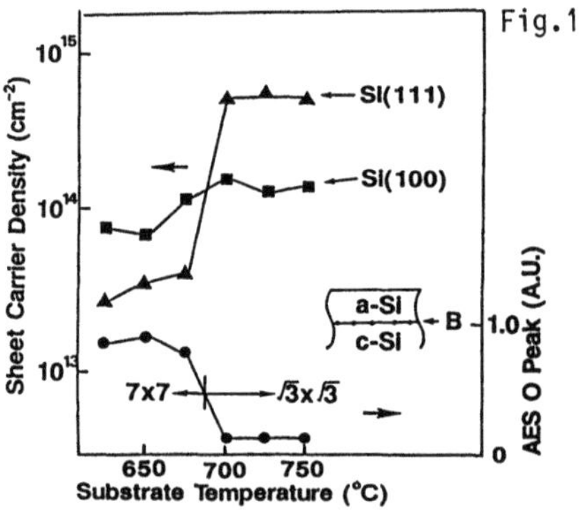

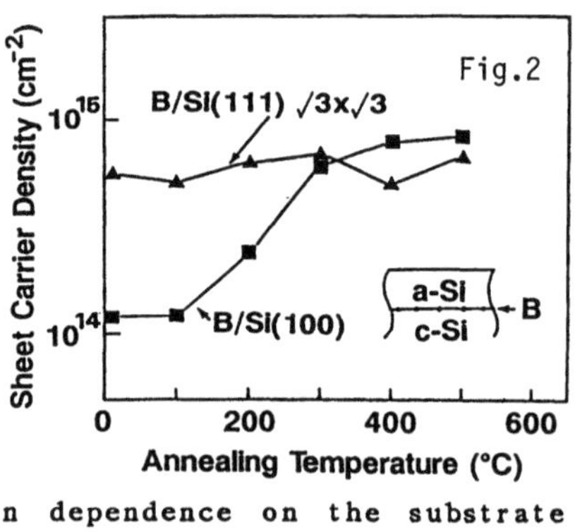

Figure 1. Sheet carrier concentration dependence on the substrate temperature at which HBO_2 was deposited for Si(111) (▲) and Si(100) (●). Also indicated here are AES peak height for oxygen and the RHEED pattern, both of which were observed just after HBO_2 deposition.

Figure 2. Sheet carrier concentration dependence of a-Si/B/c-Si on annealing temperature in N_2 atmosphere. Annealing time at each temperature was 10 min.

dependence on annealing temperature. Annealing time at each temperature was 10 min. While carrier concentration for Si(111) did not change with temperature, that for Si(100) began to increase at 200 °C and reached the level of Si(111) at approximately 400 °C. Robinson et al. have reported that an a-Si overlayer on a clean Si surface started to recrystallize at about 200°C.[7] For (100), then, B may be considered to have been activated because of such an a-Si overlayer-recrystallization occurring in the annealing process. The difference in activation efficiency between (100) and (111) strongly suggests that B is activated at the a-Si/Si(111) interface because of the $\sqrt{3}x\sqrt{3}$ structure formation.

On Si(111) at substrate temperature above 700 °C, oxygen atoms are desorbed and B forms a $\sqrt{3}x\sqrt{3}$ structure. Sheet carrier density is sufficient because B atoms replace B_5-site Si atoms and are activated. In the case of Si(100), B atoms are considered to exist at the adatom sites. B atom sites on the surface are not unique compared with B_5-site on Si(111) where B atoms occupy the exact lattice sites. Thus, B atoms are not activated on Si(100) with a-Si overlayer fabrication.

Hole mobility was 5-30 cm^2/Vs at room temperature and there was little difference between Si(111) and Si(100). At 77K, sheet carrier density decreased by about two orders. This was probably because the holes moved from the interface into the a-Si overlayer and were trapped at a deep level in the a-Si. Thus, to obtain the electrical conduction at the two-dimensional $\sqrt{3}x\sqrt{3}$ interface, it is necessary to contain holes within the $\sqrt{3}x\sqrt{3}$ plane. Then, we tried the substantial preservation of the $(\sqrt{3}x\sqrt{3})$B structure beneath an epitaxial Si cap. This was achieved by the deposition of B on a Ge_xSi_{1-x} substrate. The $(\sqrt{3}x\sqrt{3})$B structure was also formed on a $(5x5)Ge_xSi_{1-x}$ surface. In the Si/$(\sqrt{3}x\sqrt{3})$B/Ge_xSi_{1-x} structure, the hetero-interface coincides with the $(\sqrt{3}x\sqrt{3})$ two-dimensional plane. Thus, holes can be contained within the $(\sqrt{3}x\sqrt{3})$ plane by the band bending formed at the hetero-interface.

Figure 3 shows the dependence of the critical B coverage (θ_B) on the Ge concentration (x) in the substrate layer. At this coverage the (7x7) pattern disappears and only the $(\sqrt{3}x\sqrt{3})$ pattern is visible. θ_B is 0.1 monolayer (ML) ($1ML=7.8x10^{14}cm^{-3}$) when B atoms are deposited on the (7x7) Si(111) clean surface.

θ_B increases linearly with increasing Ge concentration (x) in the substrate layer until x=0.6, and saturates at 0.3ML. As Ge atoms are about 4% larger than Si atoms, compressive strain is produced in the Ge_xSi_{1-x} alloy surface. The increased θ_B on the Ge_xSi_{1-x} surface suggests that the compressive strain produced by the presence of Ge compensates for the tensile strain produced by $(\sqrt{3}x\sqrt{3})$ B formation. Stress compensation in epitaxial Si films by the simultaneous doping of a counterbalancing atom has been reported by several authors.[9)-10)] In the case of B and Ge, it has been reported that lattice defects in heavily B doped layers were reduced because the stress in the heavily doped layer was compensated by the presence of Ge .[11] The same mechanism may also be operating in the $(\sqrt{3}x\sqrt{3})$B case.

If the interface strain produced by B obstructs the burial of the $(\sqrt{3}x\sqrt{3})$ structure at the interface between epitaxial layer and substrate, the $(\sqrt{3}x\sqrt{3})$ structure on the Ge_xSi_{1-x} layer, with less strain than that on Si(111), should be better preserved at the interface. Figure 4 shows diffraction profiles from samples of 100Å epitaxial Si/$(\sqrt{3}x\sqrt{3})$B/50Å$Ge_{0.4}Si_{0.6}$/Si(111) and 100Å epitaxial Si/$(\sqrt{3}x\sqrt{3})$B/Si(111) structures. Diffraction peaks were observed at the

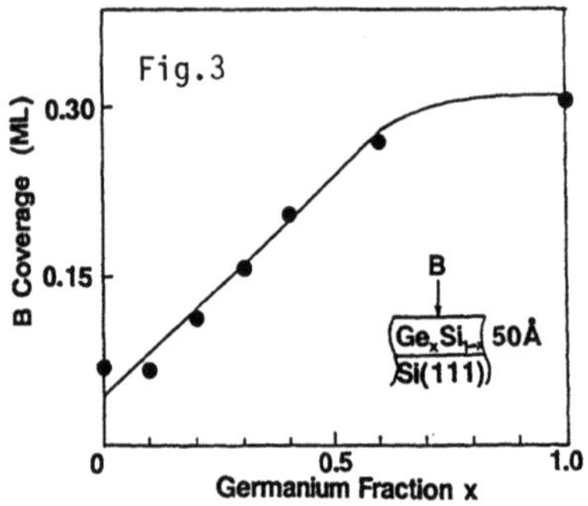

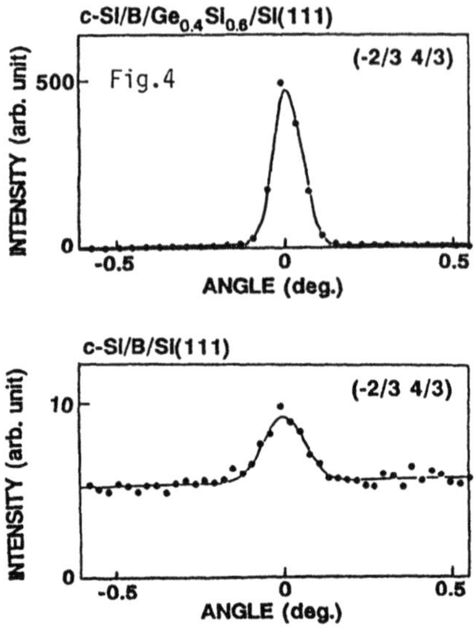

Figure 3. Dependence on the substrate Ge concentration (x) of the critical B coverage, at which the (7x7) pattern disappears and only the ($\surd 3 \times \surd 3$) pattern is visible. (1ML=$7.8\times10^{14}cm^{-2}$)

Figure 4. Diffraction profiles from samples of 100Å epitaxial Si/($\surd 3 \times \surd 3$)B/50Å$Ge_{0.4}Si_{0.6}$/Si(111) and 100Å epitaxial Si/($\surd 3 \times \surd 3$)B/Si(111) structures. Diffraction peaks were observed at the (-2/3,4/3) fractional reciprocal lattice points. The amount of B coverage in each case was 0.3ML, and an epitaxial overlayer was grown at 300°C.

(-2/3,4/3) fractional reciprocal lattice points. The amount of B coverage in each case was 0.3ML, and an epitaxial overlayer was grown at 300°C. The RHEED pattern of each sample was (7x7), somewhat streaky and diffuse. The intensity of the diffraction peak of ($\surd 3 \times \surd 3$)B on Ge_xSi_{1-x} was fifty times larger than that on Si(111). This shows that the ($\surd 3 \times \surd 3$) structure is better preserved on Ge_xSi_{1-x} than on Si(111) after epitaxial overlayer growth at 300°C. We considered that ($\surd 3 \times \surd 3$) structure between the Si epitaxial over layer and Si(111) is destroyed because of the large strain produced by B which replaces Si.

4. Conclusion

The electric activation efficiencies for a-Si/B$\surd 3 \times \surd 3$ /Si(111) and a-Si/B/Si(100) systems were measured before and after annealing by Hall measurement. The difference in the activation efficiencies between Si(100) and Si(111) strongly suggests that boron atoms are activated at the a-Si/Si(111) interface because of the $\surd 3 \times \surd 3$ structure formation. We found that the B$\surd 3 \times \surd 3$ structure was also formed on a (5x5)Ge_xSi_{1-x} surface. A Si epitaxial over-layer can be grown on the ($\surd 3 \times \surd 3$) B/50Å $Ge_{0.4}Si_{0.6}$/Si(111) structure at a growth temperature of 300°C while preserving a large fraction of ($\surd 3 \times \surd 3$)B structure due to strain compensation.

Acknowledgments

The authors wish to thank Dr. H. Ogawa and N. Aizaki for their suggestions and encouragement. The continual and helpful suggestions about the MBE system by J. Sakai of ANELVA Corp. are also gratefully acknowledged.

References

1)J.J.Lander and J.Morrison, Surface Sci.2,553(1964).
2)D.Bolmont, P.Chen,C.A.Sebenne and F.Proix, Surface Sci.137,280(1984).
3)H.Hirayama, S.Baba and A.Kinbara, J.J.Appl.Phys.25,L452(1986).
4)T.Tatsumi, H.Hirayama and N.Aizaki, Appl.Phys.Lett.50,1234(1987).
5)H.Hirayama, T.Tatsumi and N.Aizaki, Surface Sci.193,L47(1988).
6)K.Akimato, J.Mizuki, I.Hirasawa, T.Tatsumi, H.Hirayama,N.Aizaki, and J.Matsui, Extended Abstracts of the 19th Conference on Solid State Devices and Materials(Business Center for Academic Societies,Tokyo, 1987),p.463.
7)R.L.Headrick, L.C.Feldman, and I.K.Robinson, Appl.Phys.Lett.55,442(1989).
8)T.Tatsumi, I.Hirosawa, T.Niino, H.Hirayama, and J.Mizuki, unpublished.
9)J.Nisizawa, T.Terasaki, K.Yagi, and N.Miyamoto, J. Electrochem. Soc. 122,664 (1975).
10)H.J.Herzog, L.Csepregi, and H.Seidel, J. Electrochem. Soc. 131,2969 (1984).
11)H.Hirayama, T.Tatsumi, and N.Aizaki Appl.Phys.Lett.52,1335(1988).

Part IV

Interface Ordering

Structure of Silicide Surfaces and Interfaces

D.R. Hamann

AT&T Bell Laboratories, Murray Hill, NJ 07974, USA

Abstract. Atomic ordering at the interfaces of $NiSi_2$ and $CoSi_2$ with Si is discussed in terms of silicide surface structures and interface energy calculations. The structural energetics of $CaSi_2$ and its Si interface are considered. The kinematics and dynamics of ballistic electron transmission at these interfaces are discussed in relation to ballistic electron emission microscopy and potential device applications.

1. Introduction

Silicides are a technologically important class of compounds because of their ability to form strongly bonded metal contacts on silicon. Several silicides are of considerable scientific interest as well, because they grow epitaxially on silicon, forming planar, atomically perfect interfaces of well defined structure.[1] They thus provide an opportunity to develop detailed comparisons between theoretical models and experimentally realizable metal-semiconductor interfaces. The two most widely studied epitaxial silicides, $NiSi_2$ and $CoSi_2$, have the cubic fluorite structure, while the most recently discovered, $CaSi_2$,[2] has a qualitatively different structure based on hexagonal units.

2. Fluorite Silicide Interface Structures

Experimentally determining the atomic structure of an interface is difficult because the buried interface is not accessible to most structural probes. Cross-section transmission-electron-microscopy lattice imaging gave the first indications of the perfection of the fluorite silicide (111) interfaces. Combined with image simulation, it gave quantitative information on the displacement of the silicide <111> lattice planes relative to the extension of the substrate Si lattice beyond the interface.[3,4] Considering only tetrahedral coordination of all Si at the interface, Cherns et al. proposed models in which the interface metal atom was either 5-fold or 7-fold coordinated. They found that only the 7-fold model was consistent with their measured displacements for both normal (A) and reversed (B) orientations of $NiSi_2$.[3] Gibson et al. found that the $CoSi_2$ interface was only consistent with the 5-fold model.[4] The atomic geometry of these interface models is sketched in Fig. 1.

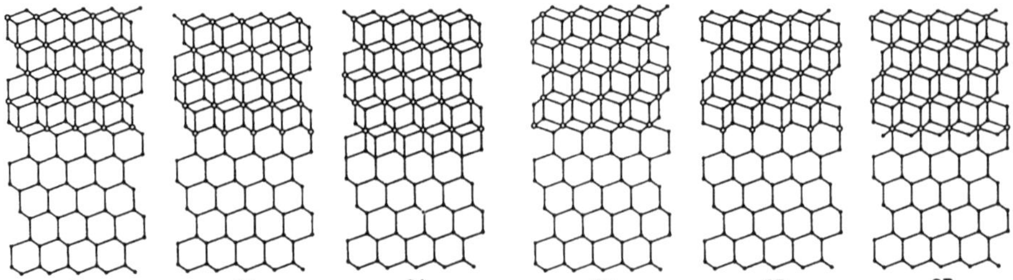

Fig. 1. Projection on the $<1\bar{1}0>$ plane of several (111) silicon-silicide interface models. The Si atoms are shown as small circles and the metal atoms as larger circles. The bonds that appear longer lie in the plane of the plot, while those that appear shorter are directed in and out of the plane.

Further clues to the interface structures can be found by considering the silicide surface structures. Low energy electron diffraction intensity analysis indicated that the structures of both the $NiSi_2$ and $CoSi_2$ (111) surfaces are ideal terminations of the bulk, with a slightly relaxed Si layer outermost.[5] The sub-surface Ni or Co is thus 7-fold coordinated. Hellman and Tung have recently shown, however, that two distinctly different structures can be formed on $CoSi_2$ (111). One corresponds to the above, but a second, Si-rich structure can be formed which is apparently energetically most stable.[6] It has been shown to correspond to the ideal surface with an added bulk-Si-like double layer.[7] This gives the Co bulk-like 8-fold coordination, and leaves all Si 4-fold coordinated, although with three rather acute bond angles of 70°. The possibility of 8-fold Co coordination at the interface was thus suggested as an issue to be examined.

The local density functional method provides a means to theoretically investigate the energies of candidate interface structures to select the most stable ones. We have used Andersen's linear augmented plane wave (LAPW) method of electronic structure calculation,[8] extended for self-consistent general potentials[9] and total valence energy,[10] for such an investigation. We applied these methods to superlattice structures, maintaining the true periodicity of the interface in the parallel direction, but introducing an artificial periodicity in the perpendicular direction, alternating slabs of Si and MSi_2. Since each unit cell of the superlattice contains two interfaces, the interface energy is half the supercell energy minus half the sum of the energies of a number of unit cells of each bulk material corresponding to the thickness of the respective slabs.[11]

For $NiSi_2$, interface structures 7A, 7B, 8B, 8A, and 5A have energies of 0.35, 0.39, 0.55, 0.78, and 1.52eV. For $CoSi_2$, the ordering is changed, and structures 8B, 8A, 7A, and 5A have energies of 0.53, 0.68, 0.86, and 2.20eV. It is clear that 5A (or, presumably B) structures aren't even remotely competitive. Only the B orientation of $CoSi_2$ is observed to grow on Si, which is consistent with the 8A-8B energy difference, 0.15 eV, which is significantly larger than the 7A-7B difference of 0.04 eV for switchable $NiSi_2$ growth.[11] The surface structure found for the Si-rich $CoSi_2$ surface corresponds to the 8A interface structure, however.[7] At the interface, 8A has an overcoordinated Si atom, while at the surface, it has a 4-coordinated Si with distorted bond angles, which appears to be energetically favorable. Recent experiments stimulated by these calculations have supported the 8-fold geometry for the $CoSi_2$ - Si interface.[12,13]

The energies alone do not give a qualitative answer to the question of why $NiSi_2$ and $CoSi_2$, which are structurally similar, have different stable interface structures. Examination of charge contour plots for the various interfaces suggests that $CoSi_2$ is more reactive at its interface than $NiSi_2$. This can be related to the existence of a "quasigap" separating bonding and antibonding states.[14] For $NiSi_2$, the Fermi level falls somewhat above the gap, into the antibonding region, while for $CoSi_2$ it falls slightly below the gap. The interface $CoSi_2$ layer can take better advantage of the extra electron made available by the 3-coordinated Si than can a $NiSi_2$ layer.

As if to prove that nature can keep ahead of our best attempts to understand it, evidence has recently been found for 7-fold Co at (111) interfaces grown by "mesotaxy,"[13] and for 8-fold Ni at a new (111) surface structure![15]

3. Structure of $CaSi_2$

The form of calcium disilicide which occurs in nature and which has been successfully grown in thin film form on the Si (111) surface,[2] is shown in Fig. 2 (a). In this structure (denoted by tr6), the stacking of trigonal Ca layers follows an AABBCC stacking sequence. Each pair of Ca layers is separated by a Si double-layer. The entire structure is hexagonal with a six-layer repeat distance. Each Ca has six Ca nearest neighbors and seven

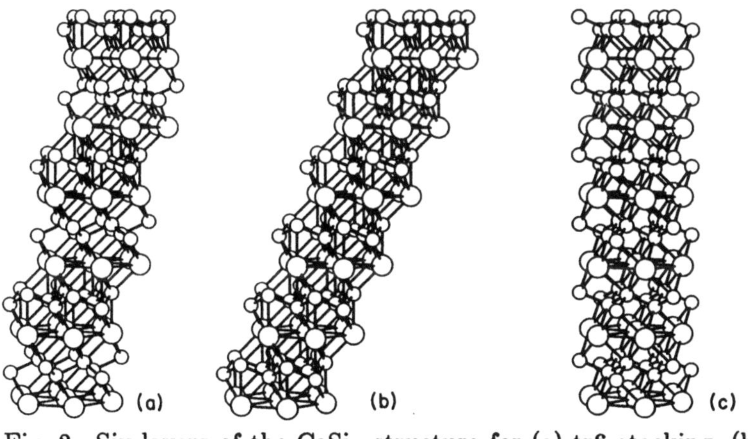

Fig. 2. Six layers of the $CaSi_2$ structure for (a) tr6 stacking, (b) tr3 stacking, and (c) h1 stacking. The larger atoms are Ca and the smaller atoms Si.

Si nearest neighbors. A second structure (denoted by tr3), which has been artificially synthesized, has a three-layer repeat distance, with Ca layers and Si double-layers both stacked in ABC sequence, as shown in Fig. 2 (b). In this structure, each Ca has six Ca nearest neighbors and eight Si nearest neighbors. It is of interest to consider a third structure (denoted by h1) similar to tr3 and tr6, which occurs naturally for $CaGe_2$ but not for $CaSi_2$. It has a simple AA stacking of alternating Ca layers and Si double-layers, as shown in Fig. 2 (c). In this last structure, each Ca atom has six Si nearest neighbors. Thus, the three structures, h1, tr6, and tr3, offer prototypes of six-, seven-, and eight-fold coordination of Ca with Si.

We have carried out LAPW calculations of the structural and electronic properties of these three polymorphs. For the tr3 structure we performed a full minimization of the total energy with respect to the two lattice constants, a and c, and the internal displacement parameter z. The values so obtained are a = 3.84, c = 16.00, and z = 0.1975. All are in excellent agreement with experiment and clearly demonstrate the quantitative reliability of the method in describing these Ca/Si based systems.[16]

The structural total energies of the relaxed tr6, tr3, and h1 structures are almost identical, the tr6 structure being approximately 10 meV per formula unit more stable than the tr3 structure, which differs in energy from the h1 structure by less than 1 meV per formula unit. We conclude that the energy differences between the three stackings are too small to be physically significant.[16] Thus, whether the coordination of Ca with Si is six, seven, or eight affects the structural energy very little.

Morar and Wittmer[2] have recently argued against eight-fold coordination of Ca with Si at the $CaSi_2$ - Si interface partly on the basis of an assumed energetic unfavorability of this coordination compared to an alternative interface structure with a stacking fault between the outermost Si double-layers. We see from the present calculation that this aspect of their argument is not valid; the energy differences between six-, seven-, and eight-fold coordination of Ca with Si are no larger than the Si stacking fault energy. Although their arguments based on electron micrographs quite convincingly demonstrate their assignment of the interface structure in the samples they have grown, it is possible that a different sample preparation could lead to six- or eight-fold coordination of Ca with Si at the interface.

4. Electron Transmission Kinematics and BEEM

The electronic structure of $NiSi_2$ and $CoSi_2$ has been calculated by LAPW and other *ab initio* methods,[17] and the bands are in good agreement with spectroscopic measurements. The utility of the silicide interfaces for device applications depends on their electron transport properties. For these highly perfect pseudomorphic interfaces, the parallel component **K** of the Bloch wave vector **k** is conserved in electron transmission. The kinematical condition for electron transmission is thus that at a given energy and **K**, energy bands exist in both materials. The perpendicular **k** components need not be the same in the two materials.

For $NiSi_2$ (111) interfaces, kinematically allowed bands exist at and near the Si conduction band minimum E_c. This minimum coincides with a gap in the $CoSi_2$ bands, however. The kinematic conditions for (111) $CoSi_2$ - Si are illustrated in Fig. 3. For energies near E_c, allowed ranges of **K** in the two materials don't overlap. The onset of overlap is at approximately E_c + 0.25eV.[18] This accounts for the near absence of transport observed in Si - $CoSi_2$ - Si metal-base transistor structures.[19] Comparable quality transistor structures have not been fabricated with $NiSi_2$.

The recently invented technique of ballistic electron emission microscopy (BEEM) allows interface transport to be probed at energies above the thermal range accessible in transistor-like structures. In BEEM, a scanning tunneling microscope is used to inject electrons into a thin overlayer, and the currents from the overlayer and the substrate are collected and measured independently.[20] The substrate current measures the interface transmission. BEEM spectra have been taken for thin layers of both $NiSi_2$ [21] and $CoSi_2$ [22] on Si. We have carried out kinematical calculations of these spectra, assuming constant transmission probability for all allowed states, and find surprisingly good agreement with the experimental spectra.[18] In particular, the calculated delayed onset of ballistic transmission for $CoSi_2$ is seen experimentally,[22] as are changes in slope associated with band thresholds for both it and $NiSi_2$.[21,22]

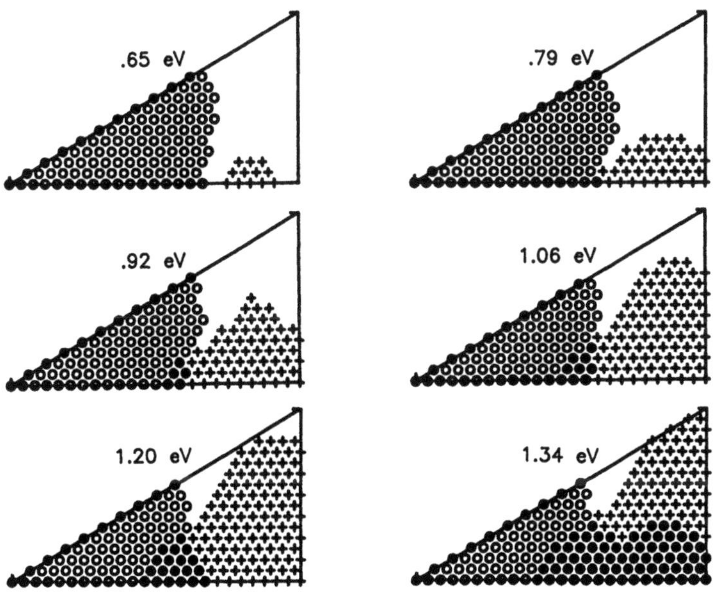

Fig. 3. Irreducible wedge of the interface Brillouin zone showing allowed **K** in Si (+), $CoSi_2$ (o), and both (•) as a function of energy above the Fermi level.

5. Electron Transmission Dynamics

Even at the most structurally perfect epitaxial interfaces, electrons can be scattered by dynamical effects arising from the change of potential encountered on passing from one material to another. A quantitative treatment of these effects is extremely important to assess the usefulness of $NiSi_2$ - Si and other interfaces in ballistic electron devices.[23]

These effects can be studied using a novel method for calculating the electronic structure of solid systems which are two-dimensionally periodic, but of arbitrary structure and infinite extent in the third dimension [24]. We divide the system by planes into a series of layers. For interface problems, we have a series of identical layers of bulk material A on the left, layers of B on the right, and an A-B interface layer in the center. We introduce a set of basis functions for each layer which have Bloch periodicity parallel to the layers with a selected parallel wave vector **K**. We then vary their coefficients to make a functional of the trial wave function stationary. The functional contains volume and surface terms. In the limit of a complete basis, it forces the wave function to satisfy Schrödinger's equation at a specified energy, with arbitrarily specified linear combinations of values and slopes on the layer boundaries.[24] We find a $2N \times 2N$ matrix operator **S** which relates N Fourier components of the constrained boundary values to those of a complementary set, and completely expresses layer's response to its neighbors.[24]

For the transmission problem, the physical boundary conditions on the wave functions are that they approach incident, reflected, and transmitted Bloch waves at $\pm\infty$. To satisfy these conditions in each bulk region, we calculate the 2N generalized Bloch states from a linear eigenvalue problem based upon its $\tilde{S}$ and the algebraic requirement that values and slopes on the two sides of the bulk layer be proportional. Appropriate propagating and decaying evanescent waves in both bulk materials are joined using $\tilde{S}$ for the interface layer to form the desired scattering states.[24] We have implemented this approach using LAPW eigenfunctions of appropriate bulk and supercell problems as our basis states, and self-consistent potentials for each type of layer from several separate bulk and supercell calculations.[10]

We carried out calculations for (111) $NiSi_2$ - Si at electron energies within a few tenths of an eV of the Si conduction band minimum (thermal carriers). Fig. 4 shows the transmission through both 7A and 7B interfaces for the **K** of the conduction band minimum (1) and several nearby **K**'s (2 and 3). All transmission probabilities are zero at the band minima for each **K**, and then increase, initially as the square root of the energy above the minimum. For an incident energy of 0.15 eV, about 50% of the incident electrons are reflected by the 7A interface and 80% by the 7B interface. Our supercell potentials give Schottky barriers which are quite close to the experimental values, and most results were calculated using the band alignments produced by these potentials. The dotted lines show the minor differences produced when we simulated the correct interface region self

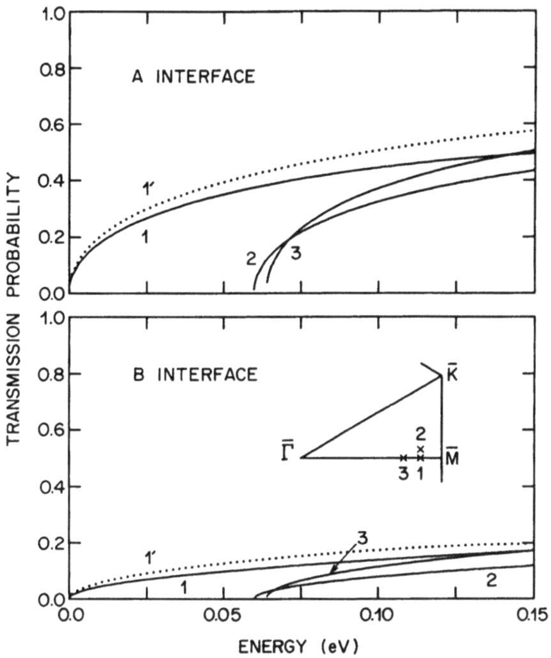

Fig. 4. Electron transmission though 7A and 7B $NiSi_2/Si(111)$ interfaces for 3 **K**'s indicated in the inset Brillouin zone segment, at energies near the conduction band minimum. The dotted curve $1'$ is explained in the text.

energy by adding a smoothly varying potential to give the experimental conduction band - Fermi level alignment.[25]

The large structural effect on the transmission was qualitatively rationalized by studying detailed differences between the electron states for 7A and 7B interfaces. For 7A, the atomic-scale details of the propagating states formed a better match, and the important evanescent states occurred with smaller amplitude than for 7B.[25] For both interface orientations, the predicted transmission is too weak to make metal-base transistors with useful gain,[23] but such device-like structures could be used to test our predictions. An A-B difference has been seen in BEEM spectra, but the detail in the data at low energies is not adequate to test our results.

One motivation for studying the electronic structure of $CaSi_2$ was the idea that this compound would have bands derived from s and p atomic orbitals. This might permit better matching of propagating wave functions, and hopefully better transmission. To our surprise, the $CaSi_2$ conduction bands turned out to be of Ca 3d character.[16] We have not yet carried out a dynamical transmission calculation for this interface, but we now expect considerable reflection.

6. Conclusions

We have shown that local density functional calculations can make useful predictions of the structure of interfaces. One of the main motivations for studying interfaces is to understand and control their electron transport properties. We see that this is now also within the reach of theoretical methods, which predict a substantial depencence on the structure and ordering at the interface.

7. References

1. R. T. Tung, J. M. Gibson, and J. M. Poate, Phys. Rev. Lett. **50**, 429 (1983).
2. J. F. Morar and M. Wittmer, Phys. Rev. B **37**, 2618 (1988).
3. D. Cherns, G. R. Anstis, J. L. Hutchinson, and J. C. H. Spence, Philos. Mag. A **46**, 849 (1982).
4. J. M. Gibson, J. C. Bean, J. M. Poate, and R. T. Tung, Appl. Phys. Lett. **41**, 818 (1982).
5. W. S. Yang, F. Jona, and P. M. Marcus, Phys. Rev. B **28**, 7377 (1983); S. C. Wu, Z. Q. Wang, Y. S. Li, F. Jona, and P. M. Marcus, Phys. Rev. B **33**,2900 (1986).
6. F. Hellman and R. T. Tung, Phys. Rev. B **37**, 10786 (1988).
7. S. A. Chambers, S. B. Anderson, H. W. Chen, and J. H. Weaver, Phys. Rev. B **34**, 913 (1986); J. Vrijmoeth, A. G. Schins, and J. F. van der Veen, Phys. Rev. B **40**, 3121 (1989).
8. O. K. Andersen, Phys. Rev. B **12**, 3060 (1975).
9. D. R. Hamann, Phys. Rev. Lett. **42**, 662 (1979).
10. L. F. Mattheiss and D. R. Hamann, Phys. Rev. B **33**, 823 (1986).
11. D. R. Hamann, Phys. Rev. Lett. **60**, 313 (1988).
12. G. Rossi, X. Jin, A. Santaniello, P. De Padova, and D. Chandesris, Phys. Rev. Let. **62**, 191 (1989).
13. C. W. T. Bulle-Lieuwma, A. F. de Jong, A. H. van Ommen, J. F. van der Veen, and J. Vrijmoeth, Appl. Phys. Lett. **55**, 648 (1989).
14. J. Tersoff and D. R. Hamann, Phys. Rev. B **28**, 1168 (1983).
15. J. E. Rowe, R. S. Becker, G. K. Wertheim, and R. T. Tung, unpublished.
16. S. Fahy and D. R. Hamann, Phys. Rev. B **41**, 7587 (1990).
17. Y. J. Chabal, D. R. Hamann, J. E. Rowe, and M. Schlüter, Phys. Rev. B **25**, 7598 (1982); D. M. Bylander, L. Kleinman, K. Mednick, and W. R. Grise, Phys. Rev. B **26**, 6379 (1982); L. F. Mattheiss and D. R. Hamann, Phys. Rev. B **37**, 623 (1988).
18. M. D. Stiles and D. R. Hamann, unpublished.
19. R. T. Tung, A. F. J. Levy, and J. M. Gibson, Appl. Phys. Lett. **48**, 635 (1986).
20. W. J. Kaiser and L. D. Bell Phys. Rev. Lett. **60**, 1406, (1988); L. D. Bell and W. J. Kaiser Phys. Rev. Lett. **61**, 2368, (1988).

21. Y. Hasegawa, Y. Kuk, R. T. Tung, P. J. Silverman, and T. Sakurai, unpublished.
22. M. Hecht, private communication.
23 A. F. J. Levi, R. T. Tung, J. L. Batstone, and M. Anzlowar, Proc. Mat. Res. Symp. Soc. Proc. **107**, 259, (1988).
24. M. D. Stiles and D. R. Hamann, Phys. Rev. B **38**, 2021, (1988).
25. M. D. Stiles and D. R. Hamann, Phys. Rev. Lett. **40**, 11349 (1989).

Structural and Magnetic Ordering at Interfaces of Metallic Superlattices

C.M. Falco and B.N. Engel

Department of Physics and Optical Sciences Center,
University of Arizona, Tucson, AZ 85721, USA

1. Abstract

We have grown a series of epitaxial Co/Pd (111) superlattices on single-crystal GaAs(110) substrates by Molecular Beam Epitaxy (MBE). Simultaneously, we have deposited polycrystalline Co/Pd multilayers on Si substrates mounted alongside the GaAs for direct comparisons of epitaxial and non-epitaxial films produced under identical conditions. The structural properties of these multilayers as determined by low and high energy electron diffraction (LEED and RHEED), low and high-angle x-ray diffraction, and Scanning Tunneling Microscopy (STM) are discussed. Vibrating Sample Magnetometry (VSM) measurements reveal significant enhancement of the perpendicular magnetic anisotropy and coercivity in the epitaxial superlattices. Implications of the structural influences on the magnetic behavior are presented.

2. Introduction

Recent interest in the magnetic and magneto-optic properties of transition metal/ transition metal multilayers has been stimulated by the discovery of perpendicular magnetism in particular systems.[1] Due to their favorable magneto-optic wavelength dependence and enhanced corrosion resistance, these materials show promise as future data storage media. However, partially due to the large variety of thin-film deposition methods and growth conditions, it has been difficult to obtain a clear understanding of the mechanisms of magnetic anisotropy in these systems. In order to create controlled and well characterized interfaces, we have grown a series of epitaxial Co/Pd (111) superlattices on single-crystal GaAs(110) substrates by Molecular Beam Epitaxy (MBE). This paper describes the growth procedure, structural characterization, and the magnetic properties of these superlattices. Comparisons are made to a series of non-epitaxial, polycrystalline multilayers that were simultaneously deposited on Si substrates.

3. Growth Procedure

To enable epitaxial growth of the Co/Pd superlattices, GaAs (110) single-crystal substrates were chosen. The GaAs wafers were In bonded to a Mo block along with Si (111) substrates. The Si substrates provided polycrystalline, UHV-evaporated multilayers for direct comparison with the epitaxial superlattices. The GaAs substrates were heated to ≈625 °C for ≈30 minutes in UHV ($< 10^{-10}$ torr) for oxide desorption and annealing. This procedure provided a clean, smooth, and highly ordered surface as determined by both reflection high and low energy electron diffraction (RHEED and LEED) and Auger electron spectrometry (AES).

To initiate epitaxial growth, a seed layer of 6 Å of Co was deposited at 0.25 Å/s from a feedback controlled electron beam evaporator onto the room temperature substrate. This Co grows epitaxially in the metastable bcc form on GaAs (110),[2] but results in a rough surface as revealed by RHEED and LEED. To heal this surface and provide a good starting point for the superlattice growth, a buffer layer of 500 Å of Pd was then deposited at 0.15 Å/s from a high temperature effusion cell. The Pd grows as fcc (111) as determined from RHEED, LEED and x-ray diffraction. Other examples of fcc/bcc growth are in the literature.[3] We found that the resulting surface could be improved by raising the substrate temperature to 200 °C for the last 200 Å of Pd growth. The resulting RHEED and LEED images show improved sharpness with a reduced diffuse background.

Following deposition of the Pd buffer layer, the substrate was cooled to room temperature and the superlattices deposited onto the rotating (10 rpm) wafers. This rotation reduces the thickness variation across the ≈5 cm wafer to < 0.5% and does not affect the quality of superlattice growth. A series with constant Pd thickness of d_{Pd} = 11 Å and varying d_{Co} was deposited. RHEED was monitored during deposition and indicated single-crystal growth throughout. The diffraction patterns displayed the familiar streaks of an ordered, atomically stepped surface.

4. Structural Characterization

In addition to using the MBE's *in situ* structural probes, the superlattice morphology and crystal structure were investigated *ex situ* with Scanning Tunneling Microscopy (STM) and x-ray diffraction. Although RHEED can provide indirect information about surface roughness, the STM gives direct real-space images of the final surface. The final surface of these superlattices was Pd, and therefore imaged well with the "in-air" STM.

A typical STM image, represented as a line scan, is shown in Figure 1. The remnants of a columnar structure are visible, with typical lateral dimensions of the individual grains ≈500 Å. These grains are very flat, with measured variations in height on the order of 2 Å (an atomic layer). These

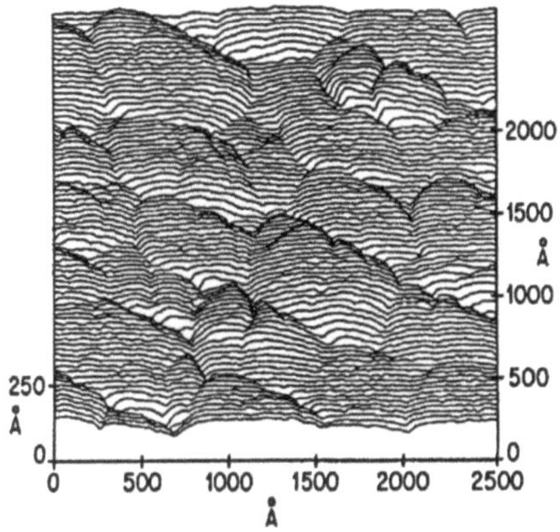

Figure 1 STM line scan of the final Pd layer (d_{Co}=4Å, d_{Pd}=11Å, N=30).

flat-topped columns are responsible for the streaked RHEED patterns seen during growth.

The crystal structure of the superlattices was investigated with x-ray diffraction in both high and low angle θ-2θ geometries. High angle rocking curves were also performed as a probe of the crystal quality and orientation. Composition modulation was confirmed in all of the structures, and a consistent determination of the bilayer period was obtained from the low-angle and high-angle superlattice peaks and Rutherford Backscattering Spectrometry (RBS). High-angle rocking curves performed on the average lattice peak of the epitaxial samples produced linewidths that vary monotonically from $0.9° < \Delta\theta < 1.8°$ as d_{Co} increases. These narrow lines confirm that these superlattices are highly oriented (111) single-crystals. In contrast, rocking curves for the polycrystalline multilayers deposited on Si reveal a (111) texture with broad linewidths on the order of 15°.

5. Magnetic Behavior

Room-temperature hysteresis curves to 14 kOe of ~ 6 x 6 mm sections of the superlattices were measured with a vibrating sample magnetometer at both 0° and 90° orientations with respect to the film plane.

5.1 Saturation Magnetization

The total measured magnetic moment divided by the Co volume in these films yields a magnetization that is significantly enhanced over that of bulk Co, especially for thin Co layers. This indicates strong Pd polarization in these films. To account for this behavior, a simple model is introduced consisting of magnetization arising from the Co layers, with an additional contribution from the polarized Pd. The total volume saturation magnetization is found

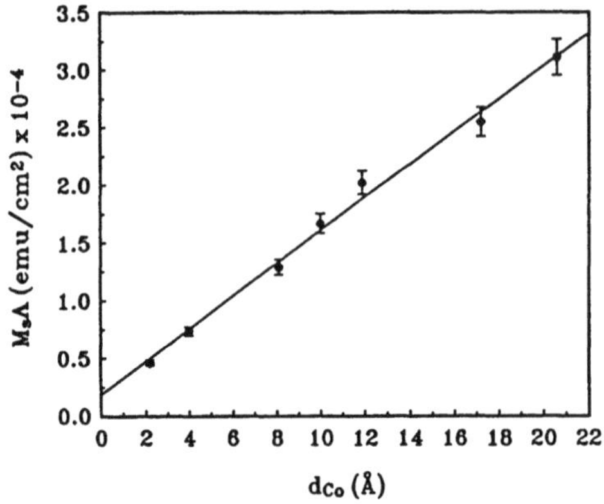

Figure 2 Total volume saturation magnetization multiplied by the bilayer period plotted versus Co layer thickness. Slope is M_{Co}, the intrinsic magnetization of the Co layers.

to follow the relation

$$M_s\Lambda = M_{Co}d_{Co} + M_{Pd}d_{Pd}^{eff}$$

where Λ is the bilayer period (d_{Co} + d_{Pd}), M_{Co} is the intrinsic Co magnetization, d_{Co} is the Co layer thickness, M_{Pd} is the effective magnetization of the polarized Pd, and d_{Pd}^{eff} is the effective polarized Pd layer thickness. Figure 2 displays M_s multiplied by Λ plotted versus the Co layer thickness for a series of superlattices with constant d_{Pd} = 11 Å. The behavior is linear and a fit of the model gives M_{Co} = 1420 ± 70 emu/cc from the slope. Within experimental error, this is identical to the quoted value for bulk, hcp Co of 1422 emu/cc. Similar behavior is seen for our polycrystalline multilayers deposited on Si. The non-zero intercept is a result of the Pd contribution. Another series of superlattices with varying d_{Pd} are in progress to uniquely determine the amount of polarized Pd (d_{Pd}^{eff}).

5.2 Coercivity

The hysteresis loop for the d_{Co} = 2.2 Å epitaxial superlattice is found to be nearly square, with a very high coercive field of H_c = 6.6 kOe. For the companion polycrystalline multilayer, the coercivity is much lower at H_c = 2.8 kOe. H_c is found to decrease monotonically with d_{Co} for both series. In Figure 3, a plot of H_c versus $1/d_{Co}$ shows a linear dependence suggestive of an anisotropy-dominated mechanism for the coercivity. However, estimates of anisotropy energy obtained assuming perfect coherent rotation yield values substantially lower than those measured. Measurements of H_c versus applied field angle are being performed in order to determine the actual spin reversal mechanism.

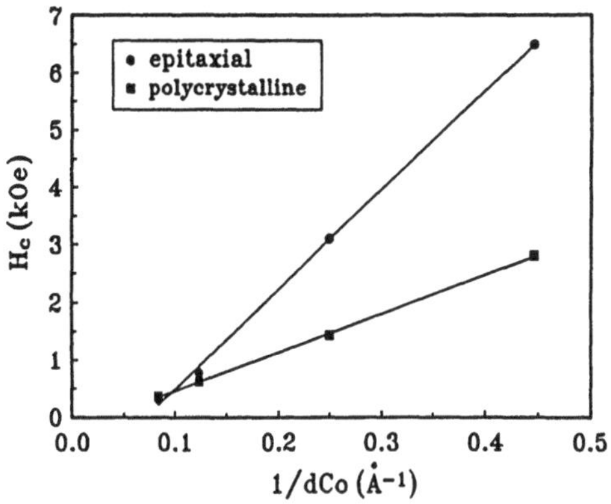

Figure 3 Coercive field versus inverse Co thickness for epitaxial and polycrystalline Co/Pd multilayers.

5.3 Uniaxial Anisotropy

The large values of coercivity for the 1 and 2 monolayer Co samples discussed above are an indication of a very strong perpendicular anisotropy in these films. Extrapolation of the 0° (in-plane) magnetization curves indicate anisotropy fields of up to 52 kOe. These films are therefore impossible to saturate in the hard direction with the field strengths available in conventional vibrating sample and torque magnetometers, making determination of anisotropy constants with conventional methods difficult.

To deduce the anisotropy constants, we have employed a technique that utilizes the magnitude and curvature of the hard-axis magnetization as a function of field and does not require saturation.[4][5] In Figure 4, the anisotropy energy per unit Co volume, K_u, multiplied by the Co layer thickness, d_{Co}, is plotted versus d_{Co} for both the epitaxial and polycrystalline multilayers. These data are seen to follow a linear relation that can be described by

$$K_u d_{Co} = (K_v^{eff} - 2\pi M_s^2) d_{Co} + 2K_s$$

where K_v^{eff} is the sum of the magneto-crystalline and magneto-elastic anisotropy energies, $2\pi M_s^2$ is the shape induced demagnetization energy, and K_s is the interface anisotropy energy. In this convention, $K_u > 0$ for a perpendicular easy-axis.

From the intercept of the linear fits to the data, the interfacial anisotropy energy, K_s, is found to be 0.6 ergs/cm^2 for both the epitaxial and polycrystalline samples. This value is comparable to the highest values reported,[6] and apparently is determined more by the deposition conditions (i.e. substrate temperature, ambient pressure, and evaporant flux rate) than by epitaxial order at the interfaces.

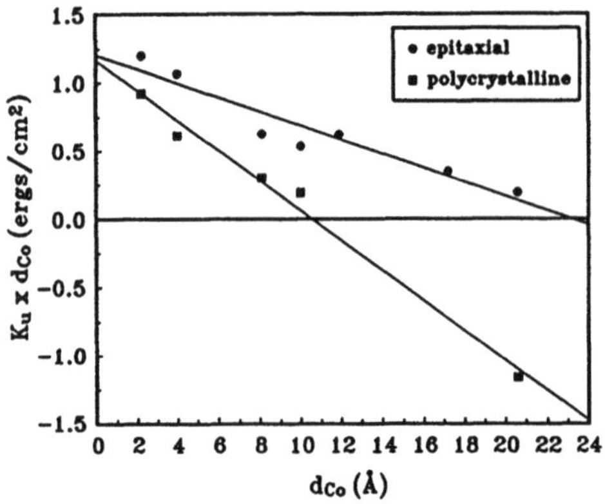

Figure 4 Uniaxial anisotropy energy multiplied by Co layer thickness versus Co layer thickness for epitaxial and polycrystalline multilayers. Crossover to in-plane easy-axis ≈ 23 Å for MBE samples.

Conversely, the effective volume anisotropy energy determined from the slopes of the linear fits is significantly higher for the epitaxial samples. The crossover point to an in-plane easy-axis for the polycrystalline samples is found to be $d_{Co} \approx 10$ Å, consistent with values reported in the literature for similarly prepared Co/Pd multilayers.[6] Extrapolation of the epitaxial curve indicates that these superlattices will retain a perpendicular ground-state for layer thicknesses as large as $d_{Co} \approx 23$ Å. However, if was found that due to demagnetizing domain effects, the perpendicular hysteresis curves became sheared for $d_{Co} > 12$ Å.

Calculation of the demagnetization energy using the bulk (and measured) Co magnetization of $M_s = 1422$ emu/cc gives an effective volume anisotropy for the epitaxial samples of $K_V^{eff} = 7.5 \pm 0.7 \times 10^6$ ergs/cc. This large value of perpendicular anisotropy can be seen to be a result of the single-crystal nature of these superlattices. As discussed earlier, x-ray diffraction rocking curves performed on the superlattice Bragg peaks display a full width at half maximum of ≈ 1°, indicating a highly oriented crystal structure with the c-axis of the Co layers perpendicular to the film plane. Thus, the magneto-crystalline anisotropy of hcp Co (6×10^6 ergs/cc for the c-axis) greatly contributes to perpendicular anisotropy in these films. The additional contribution to K_V^{eff} may be due to a magneto-elastic energy arising from the ≈9% lattice mismatch of Co and Pd. Further work is necessary to distinguish the individual contributions.

6. Summary

We have produced a series of single-crystal, Co/Pd (111) superlattices by MBE and compared their magnetic properties to a companion set of polycrystalline multilayers deposited under identical conditions onto different substrates. The magnetization behavior for both sets of samples is well described by a model of bulk-like Co layers, with an additional contribution from partially polarized Pd. The coercivity of the epitaxial samples was found to be significantly higher than their polycrystalline counterparts and varies as the inverse of the Co layer thickness. The interface contribution to the uniaxial anisotropy energy for both sets of samples was found to be 0.6 ergs/cc, and hence appears to be independent of the epitaxial order. The effective volume anisotropy contribution to K_u was found to be significantly higher for the epitaxial superlattices resulting from the highly oriented and possibly strained nature of the Co layers. In conclusion, we have shown that these epitaxial Co/Pd (111) superlattices possess an enhanced perpendicular anisotropy arising from both their high quality interfaces and single-crystal structure.

7. Acknowledgements

The authors thank Craig England and Robert Van Leeuwen for their contributions to this research and Jon Slaughter for useful discussions. This research is supported by the Optical Data Storage Center at the University of Arizona and DOE contract No. DE-FG02-87ER45297.

8. References

1. P.F. Carcia, A. Suna, D.G. Onn and R. van Antwerp, Superlattices and Microstructures **1**, 101 (1985).
2. G.A. Prinz, Phys. Rev. Lett **54**, 1051 (1985).
3. C.H. Lee, H. He, F. Lamelas, W. Vavra, C. Uher, and R. Clarke, Phys. Rev. Lett. **62**, 653 (1989)
4. W. Sucksmith and J.E. Thompson, Proc. Roy. Soc. **A255**, 362 (1954)
5. H. Zijlstra, Experimental Methods in Magnetism Vol. 2, North-Holland Publishing - Amsterdam (1967)
6. F.J.A. den Broeder, D. Kuiper, H.C. Donkersloot, and W. Hoving, Appl. Phys. A **49**, 507 (1989)

Magnetic Properties of Interfaces and Magnetoresistance in Multilayers

T. Shinjo

Institute for Chemical Research, Kyoto University,
Uji, Kyoto-fu 611, Japan

Abstract. Recent studies of the peculiar magnetoresistance (MR) properties of multilayers are surveyed. Due to the spin-dependent scattering mechanism, resistance of a multilayer increases remarkably if the magnetic components are oriented antiparallel, in comparison with the parallel (ferromagnetic) case. Such a situation is realized in Fe/Cr multilayers because of the interlayer exchange coupling, and in Cu/Co/Cu/Ni(Fe) multilayers because of the difference of anisotropy. Accordingly, large MR changes are induced in both cases. The role of the interfaces is certainly crucial to this effect, but is not yet well understood.

1.Introduction

Artificial multilayers consisting of metallic elements have been extensively studied in the last decade[1]. Since artificially structured multilayers are new materials, which do not exist in nature, it has been hoped to find multilayers exhibiting unique physical properties.

Very recently, novel MR properties have been found in magnetic multilayers consisting of magnetic and non-magnetic components. Fert's group has discovered that the MR effect of Fe/Cr multilayers is surprisingly large [2]. Their results suggest that the resistance of a multilayer is greatly enhanced if the magnetizations of adjacent layers are oriented antiparallel with respect to each other (i.e. giant antiferro- or ferrimagnetic state). Results on Fe/Cr multilayers are summarized in the next section.

The author's group demonstrated that a giant ferrimagnetic state can be realized in a multilayer including two magnetic components with different anisotropies [3]. As is described in the third section, this type of multilayer shows a large MR effect in moderate fields and therefore has a potentiality for technical applications. In both cases, the large MR effect is attributed to the spin-dependent electron scattering, and the experimental results are explained phenomenologically by this mechanism. However, the mechanism of the spin-dependent scattering from a microscopic viewpoint is not yet well understood.

2. Fe/Cr multilayers

Grünberg et al. have studied the magnetic properties of Fe/Cr/Fe(100) three-layer films and found that the magnetizations of the two Fe layers are oriented antiparallel by an exchange interaction through the intervening Cr layer [4]. Also, from the spin-polarized electron diffraction measurement by Carbone and Alvarado, the interlayer exchange interaction was confirmed to be antiferromagnetic [5]. Very recently, Krebs et al. have estimated the Cr layer thickness dependence of the interlayer coupling from ferromagnetic resonance studies on Fe/Cr/Fe(100) three-layer films [6]. The interaction becomes strongest when the Cr layer thickness is about 15Å.

MR properties of Fe/Cr multilayers were reported by Baibich et al.[2]. They prepared (100) oriented epitaxial Fe/Cr multilayers on GaAs substrates, and observed that the resistance of a [Fe(30Å)/Cr(9Å)]×60 decreased by 46% at 4.2K by applying an external field of 20 kOe. The reproducibility of this "giant" magneto-resistance effect was confirmed by the present author's group[7], where it was demonstrated that the large MR effect occurs in Fe/Cr multilayers prepared on both MgO(100) and glass substrates. In order to clarify the magnetic structure in these materials, neutron diffraction studies were conducted. For a sample with the nominal structure of [Fe(27Å)/Cr(12Å)]×30, the 1/2-order Bragg peak was successfully observed[8]. The 1st-order peak arises from the compositional modulation, and is always observable when the multilayer structure is a periodic one. The wavelength in this case is the sum of a single Fe layer thickness(27Å) and a Cr one(12Å). However, the wavelength of the 1/2-order peak is twice the compositional modulation, and this peak can be observed only if the magnetizations of adjacent Fe layers are oriented antiparallel. The observation of the 1/2th-order neutron diffraction peak therefore is direct evidence for the giant antiferromagnetic state.

Since the 1/2-order peak is entirely due to magnetic diffraction, it should disappear if the magnetizations are fully aligned parallel by applying a sufficiently strong external field. In Fig.1, the field dependence of the 1/2th-order peak intensity is shown in comparison with those of magnetization and the MR. Qualitatively, these three curves have consistent field dependences. It is apparent that the resistance is decreased with the change of the magnetic structure from antiparallel to parallel. Independently, Barthelemy et al.[9] also applied the neutron diffraction technique to an Fe/Cr multilayer and observed a similar field dependence of the 1/2th-order peak as the result in Fig.1

From the results on Fe/Cr multilayers, two questions have been raised. One is the mechanism of spin-dependent scattering, and the other is the origin of the strong antiferromagnetic coupling between Fe layers across a Cr layer. Very recently, Parkin et al. have prepared Fe/Cr multilayers over a wide thickness range and found that the interlayer coupling strength oscillates with the Cr layer thickness [10]. Oscillatory features also were

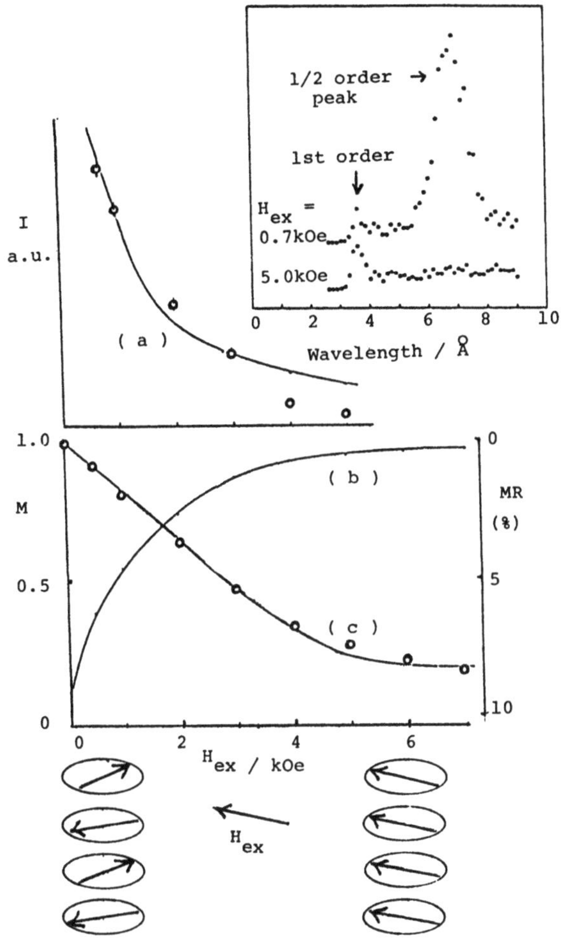

Fig.1. External field dependences of (a) 1/2th–order neutron diffraction peak intensity (the inset is the time–of–flight polarized neutron diffraction peak profile), (b) magnetization, and (c) MR effect, measured at 300K. The sample was a [Fe(27Å)/Cr(12Å)]×30 prepared on a glass substrate.

observed for Co/Ru and Co/Cr multilayers, which suggests a similarity to the Rudermann–Kittel–Kasuya–Yosida(RKKY) interaction.

Apparently, the magnetic properties of a thin Cr layer sandwiched by Fe layers cannot be understood by analogy to bulk Cr metal. Theoretically, using a tight–binding band calculation, Hasegawa has studied the electronic structures of the ferromagnetic and giant antiferromagnetic Fe/Cr multilayers[11]. However, interactions between two magnetic layers across non–magnetic layers have not yet been systematically investigated. This should be a subject to be challenged both experimentally and theoretically in the near future.

3. Cu/Co/Cu/Ni(Fe) multilayers

If a giant antiferro- or ferrimagnetic state is realized in some other type of multilayer, it is of great interest to compare the MR properties with those of Fe/Cr films. The author's group has succeeded in producing such a condition in a multilayer consisting of two magnetic components with different anisotropies [3]. The nominal structure of the multilayer is [Cu(50Å)/Co(30Å)/Cu(50Å)/$Ni_{80}Fe_{20}$(30Å)]×15. The sample was prepared by successive vacuum deposition on a glass substrate.

Figure 2(a) shows a hysteresis curve of this sample measured by a vibrating sample magnetometer (VSM), with applied magnetic field in the in-plane direction. In Fig. 2(b), the magnetization curves of the respective components are shown (the case of increasing field

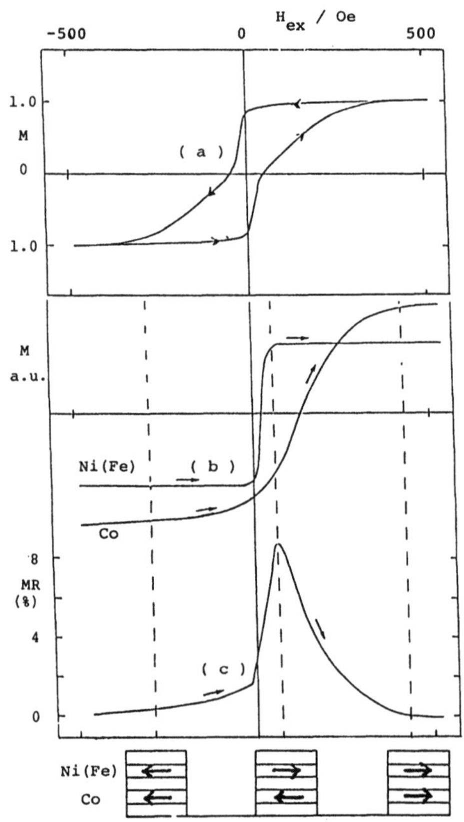

Fig.2. (a)Magnetic hysteresis curve of [Cu(50Å)/Co(30Å])/Cu(50Å)/$Ni_{80}Fe_{20}$(30Å)]×15.
(b) Individual magnetization processes of the two magnetic components inferred from the curve (a). Only the case of increasing field is shown.
(c)Observed MR effect as a function of increasing of field.
All results were measured at 300K [3].

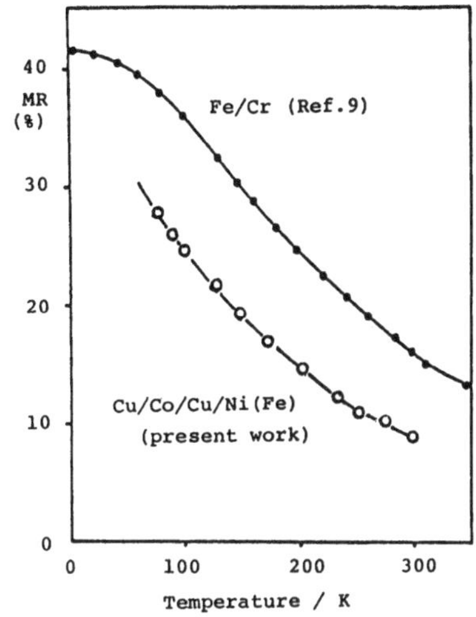

Fig.3. Temperature dependence of the MR change in the field–induced giant ferrimagnetic multilayer, which is expressd as $\Delta\rho/\rho(H=500\ Oe)$ [12]. For comparison, the result on Fe/Cr from Ref.9 is also shown. For the Fe/Cr case, the MR ratio is expressed as $[\rho(H=0)-\rho(Hsat)]/\rho(H=0)$.

only). These curves were inferred from the curve of Fig. 2(a). Actually measured is the sum of the two magnetizations, but this can be decomposed unambiguously into two smooth curves. The magnetization of the soft component (NiFe alloy layer) rotates in weak fields, below 50 Oe, while that of the hard component (Co) is not saturated up to about 400 Oe. Thus, an artificial giant ferrimagnetic state is realized in the region between these two coercive fields, and the ferrimagnetic order degree becomes the highest at about 80 Oe.

The MR curve for increasing field is shown in Fig.2(c). It can be seen that the resistance has remarkably increased in the giant ferrimagnetic region. Thus, there is no doubt that the spin–dependent scattering enhances the resistance significantly when the orientations of magnetization in the respective layers are not parallel. We can express the MR ratio as $\Delta\rho/\rho_F$, where ρ_F is the resistance in the fully ferromagnetic state. When the value of ρ at 3,000 Oe is taken as ρ_F, the observed maximum MR effect at 300K is 9.9%.

In Fig. 3, the observed MR effect of the Cu/Co/Cu/Ni(Fe) multilayer is plotted as a function of temperature, and compared with a typical result on Fe/Cr [9]. On decreasing the temperature, the effect increases remarkably. It is of interest that the profiles of the two curves are rather similar.

4. Discussion

A large enhancement of resistance occurs in the field–induced giant ferrimagnetic multilayer, Cu/Co/Cu/Ni(Fe), as well as the case of Fe/Cr multilayers. Thus, it is certain that a large enhancement of resistance is caused by a non–parallel alignment of magnetizations, but does not depend on the strength of interlayer magnetic coupling. In other words, seemingly, the nature of Cr magnetism in Fe/Cr multilayers is not essential for the MR properties.

The increase of the resistance depends on the relative angle of magnetizations of adjacent layers, but only slightly on the angle between the direction of the current and of the magnetization. The MR effect in a classic definition (the anisotropy of conductance with respect to the direction of magnetization) is negligibly small. Since multilayers are parallel circuits consisting of ferromagnetic and non–magnetic layers, the classic MR effect in each ferromagnetic layer only slightly affects the total MR value of the multilayer. The mechanism of the present MR effect is entirely different from that of the classic MR effect.

The observed MR ratio changes described above are much larger than those caused by the classic MR effect (the maximum of which so far known is only about 6%). Such a large MR effect is attractive for technical applications. However, the magnetic field necessary for the Fe/Cr system is too high for practical uses. In contrast, for the Cu/Co/Cu/Ni(Fe) multilayer, a much weaker field, below 50 Oe, is sufficient to produce a significantly large resistance change at room temperature. Therefore, it is very hopeful to use such multilayers as MR sensors. Desirable MR properties for technical uses may be obtained by tailoring the multilayered structures.

It is too early to speculate about the attainable largest MR effect at 300K, but a hint may be available from Fig.3. If the background resistance can be kept low, the magnetic contribution might reach about 20%.

The importance of spin–dependent scattering is thus recognized. Dupas et al. have prepared Au/Co/Au/Co/Au five–layer samples with different Co layer thicknesses, and observed an enhancement of resistance when the external field induced an antiparallel state[13]. Dieny et al. have prepared a Ni(Fe)/Cu/Ni(Fe)/Fe(Mn)/Ag five–layer sample, where the anisotropies of Ni(Fe) layers were differentiated by attaching an Fe(Mn) layer on one Ni(Fe) layer as an anchor.[14]. The difference of the resistance between the parallel and antiparallel magnetic states was clearly observed, and they named the phenomenon the spin valve effect.

Results on such systems having two magnetic layers are useful when trying to understand the mechanism of spin–dependent scattering. However, experimentally it is evident that the relative MR change of a multilayer is much larger than that of a few layer film. According to a very recent report by Petroff et al., to a certain degree, the roughness of the interfaces is an essential condition for the very large MR effect in Fe/Cr multilayers [15]. Further studies are required to determine whether the large MR effects in multilayers are merely due to interface roughness in the actual multilayers, or there is an intrinsic mechanism that results in the giant ferri- or antiferromagnetic multilayer structure enhancing the MR change. Using the Fuchs-Sondheimer theory, phenomenological explanations were presented [16,17]. However from a microscopic viewpoint, the mechanism of the spin-dependent scattering has not yet been clarified. Further theoretical studies will be required to answer to this question.

Acknowledgements

The author thanks S.Araki, N.Hosoito, K.Mibu, H.Yamamoto. H. Dohnomae and T.Okuyama for collaborations, and Professor C.M.Falco for reading manuscript. The majority of the author's work presented here was supported by the Grant-in-Aid for Scientific Research from Ministry of Education, Science and Culture.

References

[1] See for example, T.Shinjo and T.Takada eds., Metallic Superlattices(Elsevier, Amsterdam, 1987).

[2] M.N.Baibich, J.M.Broto, A.Fert, F.Nguyen Van Dau, F.Petroff, P.Etienne, G.Creuzet, A.Friederich and J.Chazelas, Phys.Rev.Letters 61 (1988) 2472.

[3] T.Shinjo and H.Yamamoto, J.Phys.Soc.Jpn.59 (1990) 3061.

[4] P.Grünberg, R.Schreiber, Y.Pang, M.B.Brodsky and F.Sowers, Phys.Rev.Letters 57 (1986) 2442.

[5] C.Carbone and S.F.Alvarado, Phys.Rev.B 36(1987) 2443.

[6] J.J.Krebs, P.Lubitz, A.Chaiken and G.A.Prinz, Phys.Rev.Letters 63 (1989) 1645.

[7] S.Araki and T.Shinjo, Jpn.J.Appl.Phys.29 (1990) L621.

[8] N.Hosoito, S.Araki, K.Mibu and T.Shinjo, J.Phys.Soc.Jpn. 59 (1989) 1925.

[9] A.Barthelemy, A.Fert, M.N.Baibich, S.Hadjouji, F.Petroff, P.Etienne, R.Cabanel, S.Lequien, F.Nguyen Van Dau and G.Creuzet, J.Appl.Phys. 67 (1990) 5908.

[10] S.S.P.Parkin, N.More and K.P.Roche, Phys.Rev.Letters 64 (1990) 2304.

[11] H.Hasegawa, Phys.Rev. B 42 (1990) 2368. See also his paper in this volume.

[12] H.Yamamoto, T.Okuyama, H.Dohnomae and T.Shinjo, in preparation.

[13] C.Dupas, P.Beauvillain, C.Chappert, J.P.Renard, F.Trigui, P.Veillet, E.Velu and D.Renard, J.Appl.Phys. 67 (1990) 5680.

[14] B.Dieny, V.S.Speriosu, S.S.P.Parkin, B.A.Gurney, D.R.Wilhoit and D.Mauri, preprint.

[15] F.Petroff, A.Barthelemy, A.Hamzic, A.Fert, E.Etienne, S.Lequien and G.Creuzet, to be published in J.Magn. & Magn.Mater.

[16] R.E.Camley and J.Barnas, Phys.Rev.Letters 63 (1989) 664.

[17] P.M.Levy, K.Ounadjela, S.Zhang, Y.Wang, C.B.Sommers and A.Fert, J.Appl.Phys.67 (1990) 5914.

Electronic Structures and Magnetism of Ferromagnetic/Antiferromagnetic Multilayers

H. Hasegawa

Department of Liberal Arts, Tohoku Gakuin University,
Ichinazaka, Izumi-ku, Sendai 981-31, Japan

Abstract. Realistic tight-binding band calculations have been performed for $(X)_3/(Cr)_m/(X)_3$ sandwiches by changing X = Fe, Co and Ni and $m = 4$ and 5. The moment distributions on the X and Cr layers have been determined self-consistently. In all the cases investigated, we have obtained locally stable, multiple solutions, in which magnetic moments on the successive X layers couple ferromagnetically and antiferromagnetically across intervening Cr layers. Implications of our calculation for the magnetoresistance recently observed in Fe/Cr and Co/Cr multilayers are discussed.

1. Introduction

Multilayers consisting of ferromagnetic and antiferromagnetic layers have attracted both scientific and technological interest. One current topic in this field is the giant magnetoresistance. Baibich et al. [1] found that the resistivity of Fe/Cr multilayers is reduced by 46 % when an external field of 20 kOe is applied. Various experiments such as light scattering [2,3], spin-polarized low-energy diffraction (LEED) [4], neutron diffraction [5] and magnetization measurements [3,6,7] show that total magnetic moments on successive Fe layers couple antiferromagnetically across intervening Cr layers. This antiferromagnetic coupling between Fe layers is expected to play an essential role in the observed large magnetoresistance [8,9]. When an external field is absent or small, the coupling between successive Fe layers is antiferromagnetic and the spin-dependent scattering at interfaces leads to the large resistivity. When a strong field is applied, on the other hand, Fe moments order ferromagnetically and the resistivity is reduced because of the vanishing of the spin-dependent scattering. This antiferromagnetic Fe-Fe coupling cannot be, however, understood within the localized (Heisenberg or Ising) model. When the localized model is applied to the Fe/Cr multilayer, the coupling between the Fe layers is ferromagnetic (FM) or antiferromagnetic (AFM) depending on whether the number of Cr layers is odd or even [10]. This is inconsistent with the experimental results showing the AFM coupling when the Cr-layer thickness is less than 20 – 30 Å [6,7], though in Fe/Cr superlattices with thicker Cr layers an oscillation in the Fe-Fe couplings with a period of 18 – 20 Å was reported [11].

Quite recently, Levy et al. [9] and Hasegawa [12] made band calculations for Fe/Cr multilayers. It was pointed out that the antiferromagnetic coupling is conceivable even when the number of Cr layers, m, is *odd*. In this case the energy cost due to the unfavorable moment distributions in Cr layers is compensated by *changing* the magnitude of Cr moments [9,12]; the itinerant electron system has an extra degree of freedom varying the magnitudes of magnetic moments while they are fixed in the localized model. It was also shown [12] that the calculated interlayer exchange interaction, J, between Fe layers is positive for the nearest-neighbor layers as expected, but negative at $1 \leq m \leq 6$. This is consistent with the observed Cr-thickness dependence of J [6,7].

Since the discovery of the magnetoresistance in Fe/Cr multilayers, a similar phenomenon has been found in many transition-metal multilayers such as Co/Cr [11], Co/Ru [11], Co/Au [13] and Co/Cu/NiFe/Cu [14]. Among them, we pay attention to a Co/Cr multilayer, which is isomorphic with a Fe/Cr multilayer having been discussed so far. It was reported that an applied field reduces the resistivity of the Co/Cr multilayer by about 2.5 % [11], which is much smaller than the 46 % reported for a Fe/Cr multilayer [1]. Now questions are raised. Why is the magnetoresistance in Fe/Cr multilayers much significant than that in Co/Cr multilayers? Does the magnetoresistance in Co/Cr multilayer arise from the AFM coupling between Co moments, just as in the Fe/Cr multilayers, although no experimental evidence has been reported yet? It is interesting to make a theoretical study on the similarities and differences between Fe/Cr and Co/Cr multilayers. This is the purpose of the present paper. We have calculated the electronic and magnetic structures of $(X)_3/(Cr)_m/(X)_3$ sandwiches by changing X = Fe, Co and Ni and $m = 4$ and 5.

The outline of the present paper is as follows: In the next section, we briefly explain the model and computation method adopted in our calculation. The calculated moment distributions are presented in section 3. The final section is devoted to supplementary discussions.

2. Model and Computational Method

We employed the model and computational method which were adopted in our previous calculations (for details see Refs. 12, 15, 16). We assumed that a multilayer whose configuration is given by $(Cu)_2/(Cr)_2/(X)_3/(Cr)_m/(X)_3/(Cr)_2/(Cu)_2$, is put on a semi-infinite Cu substrate with bcc [001] surface. The layer parallel to the interfaces is assigned by the index, n. This sandwich is assumed to be described by the realistic tight-binding d-band Hamiltonian, which includes the band parameters such as core potentials, two-center transfer integrals and electron-electron interactions. We adopted the transfer integrals after the canonical band theory [17]. Core potentials were chosen such as to preserve the local charge neutrality. We employed the electron-electron interactions such that the calculated moments for pure elements are consistent with the ground-state magnetic moments which are observed experimentally or calculated theoretically. The band parameters used in our calculations have been reported in Refs. 12, 15, 16.

The averages of local magnetic moments and number of electrons on the n-th layer, M_n and N_n, were calculated self-consistently by an iterative method. By adopting a variety of initial, trial solutions, we repeated our calculations looking for solutions which are locally stable in the configuration space.

3. Calculated Results

3.1 Fe/Cr/Fe Sandwiches

First we discuss the results of Fe/Cr/Fe sandwiches. We obtained the FM and AFM solutions, both of which are numerically stable. Figure 1(a) and (b) show the calculated moment distributions on the inner Fe/Cr/Fe layers with AFM and FM configurations, respectively; moment distributions on the outer Cr and Cu layers are not shown [12]. The substrate Cu layers have no local moments while interfacial Cu layers are slightly spin polarized because of the molecular field arising from adjacent Fe layers. The calculated moment distributions on Fe and Cr layers have the following features:

(i) In both the AFM and FM solutions, the Fe regions are ferromagnetic while the Cr regions are antiferromagnetic.

(ii) Magnetic moments on the central and interfacial Fe layers are 2.5 and 1.5 μ_B, respectively, both in the FM and AFM solutions.

(iii) Magnitudes of Cr moments in the AFM solutions for $m = 4$ and in the FM solutions for $m = 5$ are about 1.0 μ_B, which is enhanced compared with sublattice moment of 0.6 μ_B of bulk Cr. In these cases, the spin density wave (SDW) on Cr layers is compatible with the moment distribution on Fe layers. On the other hand, in the other cases where the Cr SDW is incompatible with Fe moments, Cr moments become very small.

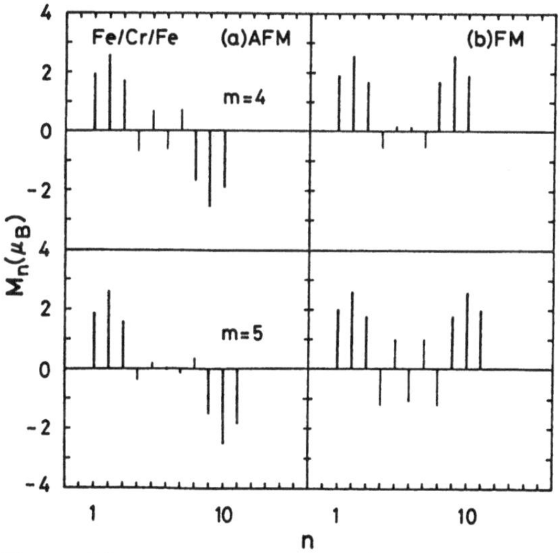

Fig. 1 Magnetic-moment distributions against the layer index, n, of (a) AFM and (b) FM solutions in $(Fe)_3/(Cr)_m/(Fe)_3$ sandwiches. Local moments on Fe and Cr layers are expressed by bold and thin solid lines, respectively.

(iv) The moment distribution in the AFM solution for odd m (= 5) has a node at the central Cr layer.

(v) The coupling between interfacial Fe and Cr moments is antiparallel in the AFM and FM solutions.

Our results for $(Fe)_3/(Cr)_5/(Co)_3$ sandwiches are similar to the calculations for $(Fe)_3/(Cr)_5$ superlattice [9,12]. The enhanced moments on Fe and Cr layers are attributed to the two-dimensional character of the sandwiches.

3.2 Co/Cr/Co Sandwiches

Next we discuss the calculated result of the Co/Cr/Cr sandwiches, for which we again obtained locally stable AFM and FM solutions. The calculated moment distributions on the Co/Cr/Co layers with the AFM and FM configurations are plotted in Figs.2(a) and (b), respectively, which show the following features:

(i) Magnitudes of Co moments are 1.0 – 1.5 μ_B, which is reduced relative to the bulk value of $1.7\mu_B$.

(ii) When the Cr SDW is compatible with Co moments (in the AFM solution for $m = 4$ and in the FM solution for $m = 5$), the magnitude of Cr moments is 0.2 – 0.5 μ_B.

(iii) The moment distribution of the AFM solution for odd m (= 5) has a node at the central Cr layer.

(iv) The coupling between the interfacial Co and Cr moments is parallel in the FM solution for $m = 4$ and in the AFM solution for $m = 5$. In the AFM solution for $m = 4$ and in the FM solution for $m = 5$, we obtained not only the parallel but also antiparallel alignments in the interfacial Co and Cr moments, though Figs.2(a) and (b) shows only the parallel solutions.

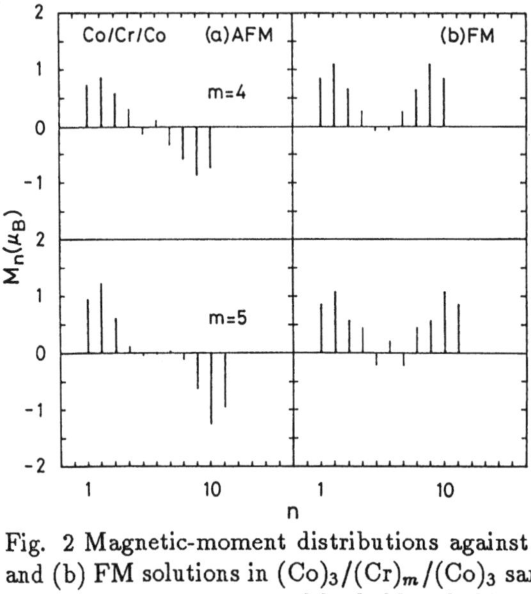

Fig. 2 Magnetic-moment distributions against the layer index, n, of (a) AFM and (b) FM solutions in $(Co)_3/(Cr)_m/(Co)_3$ sandwiches. Local moments on Co and Cr layers are expressed by bold and thin solid lines, respectively.

Thus, we obtained much more stable solutions in Co/Cr/Co sandwiches than in Fe/Cr/Fe sandwiches. First-principles calculations for $(Co)_3/(Cr)_5$ superlattice [18] yield the FM solution similar to ours for the relevant sandwich.

3.3 Ni/Cr/Ni Sandwiches

The calculated moment distributions in Ni/Cr/Ni sandwiches are plotted in Figs.3(a) and (b), which show the stable AFM and FM solutions, respectively. They have the following features:

(i) Although the Cr regions are antiferromagnetic, the Ni regions are not ferromagnetic, having peculiar moment distributions.

(ii) Magnitudes of Ni moments are less than 0.3 μ_B, which is smaller than the bulk value of 0.6 μ_B.

(iii) Magnetic moments on Cr layers in the FM solution for $m = 4$ and in the AFM solution for $m = 5$ are 1.2 – 1.8 μ_B, which are much enhanced compared with the bulk value of 0.6 μ_B. In the other cases, the magnitudes of Cr moments are less than 0.2 μ_B.

(iv) The coupling of the interfacial Ni and Cr moments is parallel in all cases.

In Ni/Cr/Ni sandwiches Ni moments are nearly quenched because of the influence of Cr layers.

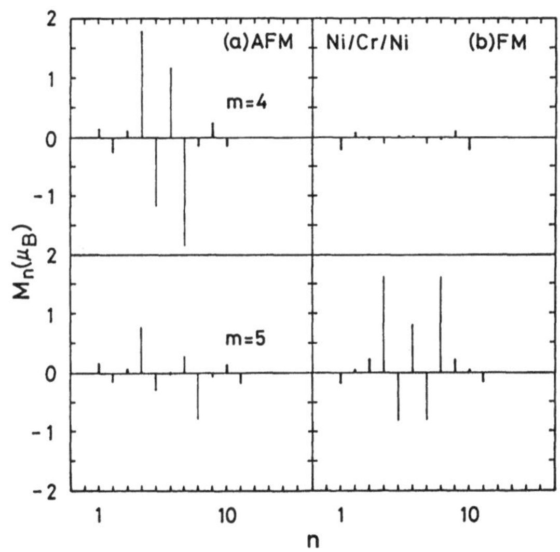

Fig. 3 Magnetic-moment distributions against the layer index, n, of (a) AFM and (b) FM solutions in $(Ni)_3/(Cr)_m/(Ni)_3$ sandwiches. Local moments on Ni and Cr layers are expressed by bold and thin solid lines, respectively.

4. Discussions

We have discussed the electronic and magnetic structures of $(X)_3/(Cr)_m/(X)_3$ sandwiches with X = Fe, Co and Ni and $m = 4$ and 5. In all the cases investigated, we have obtained the AFM and FM solutions, both of which are locally stable in the configuration space. It has been shown that the AFM solutions can be stable even when the number of intervening Cr layers is *odd* within the itinerant electron model [9,12]. This is realized by the characteristics of the itinerant model having an extra degree of freedom changing the magnitude of local moments. We notice that when the number of Cr is odd, the SDW is induced by applying an external field (*the field-induced SDW*), whose observation by experiments such as the neutron diffraction would be interesting.

In order to establish which of the FM and AFM solutions is more stable, it is necessary to make a comparison of their ground-state energies. We have tried it, but could not determine the stabler solutions because the energy difference is too small compared with our computational accuracy. Note that the energy difference is expected to be of the order of the saturation field, H_s, whose typical value is 10 kOe ($\sim 10^{-3}$ eV), while the total d-band energy is of the order 10 eV. In order to bypass this difficulty, we have calculated the interlayer exchange interaction, $J_{nn'}$, between n- and n'-th layers in Fe/Cr superlattices as a function of m, the number of intervening Cr layers [12]. The exchange interaction $J_{nn'}$ is positive for the nearest-neighbor pair ($m = 0$) as expected. It is nearly vanishing for $m = 1$, and becomes negative for $2 \leq m \leq 6$, which suggests that the AFM solutions become more stable than the FM solutions. This is consistent with the experimental result showing antiferromagnetic Fe-Fe coupling in Fe/Cr multilayers [6,7], although for far neighboring layers, a RKKY-type oscillation in J with a period of 18 – 20 Å has been reported [11].

We plot, in Fig.4, the X-dependence of magnetic moments at the central X layer (M_{XC}), those at the interfacial X layer (M_{XI}) and those at the interfacial Cr layer (M_{CrI}). This shows the trend in magnetic properties of X/Cr/X

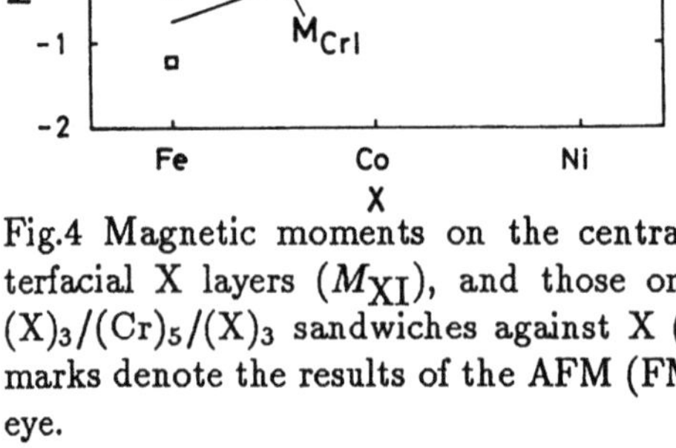

Fig.4 Magnetic moments on the central X layers (M_{XC}), those on the interfacial X layers (M_{XI}), and those on the interfacial Cr layers (M_{CrI}) in $(X)_3/(Cr)_5/(X)_3$ sandwiches against X (= Fe, Co and Ni). The filled (open) marks denote the results of the AFM (FM) solutions. The line is a guide to the eye.

sandwiches as changing X. Magnitudes of magnetic moments on the central and interfacial X layers decrease going from X = Fe to X = Ni. We note that the interfacial Cr moments are antiparallel to M_{XI} for X = Fe while they are parallel for X = Ni. In the case of X = Co, we obtained two solutions corresponding to the parallel and antiparallel alignments of the interfacial moments, as discussed in subsection 3.2. Figure 4 clearly shows that magnetic properties are very sensitive to a chice of X atom in the three sandwiches.

It was reported that the magnetoresistance is large (46 % reduction) in Fe/Cr multilayers [1] while it is fairly small (2.5 % reduction) in Co/Cr multilayers [11]. Our calculation might explain the difference in the magnetoresistance between Fe/Cr and Co/Cr multilayers as follows: According to the existing theories [8,9], the magnetoresistance is expected to arise from the spin-dependent scatterings at the interfaces. The spin-dependent random scattering potential, V, at the interface between X and Cr layers is given by $V = (U/2)\delta M$ where $U = U_{\mathrm{X}} \cong U_{\mathrm{Cr}}$ and $\delta M = \mid M_{\mathrm{XI}} - M_{\mathrm{CrI}} \mid$. Our calculation shows that in the Co/Cr/Co superlattice, we get $\delta M = 0.1\mu_B - 0.8\mu_B$ because $M_{\mathrm{CoI}} = 0.5\mu_B - 0.6\mu_B$ and M_{CrI} = - $0.2\mu_B - 0.4\mu_B$. On the other hand, in the case of the Fe/Cr/Fe sandwich, $M_{\mathrm{FeI}} = 1.6\mu_B - 1.8\mu_B$ and M_{CrI} = - $0.4\mu_B$ to - $1.2\mu_B$ lead to $\delta M = 2.0\mu_B - 2.8\mu_B$, which is much larger than that in the Co/Cr/Co sandwich. Since the resistivity due to the spin-dependent scattering is nearly proportional to V^2, it is probable that this difference in δM explains the difference in the observed magnetoresistance of Fe/Cr and Co/Cr multilayers.

References

[1] M. N. Baibich, J. M. Broto, A. Fert, F. Ngugem Van Dau, F. Petroff, P. Eitenne, G. Creuzet, A. Friedrich, and J. Chazelas, Phys. Rev. Lett. **61**, 2472 (1988).

[2] P. Grünberg, R. Schreiber, Y. Pang, M. B. Brodsky, and H. Sower, Phys. Rev. Lett. **57**, 2442 (1986)

[3] F. Saurenbach, U. Walz, L. Hinchey, P. Grünberg, and W. Zinn, J. Appl. Phys. **63**, 3437 (1988)

[4] C. Carbone and S. F. Alvarado, Phys. Rev. B **36**, 2433 (1987).

[5] N. Hosoito, S. Araki, K. Mibu, and T. Shinjo, J. Phys. Soc. Jpn. **59**, 1925 (1990).

[6] F. Nguyen Van Dau, A. Fert, P. Eitenne, M. N. Baibich, J. M. Broto, J. Chazelas, A. Friederich, S. Hadjoudj, H. Hurdequint, J. P. Radoules, and J. Massies, J. Phys. (Paris) Colloq. **49**, C8-1633 (1988).

[7] J. J. Krebs, P. Lubitz, A. Chaiken, and G. A. Prinz, Phys. Rev. Lett. **63**, 1645 (1989).

[8] R. E. Camley and J. Barnas, Phys. Rev. Lett. **63**, 664 (1989).

[9] P. M. Levy, K. Ounadjela, S. Zhang, Y. Wang, C. B. Sommers, and A. Fert, J. Appl. Phys. **67**, 5914 (1990).

[10] L. L. Hinchey and D. L. Mills, Phys. Rev. B **33**, 3329 (1986).

[11] S. S. P. Parkin, N. More, and K. P. Roche, Phys. Rev. Lett. **64**, 2304 (1990).

[12] H. Hasegawa, Phys. Rev. B **42**, 2368 (1990); *ibid.* (to be submitted).

[13] E. Velu, C. Dupas, D. Renard, J. P. Renard, and J. Seiden, Phys. Rev. B **37**, 668 (1988).

[14] T. Shinjo and H. Yamamoto, J. Phys. Soc. Jpn. (to be published.)

[15] H. Hasegawa, J. Phys. F **16**, 1555 (1986); Surf. Sci. **182**, 591 (1987); *Magnetic Properties of Low Dimensional Systems II*, ed. by L. M. Falicov and J. L. Moran-Lopez (Springer, Berlin, 1990).

[16] H. Hasegawa and F. Herman, Phys. Rev. B **38**, 4863 (1988).

[17] D. G. Pettifor, J. Phys. F **7**, 613 (1977).

[18] F. Herman, P. Lambin and O. Jepsen, Phys. Rev. B **31**, 4394 (1985).

Energetics of Metal–Metal Interfaces

M. Weinert, R.E. Watson, J.W. Davenport, and G.W. Fernando
Department of Physics, Brookhaven National Laboratory,
Upton, NY 11973, USA

Understanding and controlling the formation and growth modes of metal-metal interfaces is a challenge for both experiment and theory. Although the kinetics are still difficult to treat theoretically from first principles, the equilibrium thermodynamics of transition-metal–transition-metal adlayer adhesion can be investigated using total energy electronic structure calculations. These calculations make it possible to disentangle the various bulk and surface contributions that control adlayer and multilayer formation, as well as providing thermodynamic values that are difficult or impossible to determine experimentally. The various terms and their physical basis are illustrated using the Ag–Nb and Pd–Nb systems as examples. The relationship between spectroscopic data such as core level shifts and the energetics of metal–metal adhesion is also discussed.

1. Introduction

The study of the growth of metallic adlayers and multilayers has a long history, but it is only rather recently that experimentally well characterized samples can be produced. These new artificial multilayer materials are of interest because of their possibly modified electronic and mechanical properties. From a practical point of view, it is important to understand the factors which determine the growth behavior when one metal is deposited on another. Over the years, a rudimentary understanding [1–3] of the basic thermodynamics of adsorption and the growth process has been developed. What has been missing, however, is a first principles theoretical determination of the various factors and their interplay. In addition, such a theoretical study can provide values of thermodynamical quantities such as heats of adsorption and surface energies which are notoriously difficult to measure accurately and test the applicability of the values derived from elemental systems to monolayer coverages. In this paper [4], we will dicuss the energetics and bonding of planar transition metal systems in an attempt to gain new insight into the mechanism of interface and multilayer formation.

Eds.: A. Yoshimori · T. Shinjo · H. Watanabe

In order to illustrate the physics, we consider the Pd-Nb and Ag-Nb alloy systems. Besides their intrinsic experimental interest, these two systems provide a juxtaposition of two different types of bulk phase diagrams. The Pd-Nb bulk phase diagram displays strong compound formation, whereas for Ag-Nb there is complete immiscibility and no intermediate compounds. Neither of these types of bulk phase diagrams favor "wetting" (pseudomorphic growth). Strong compound formation favors interdiffusion, while immiscibility favors similiar nearest neighbors, i.e., clustering. Experimentally [5], however, Pd forms thick epitaxial layers on Nb(001), a single commensurate layer on Nb(110), and Ag grows epitaxially on Nb(110) for at least one layer.

The basic thermodynamic framework in which we discuss the energetics is, not surprisingly, due to Gibbs [1–3]. Consider a crystalline substrate of element B covered with a thick epitaxial layer of element A. The heat of adsorption (adhesion) of such a layer is

$$\Delta E = \gamma_A - \gamma_B + \xi + \Delta E_{str} , \tag{1}$$

where γ_A, γ_B are the surface energies of the elemental solids, ξ is the energy of bonding at the interface, and ΔE_{str} is the structural (strain) energy cost required to deform A to be in registry with the substrate B. The structural and surface energies individually are always positive since the bulk solids exist, while ξ may have either sign depending on whether the chemical interaction is bonding (negative) or non-bonding (positive). For $\Delta E < 0$, layer-by-layer growth (Frank-van der Merwe) is stable compared to clustering (Volmer-Weber). [Since $\Delta E_{str} > 0$ scales with the number of A atoms, at some point ΔE must become positive; more accurately, $\Delta E < 0$ describes Stranski-Krastanov growth.] For thin adlayers, ξ may depend on the number of layers. In the case of a multilayer system, there are no free surfaces, but two interfaces per unit cell, and Eq. (1) becomes

$$\Delta E = 2\xi + \Delta E_{str} . \tag{2}$$

The decomposition given in Eqs. (1) and (2) is in principle exact, but is useful only if the terms so defined are transferable.

In this paper, we determine the various energy contributions for a) bulk ordered alloys, b) bulk multilayers, c) elemental surfaces, and d) adlayers on surfaces using local density total energy electronic structure methods [4,6]. From these results, we will be able to draw conclusions and make predictions concerning the growth modes of adlayers and multilayers. In addition, we discuss the relationship between spectroscopic data and adhesion energies.

2. Bulk Alloys and Multilayers

The bulk calculations were all done using the Linearized Augmented Slater-Type Orbital method, while the adlayer and surface calculations used the Full-potential Linearized Plane Wave method (FLAPW) [6]. Optimization of volumes and internal coordinates were performed for some sets of calculations, but a complete optimization was not attempted in all cases. The possible changes in the energies are such that they will not effect our conclusions or even most of the values to the quoted number of digits.

The distortion energies were obtained in a straighforward manner by determining the energy cost of a given distortion in the elemental bulk system compared to the bulk groundstate crystal structure. These energies are related to various elastic constants of the materials. Ordered alloy calculations were done for various stoichiometries and crystal structures, including CsCl, CuAu I, $MoSi_2$, $MoPt_2$, Cu_3Au, and $TiAl_3$. These alloy calculations set the energy scale of the interactions and make contact with the known bulk phase diagram. The Pd-Nb system favors Pd-rich alloys with calculated heats of formation on the order of ~0.5 eV/atom. This skewing of the heats arises because the hybridization of the Nb states into the predominately Pd-like "*d* bands" causes the Pd *d* bands to fill and "float" free of the Fermi level, thus increasing their effectiveness in bonding. The calculations are consistent with the low temperature bulk phase diagram, correctly predicting the observed line compounds. Bulk superlattices ranging from Pd_2Nb_2 to Pd_6Nb_6, with the in-plane spacing constrained to that of Nb(001), all have negative heats of formation. Even for the thickest multilayers considered, the heats have not yet reached the asymptotic value of $\Delta E_{str}/2$ for Pd on Nb(001). For Pd-Nb(110) multilayers, on the otherhand, the heats of formation become positive already for Pd_3Nb_3 multilayers. That the (001) stacking is more stable than the (110) is due to two rather simple factors: 1) the distortion energy of Pd to fit the Nb(001) lattice is half of that for the Nb(110); and 2) the (001) face has more unlike nearest neighbors (and more bonding) than the (110) face. The heats for the alloys and ξ (per atom) are comparable, demonstrating that Pd-Nb multilayers can compete energetically with alloy formation.

For Ag-Nb, on the otherhand, both alloy and multilayer heats are positive, consistent with the bulk phase diagram. Note that positive heats do not mean non-bonding between Ag and Nb, but rather that the energy gain from Ag-Nb bonding is smaller than the cost of breaking the stronger Nb-Nb bond. At best such multilayers are metastable and will exist only if the kinetics prevent equilibrium from being reached.

The bulk multilayer results suggest that sharp bulk multilayers of Pd-Nb(001) several layers thick are possible, but only one or two epitaxial layers of Pd on Nb(110) are stable, and for Ag-Nb clustering always occur. Although some of the experimental results can be rationalized from the bulk calculations, some major discrepancies still remain. A complete description of the growth and energetics requires a discussion of the surface terms.

3. Clean Surfaces and Adhesion

Surface energies are extremely difficult to measure and, consequently, difficult to find reliable data for. Calculations are now at the stage where such basic thermodynamic data as surface energies and heats of formation can be computed at least as accurately as they can be measured. The calculated surface energies of Nb, Ag, and Pd are comparable to the other energies in the problem and show large differences between various elements (Nb: $\sim$3.0; Pd: $\sim$2.3; Ag: $\sim$1.3 J/m^2). Between different crystallographic faces of the same elements, however, the variations (per unit area) are only $\sim$10%, with the closer packed faces having the smaller surface energies. This near constancy of γ also holds for distorted structures. For example, if a thick Pd(001) film is tetragonally distorted to have an in-plane lattice constant equal to that of Nb(001), and if the distortion energy of Pd is accounted for, the surface energy per unit area is virtually unchanged. These results justify the use of elemental γ's even for the case of large lattice mismatches and demonstrate the transferability of these terms. (Note that surface energies per surface atom show large variations between faces and are not transferable.) In addition, these results imply that reasonably spherical bulk crystals can be grown.

The differences in surface energies favor the wetting of Ag and Pd on Nb. The calculated adhesion energies of monolayers of Ag (~ -0.5 J/m^2) or Pd (~ -1.3 J/m^2) on Nb(001) and Nb(110) are all bonding, with Pd more bound than Ag. Although the surface energies are transferable for thick layers, at monolayer coverages there is a strong interplay between interface bonding and surface energies. We define the effective surface energy of a monolayer of metal A on B as

$$\gamma(A/B) = \Delta E + \gamma(B) - \xi - \Delta E_{str}\,. \tag{3}$$

The effective γ's [4] can be used to derive a criterion for wetting of a second layer. First, assume that after the first layer the interface does not alter the energetics, i.e., $\xi \approx 0$ and $\gamma(A_n/B) = \gamma(A)$ for $n > 1$. (These assumptions are only approximately true.) Then the second layer adhesion energy will be

$$\Delta E^{(2)} = -\gamma(A/B) + \gamma(A) + \Delta E_{str}\,. \tag{4}$$

A necessary condition for wetting is $\gamma(A/B) > \gamma(A)$. The calculations predict second layer wetting only for Pd on Nb(001), which is consistent with all the experimental data and can be traced to the fact that ξ is the most binding for the (001) face of Nb.

The surface terms will clearly hinder the wetting of Nb on either Ag or Pd. For Nb on Ag(001) and Pd(001), further calculations [7] confirm the predictions [4] that Nb on Pd will form Pd-rich surface compounds and that Nb on Ag will cluster. In particular, for Nb/Pd(001), the wetting energy of a monolayer of Nb on the surface is $\Delta E \approx -0.0$ J/m^2, but surface compound formation has $\Delta E \approx -2.0$ J/m^2. This energy is even larger than Pd on Nb. Thus while it is possible to grow sharp interfaces of Pd on Nb, the converse is not possible using vapor deposition.

The case of Nb on Ag shows even more interesting behavior. Surface wetting is extremely unfavorable (2.5 J/m^2) and segration into Nb clusters is indicated. These clusters, however, will be buried in the surface and will be covered by a monolayer of Ag in order to reduce the effective surface energy. Since the gain in energy is so significant, it is unlikely that kinetics will change this conclusion. The growth mode we propose is that as Nb is deposited on the surface, small cluster of Nb will diffuse beneath the surface. These clusters, of perhaps only one or two atoms, will aggregate into larger clusters. The Ag will "flow" over these clusters to maintain a monolayer of Ag on the outer surface. Only if the kinetics are altered significantly by, for example, deposition temperature or rate, will Nb be found at the surface.

4. Spectroscopy and the Energetics of Adhesion

Since thermodynamic data such as heats are difficult to directly measure, other more indirect methods to obtain values are desirable. The most obvious method is to use spectroscopic data. In a one-electron picture, it is possible to relate [8] the heat of formation to the changes in the valence band photoemission and core levels binding energies between the elements and the compounds, i.e.,

$$\Delta E \approx \sum_i N_i \left[\Delta\epsilon_i^v - \Delta\epsilon_i^c\right]\,, \tag{5}$$

where $\Delta\epsilon_i^v$ is the change in the center of gravity of the valence band local density of states between the compound for the ith constituent and $\Delta\epsilon_i^c$ is the corresponding core level shift. The core levels are used to reference the energies assuming that the core levels do not play any role in the bonding. From the calculations, we can determine all the centers of gravity of the local density of states and the core level shifts needed to evaluate Eq. (5). There is [7] reasonable agreement, typically within $\sim$ 0.3 eV for Pd and Ag on the Nb surfaces. On the otherhand, for Nb on the Ag and Pd surfaces, errors of more than 1 eV can occur. This breakdown occurs since the implicit assumption [9] that the onsite density and potential ("double counting") contributions to the total energy are the same in the adlayer and the bulk is no longer valid; generally this problem is worst for metals with more holes in the d bands. To apply this method experimentally requires the ability to resolve local valence density of states and also a high accuracy in determining the core level shifts. For systems not in the split band limit, obtaining ground state local density of states will be very difficult.

A different approach [10] is based on Born-Haber cycles and the so-called Z^* approximation in which a core-hole excited atom of atomic number Z is treated as a Z+1 impurity. The only spectroscopic data used are the core level binding energies. This method does not give absolute numbers, but rather differences between materials with atomic numbers Z and Z+1; in addition, the energetics are related to the core-hole excited final state. The trends for differences in surface energies are reasonable, and for the case of Pd and Ag for which we have all the numbers, the calculated values using this approximation are accurate to 0.1 eV; using semi-empirical values for γ, the agreement in other cases is between 0.5 and 1.0 eV [7]. Obtaining the trends for differences in γ is not particularly significant since any simple model taking into account the inverted parabolic shape of the cohesive energy will get the trend correctly.

Extending this framework to adhesion [11], the core level shifts (ΔE_c) between the interface and the free surface are equal to the differences in adhesion energy between monolayers of atomic number Z+1 and Z on substrate B, i.e.,

$$\Delta E_c = \Delta E(Z{+}1/B) - \Delta E(Z/B)\,. \tag{6}$$

While the spectroscopic-based differences agree with the calculated ones to within a few tenths of an eV, there can be a spread of another few tenths of an eV resulting from the choice of the reference surface core level. The results (and underlying assumptions) work best for lattice matched materials.

The comparison of the "spectroscopic" determination of the adhesion energies and the calculated ones is encouraging at the level of several tenths of an eV. However, it is important to note that the comparisons are for idealized cases in which only the basic assumptions of the methods are tested, not the accuracy to which quantities can be measured. In an actual experiment, final state correlation effects, resolution problems, reconstructions and relaxations, and sample quality will introduce further possible sources of errors. These spectroscopic methods, while helpful, are not a substitute for more traditional thermodynamical measurements.

5. Summary

From our investigations of the energetics, we deduce conditions that are favorable for multilayer formation: 1) the surface energies are approximately equal, so that there is no bias of A on B versus B on A. 2) ΔE_{str} is small (good lattice matching), and 3) heats are binding but moderate to avoid interdiffusion. Neither of the systems considered here is a good candidate for making ideal multilayers because of the difficulty of depositing Nb: for example, Pd-Nb multilayers would be expected to have one sharp (Pd on Nb) and one diffuse (Nb on Pd) interface. For systems with smaller energy differences than here, electrochemical deposition techniques may be able to alter the energy balance, thereby making other growth modes possible. Finally, while spectroscopic data do yield information concerning the energetics of adhesion, the errors involved in the experiments and the underlying approximations can often be of the same order as the effects that are being studied.

Acknowledgments

This work work was supported by the Division of Materials Sciences, Office of Basic Energy Research, U.S. Department of Energy under Contract No. DE-AC02-76CH00016 and a grant of computer time at the National Energy Research Supercomputer Center (NERSC).

References

1. J.W. Gibbs, *Collected Works, Vol. 1*, pp. 55–372 (Longmans, NY, 1931).
2. D. Turnbull, *Impurities and Imperfections*, pp. 121–144 (American Society for Metals, Cleveland, 1955); Solid State Physics **3**, 226 (1956).
3. E. Bauer, Z. Krist. **110**, 372, 395 (1958).
4. Some of this work has previously been discussed in M. Weinert, R.E. Watson, J.W. Davenport, and G.W. Fernando, Phys. Rev. B**39** 12585 (1989).
5. M.J. Sagurton, M. Strongin, F. Jona, and J. Colbert, Phys. Rev. B**28**, 4075 (1983); M.W. Ruckman and L.Q. Jiang, Phys. Rev. B**38**, 2959 (1988); G. Kellog (unpublished); J. Tranquada and P.D. Johnson (unpublished).
6. M. Weinert, E. Wimmer, and A.J. Freeman, Phys. Rev. B**26**, 4571 (1982); J.W. Davenport, M. Weinert, and R.E. Watson, Phys. Rev. B**32**, 4876 (1985).
7. M. Weinert, R.E. Watson, J.W. Davenport, and G.W. Fernando (to be published).
8. P. Oelhafen, J. Phys. F**11**, L41 (1981).
9. M. Weinert, R.E. Watson, and J.W. Davenport, Phys. Rev. B**32**, 2115 (1985).
10. B. Johansson and N. Mårtensson, Phys. Rev. B**21**,4427 (1980); W.F. Egelhoff, Phys. Rev. Lett. **50**, 587 (1983).
11. N. Mårtensson, A. Stenborg, O. Björneholm, A. Nilsson, and J.N. Andersen, Phys. Rev. Lett. **60**, 1731 (1988).

Structure at the Yb/GaAs Interface Studied by Anomalous X-Ray Scattering

K. Akimoto, K. Hirose, J. Mizuki, I. Hirosawa, T. Mizutani, and J. Matsui

Fundamental Research Laboratories, NEC Corporation,
34 Miyukigaoka, Tsukuba, Ibaraki 305, Japan

Two kinds of 4×1 superstructures are found for Yb/GaAs(001) contacts with different Yb thicknesses. Different Schottky-barrier height values are also found for these samples. To clarify the role of Yb atoms in the 4×1 structure, an X-ray anomalous dispersion effect has been used in the grazing-incidence X-ray diffraction geometry. Diffraction intensities measured in the X-ray energy range near the Yb L_{III} absorption edge show a characteristic variation, which reflects the energy dependence of the real part of the anomalous scattering factor (f') for Yb atoms. As a result, for all the samples, Yb atoms are found to order in 4×1 structures. By comparison of the calculated diffraction intensities with observed intensities, two (4×1) structural models are found for the thin and thick Yb-layer samples, respectively.

1. Introduction

Metal-semiconductor contacts are of both technological and scientific interest, and have been extensively studied for over thirty years. The study of metal-semiconductor interface structures is important in understanding Schottky-barrier formation. The atomic structure at a buried interface was directly observed for metal-semiconductor systems only recently [1-3]. The superstructure was found for the Al/GaAs interface [4]. On the other hand, anomalous Schottky-barrier height changes were found for Al/GaAs contacts by inserting a rare-earth metal (Sm, Dy, Yb) interlayer [5].

This paper shows the existence of two kinds of 4×1 superstructures at Yb/GaAs(001) interfaces observed by X-ray diffraction. Different Schottky-barrier height values are also found for these samples. For all the samples, Yb atoms are found to order in the 4×1 structures by measuring the diffraction intensities in the X-ray energy range near the Yb L_{III} absorption edge. By comparison of the calculated diffraction intensities with observed ones, two (4×1) structural models are found for the thin and thick Yb-layer samples, respectively.

2. Experimental

The sample used in this study consisted of a 90 Å - thick Al caplayer, a very thin Yb interlayer and a Si-doped ($1\text{-}2\times10^{17}$ cm^{-3}) GaAs epitaxial layer. The epitaxial GaAs layer was grown under As-stabilized conditions at 600 °C on (001) Si-doped GaAs substrates by molecular beam epitaxy (MBE) techniques. Yb and then Al were sequentially deposited at 30 °C in an MBE growth chamber. Before the deposition of the Yb layer, the surface structure of GaAs(001) was a superposition of 4×1 and 1×6 structures. Six samples were prepared for the present study. The thicknesses of the Yb interlayers were 3, 4, 8, 16, 20, and 32 Å, respectively.

The interface atomic structure was studied using a grazing-incidence X-ray diffraction technique. In this geometry, the grazing-incidence angle is set just above the critical angle of total reflection. The experiments were carried out with synchrotron radiation, partly at the Photon Factory in Japan and partly on beamline x22c at the National Synchrotron Light Source in the U.S.A.

3. Results and discussion

Grazing-incidence X-ray diffraction measurements revealed that the interface between Yb and GaAs(001) has two different kinds of 4×1 superstructures [6]. Figures 1(a) and 1(b) show partial Patterson functions (partial pair-correlation functions) calculated from observed structure factors for fractional-order reflections. For the sample with relatively thin (3 Å) Yb-layer, the Patterson map is shown in Fig. 1(a). On the other hand, for the sample with a relatively thick (20 Å) Yb-layer, the Patterson map is shown in Fig. 1(b). The Two Patterson maps are obviously different from each other. The difference between the two Patterson maps appears mainly in the central peak of the 4×1 lattice in Fig. 1(b). Because this intensity is very large, this Patterson peak may come from the correlation between Yb atoms (with a large structure factor). However, at this moment, it is not clear that Yb atoms are ordered in the 4×1 structure. There are at least three possibilities. First, only for the thick Yb-layer sample do Yb atoms participate in the formation of the 4×1 structure. Second, for both (thin and thick) Yb-layer samples, Yb atoms participate in the formation of the 4×1 structure. Third, for both samples, Yb atoms are not involved in the formation of the 4×1 structure. To determine which one is correct it is necessary to determine the real-space structure.

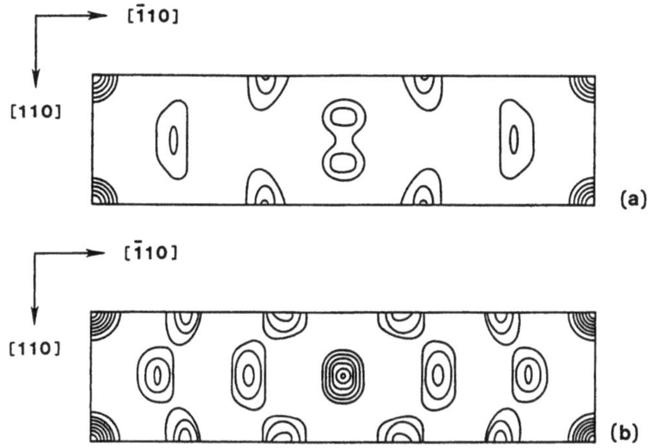

Fig. 1 Partial Patterson function calculated from the observed structure factors. Positive contour levels above zero are shown. (a) Sample with 3 Å - thick Yb layer, and (b) with 32 Å - thick Yb layer.

For that purpose, an X-ray anomalous dispersion effect has been used. The diffraction intensity was measured in the X-ray energy range near the Yb L_{III} absorption edge. Figure 2(a) shows diffraction intensity change at the ($\bar{6}/4$ 0) reciprocal-lattice point for the thick (32 Å) Yb-layer sample. We refer to (1 $\bar{1}$ 0) and (1 1 0) indices as ($\bar{1}$ 0) and (0 1), respectively. The diffraction intensity change shows a dip at the X-ray energy of the Yb L_{III} absorption edge (8944 eV). This dip corresponds to the change of f' (real part of the anomalous scattering factor) near the Yb L_{III} absorption edge. Figure 2(b) shows the results of theoretical calculations of f' for Yb atoms [7]. In this energy range (8850-9050 eV), only the Yb absorption edge exists. The Ga, As, or Al absorption edge doesn't exist in this energy range. So, this results clearly indicates that diffraction intensity at the ($\bar{6}/4$ 0) fractional-lattice point involves Yb atoms. Four samples have been measured. The thicknesses of the Yb layer were 4, 8, 16, and 32 Å, respectively. Figures 3(a), 3(b), and 3(c) show the diffraction intensity changes at the ($\bar{6}/4$ 0), ($\bar{5}/4$ 1) and ($\bar{3}/4$ 1) reciprocal-lattice points, respectively. Yb anomalous dispersion effects are detected in all samples. So it is concluded that in all the samples, Yb atoms are ordered in a 4×1 unit cell.

Recently, the c(8×2) model of the structure of the GaAs(001) surface was reported by studies of scanning tunneling microscopy [8]. This model consists of two Ga dimers and two missing dimers, mainly. The 4×1 structure is considered to be an incompletely ordered c(8×2) structure. This model has been adopted as the structure of the surface before the

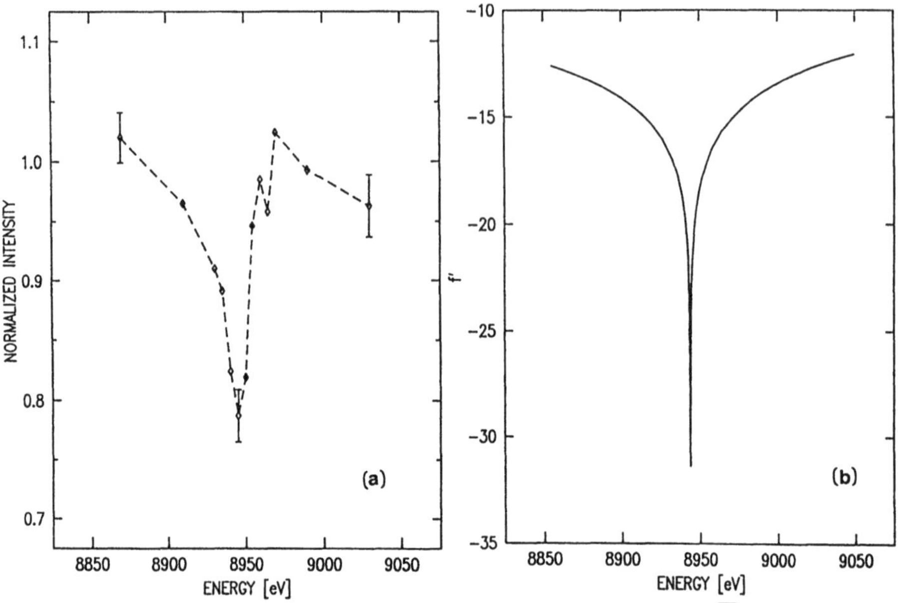

Fig. 2 (a) Observed diffraction intensity change at the ($\overline{6}$/4 0) reciprocal-lattice point in the X-ray energy range of the Yb L_{III} absorption edge for the thick (32 Å) Yb-layer sample, and (b) theoretical change of f' for Yb atoms in the same X-ray energy range.

deposition of the Yb layer. Among many interface structure models, the structure models that give the best overall agreement with the observed data are shown in Fig. 4. The present results give only information about in-plane structures. So, the positions of each atom in the depth direction are unknown. The structure models in Figs. 4(a) and 4(b) are for the thin Yb-layer sample and the thick Yb-layer sample, respectively. Both structures contain Yb dimers. The model for the thin Yb-layer sample contains one set of Yb dimers along the [1 1 0] direction and has 3 Yb atoms per unit cell. The in-plane distance of the Yb-Yb dimer bonds is about 70 % of a side length of GaAs 1×1 unit cell. The model for the thick Yb-layer sample contains two sets of Yb dimers along the [$\overline{1}$ 1 0] direction and has 4 Yb atoms per unit cell. The in-plane distance of the Yb-Yb dimer bonds is about 80 % of a side length of GaAs 1×1 unit cell. Figures 5(a) and 6(a) are structure factor maps calculated from the structure of Figs. 4(a) and 4(b), respectively. In the calculation, it is assumed there are no Ga dimers. Figure 5(b) shows the observed structure factor map for the thin (3 Å) Yb-layer sample and Fig. 6(b) shows the observed one for the thick (32 Å) Yb-layer sample. The radii of the filled circles were defined as being proportional to the observed

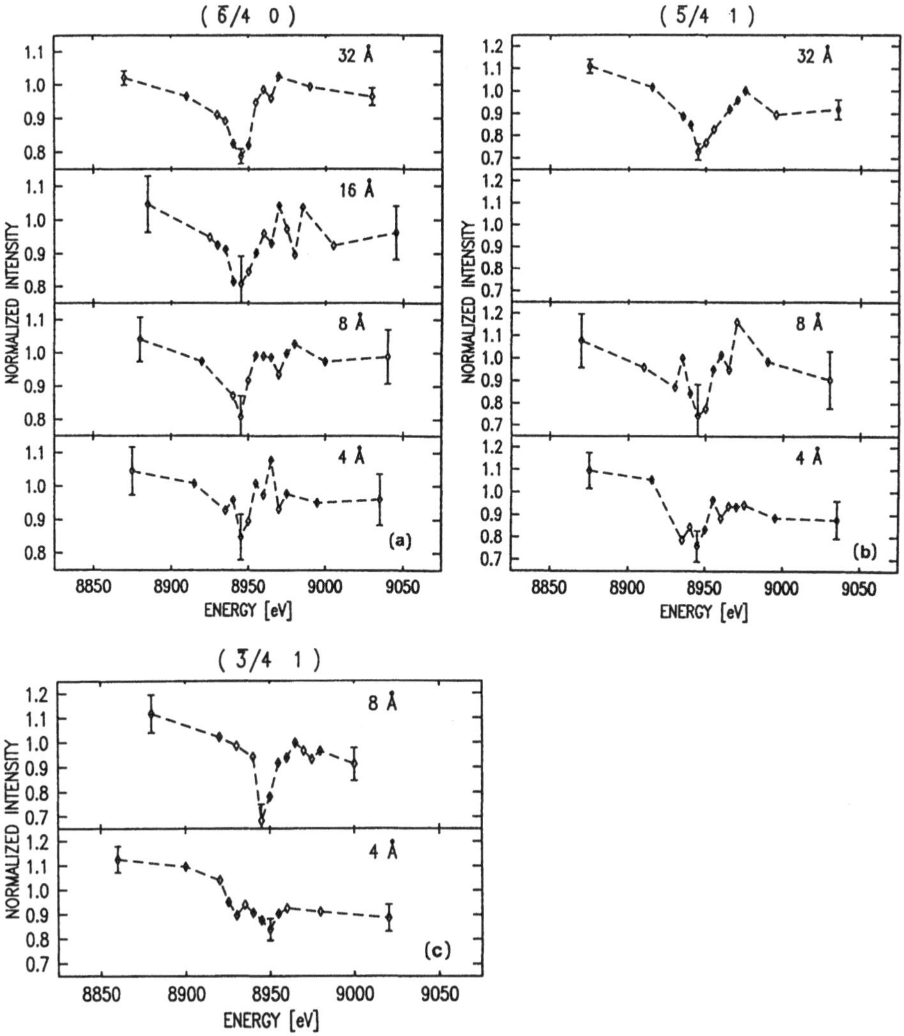

Fig. 3 Diffraction intensity changes for the various thick (4, 8, 16, and 32 Å) Yb-layer samples. (a) At the ($\bar{6}$/4 0) reciprocal-lattice point, (b) at the ($\bar{5}$/4 1) point, and (c) at the ($\bar{3}$/4 1) point.

structure factor with correction for the polarization, Lorentz factor, and the variation of the active sample area. The star marks show the fundamental-lattice points [(1 0), (0 1), etc.]. At the reciprocal lattice points indicated by the open circles, no diffraction peaks were observed. The calculated structure factors shown in Figs. 5(a) and 6(a) give similar values to the observed ones shown in Figs. 5(b) and 6(b), respectively. The R-factors (reliability factors) for the calculated structure factors

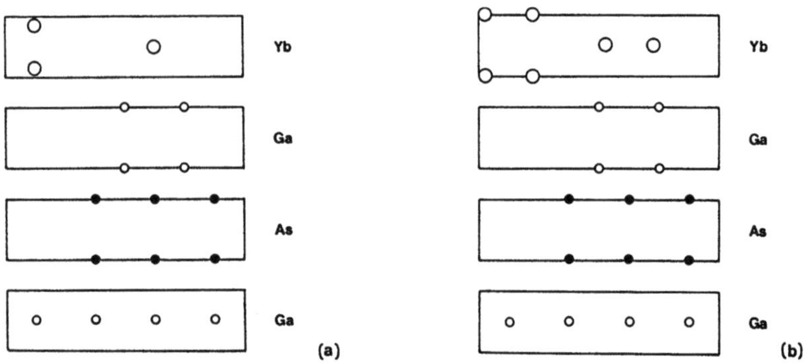

Fig. 4 In-plane structural models for the Yb/GaAs interface. (a) For the thin Yb-layer sample, and (b) for the thick Yb-layer sample.

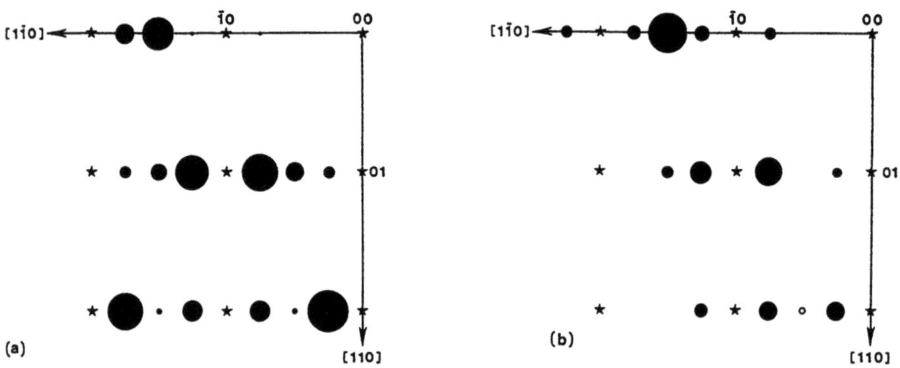

Fig. 5 (a) Calculated structure factors for the thin Yb-layer sample using the structure model in Fig. 4(a). (b) Observed structure factors for the thin (3 Å) Yb-layer sample.

shown in Figs. 5(a) and 6(a) are 32 % and 34 %, respectively. (Refinement of the Ga and As atom positions has not yet been done.) If no GaAs substrate structures are taken into account in the calculation, the calculated structure factors do not give similar values to the observed ones.

Schottky-barrier height values for the thin Yb-layer sample are 0.75-0.76 eV. On the other hand, those for the thick Yb-layer sample are 0.82-0.84 eV, which are close to those for ideal rare-earth-metal/n-GaAs contacts. Note that the Schottky-barrier height values for the thick Yb-layer sample are larger than that for the thin Yb-layer sample by about 80 meV. At this moment, we can only simply say different interfacial structures give different Schottky-barrier height values. However, the detailed atomic interfacial structures found by this study will make possible precise calculations of the relation between structures and electronic properties.

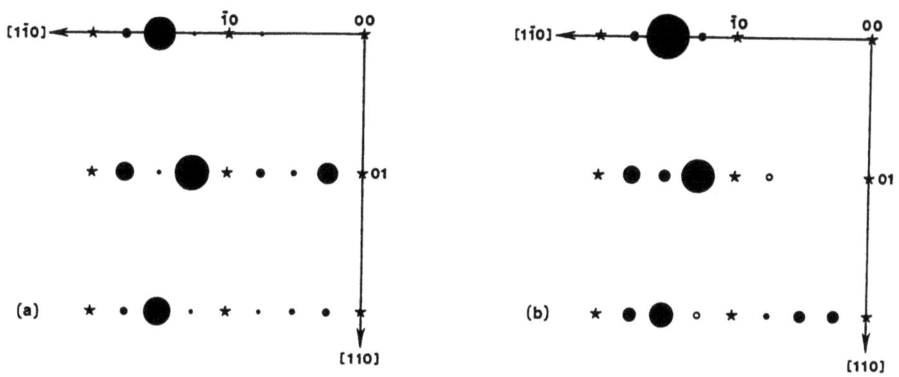

Fig. 6 (a) Calculated structure factors for the thick Yb-layer sample using the structure model in Fig. 4(b). (b) Observed structure factors for the thick (32 Å) Yb-layer sample.

Acknowledgements

The authors would like to express their deep appreciation to Drs. D. Gibbs, B. M. Ocko, and S. K. Sinha of Brookhaven National Laboratory for their research collaboration at the National Synchrotron Light Source. The work performed at Brookhaven National Laboratory is supported by the U. S. Department of Energy under Contract No. DE-AC02-76CH00016. The cooperation of the members of the Photon Factory is also gratefully acknowledged.

References

1. E.J. van Loenen and J.F. van der Veen, in *Layered Structures, Epitaxy, and Interfaces*, edited by J.M. Gibson and L.R. Dawson, Materials Research Symposia Proceedings, Vol. 37 (Materials Research Society, Pittsburgh, 1985), p.361.
2. K. Akimoto, T. Ishikawa, T. Takahashi, and S. Kikuta, Jpn. J. Appl. Phys. **24**, 1425 (1985).
3. K. Akimoto, J. Mizuki, I. Hirosawa, and J. Matsui, Appl. Surf. Sci. **41/42**, 317 (1989).
4. J. Mizuki, K. Akimoto, I. Hirosawa, K. Hirose, T. Mizutani, and J. Matsui, J. Vac. Sci. Technol. **B6**, 31 (1988).
5. K. Hirose, H. Tsuda, and T. Mizutani, J. Appl. Phys. **64**, 6575 (1988).
6. K. Hirose, K. Akimoto, I. Hirosawa, J. Mizuki, T. Mizutani, and J. Matsui, Phys. Rev. B**39**, 8037 (1989).
7. S. Sasaki, KEK Report, Vol. 88-14 (National Laboratory for High Energy Physics, Tsukuba, 1989).
8. D.K. Biegelsen, R.D. Bringans, J.E. Northrup, and L.-E. Swartz, Phys. Rev. **B41**, 5701 (1990).

Interfacial Superstructure and Schottky-Barrier Height of Metal/GaAs Contacts

K. Hirose, K. Akimoto, I. Hirosawa, J. Mizuki, T. Mizutani, and J. Matsui

Fundamental Research Laboratories, NEC Corporation,
34 Miyukigaoka, Tsukuba, Ibaraki 305, Japan

The interfacial superstructures of metal(Al or Sb)/ GaAs(001) contacts are observed by using the grazing incident x-ray diffraction (GID) method. The Schottky-barrier heights for these contacts are also determined by using the current-voltage(I-V) or capacitance-voltage(C-V) methods. The Schottky-barrier heights for Al/GaAs contacts with and without an interfacial superstructure are almost the same. On the other hand, the Schottky-barrier heights of the Sb/GaAs contacts are found to differ by as much as 0.1 eV according to the difference in interfacial superstructure. This indicates the importance of the local electronic structure for Schottky-barrier formation at the metal/GaAs interfaces. The present results are discussed in connection with the current Schottky-barrier models.

1. Introduction

For over thirty years, metal-semiconductor Schottky-barrier (SB) has been extensively studied because of both its technological and scientific interest. However, it was not until Tung's discovery[1] in 1984 that the interfacial atomic structure was recognized as being important in determining the Schottky-barrier height(SBH). Tung found different SBHs at epitaxial $NiSi_2$/Si(111) contacts to be dependent on the atomic structure at the interfaces; a type-A interface had a SBH of 0.64 eV, while a type-B interface, where the orientation of the silicide is rotated by 180° from the orientation of type-A, had a SBH of 0.78 eV[1]. A similar observation was reported by Heslinga et al.[2] in 1990. They found a surprisingly large difference in SBH for two types of Pb/Si(111) contacts, both of which differ in the arrangement of the first layer of Pb and Si atoms at the interface. For a Si(111) (7×7)-Pb structure a SBH of 0.7 eV was determined, while the Si(111) ($\sqrt{3}\times\sqrt{3}$)R30-Pb structure revealed a SBH of 0.93 eV[2].

The $NiSi_2$/Si(111) and the Pb/Si(111) interfaces show homogeneity in the interface structures and associated electronic properties, making it possible to clarify the relationship between the atomic structures and the SBH. Thus, up to now, only these interfaces have been able to form ideal model systems for the microscopic theory of the SB formation mechanism.

However, most metal/GaAs interfaces show much more complicated interface structures caused by chemical reaction and/or diffusion. Therefore, it has been thought that the electrons traveling through different parts of the interfaces may experience different local barriers, thus giving rise to average SBHs with little information regarding the relationship between local electronic and structural properties [3]. Recently, however, we have found ordered interfacial atomic structures, or interfacial superstructures, even at such metal(Al,Yb)/GaAs interfaces [4,5], by using the GID method. This suggests the possibility that the metal/GaAs interfaces are also homogeneous enough for the local electronic structure to be recognized as being important to SB formation.

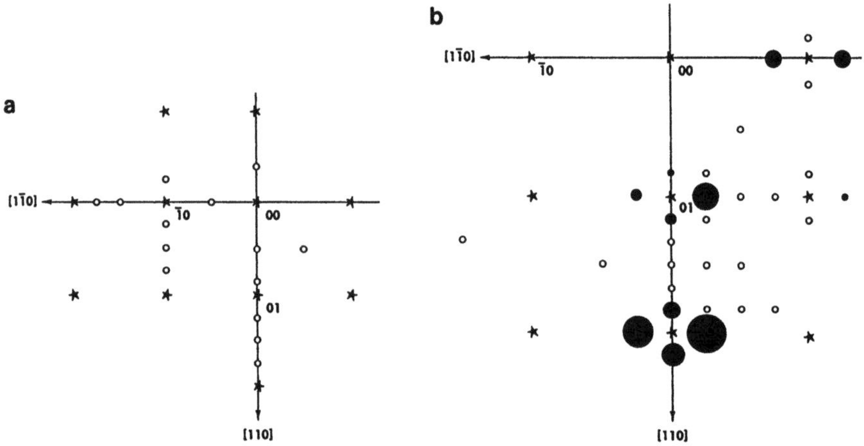

Figure 1. (a) No superstructure at the interface between Al and GaAs(2×4). (b) Observed structure factors from the 4×6 superstructure at the interface between Al and GaAs(4×6). The radius of each filled circle is proportional to the observed structure factor. At the reciprocal-lattice points indicated by open circles, no diffraction peaks were observed. Star marks are the fundamental lattice points.

the GaAs(2×4) has no superstructure, while the interface between the Al and the GaAs(4×6) has a (4×6) superstructure which has the same symmetry as the original GaAs surface. In fig.1(b), the peaks at the points only around the fundamental reciprocal lattice points were observed, but the definite peaks at the (n/4,m/6) (n≠0 and m≠0) could not be observed. These results indicate that the observed (4×6) superstructure is an overlap of the (4×1) and (1×6) domains, as is the case of the GaAs(4×6) surfaces[8].

Figures 2(a) and (b) show the observed structure factors for the interfaces between Sb and GaAs(2×4) or GaAs(4×6), respectively. The interface has a (1×4) superstructure when Sb is deposited on the GaAs(2×4) surface, which is in surprising contrast to the situation of Al, where no interfacial superstructure was observed. On the other hand, the interface has a (1×6) superstructure when Sb is deposited on the GaAs(4×6) surface. This observed superstructure is different from that observed at the interface between Al and GaAs(4×6). As a result, it was determined that the observed interfacial superstructures at both interfaces share a part of the symmetry which exists in each of the original GaAs surface structures.

Table I summarizes the SBH(I-V) and SBH(C-V) determined from I-V and C-V measurements. The observed symmetries of the interfacial superstructures are also shown. Since the SBH(I-V) and the SBH(C-V) are consistent with each other, this data is sufficiently reliable. The SBH for the Al/GaAs(2×4) is 0.82-0.83 eV which is almost the same as the SBH of 0.79-0.82 eV for the Al/GaAs(4×6). On the other hand, the SBH for the Sb/GaAs(2×4) is 0.68-0.71 eV, which differs by as much as 0.1 eV from the SBH of 0.78-0.80 eV for the Sb/GaAs(4×6). Furthermore, these values are larger than what is expected for Al and Sb from the original Schottky model[9], showing that the interface Fermi levels are pinned due to interface states at all of these interfaces.

In this paper, we describe the SBH and the interfacial superstructures of III-metal(Al) or V-metal(Sb)/GaAs contacts, each of which has two types of original GaAs surface superstructures; namely (2×4)- and (4×6)-reconstructed surfaces. The interfacial superstructures are observed at some of the interfaces, where different SBHs are measured according to the difference in interfacial superstructure.

2. Experimental

Each of the sample structures used in this study consists of either a thin Al (90 Å) or a thin Sb(40 Å) layer and a 0.7 µm thick Si-doped (2×10^{17} cm^{-3}) GaAs epitaxial layer. The Sb layers have 12 Å thick Al cap layers to prevent oxidation of the Sb surfaces. The epitaxial layers were grown on Si-doped n^+-type GaAs(001) substrates under As-stabilized conditions at 600 °C by the molecular-beam epitaxy (MBE) technique. The Al and Sb layers were deposited at room temperature in the MBE-growth chamber after a desired surface superstructure was identified on the GaAs epitaxial layer by a reflection high-energy electron diffraction (RHEED) technique. Two kinds of surface structures were prepared for both the Al and Sb layer depositions; a (2×4)-reconstructed surface (an As-rich surface) and a (4×6)-reconstructed surface (a Ga rich surface.) Hereafter, we will refer to these original GaAs surfaces as GaAs(2×4) and GaAs(4×6).

The (2×4)-reconstructed surface was obtained by simply exposing the epitaxial layer to the As molecular beam until the substrate temperature decreased to 550 °C. Slightly more complicated was the formation of the (4×6)-reconstructed surface. After the growth of the epitaxial layer, four monolayer thick Ga was deposited on the surface, once the substrate was cooled to 200 °C and the background As pressure in the growth chamber was reduced to $<10^{-9}$ Torr. After the reconstructed surface structure disappeared once, the substrate was again heated to 620 °C until the surface reconstructed structure reappeared. A clean (4×6)-reconstructed structure was obtained when the substrate temperature was decreased.

The interfacial atomic structures were studied using the GID method. The experiment was carried out with synchrotron radiation at beamline 9C, installed at the Photon Factory in Tsukuba, Japan. The 1.5 Å wavelength x-ray impinged on the samples with a 0.3° grazing incidence angle. The detailed experimental geometry has been given in a previous report[6].

In order to measure the SBHs, diodes of 500 µm in diameter were defined after Au-Ti deposition in another deposition chamber. SBHs were carefully determined [7] from both I-V and C-V measurements. The SBH(I-V) were determined from the saturation current density measured from the I-V curves. In this calculation, the tunneling current was taken into account, as was the lowering effect produced by the image force. The SBH(C-V) was determined from the intercept voltage for $1/C^2$-V curves, taking into account Fermi energy. Ohmic contacts were made with In solder at the back surfaces of the substrates.

3. Results

The structure factors given by the GID measurements for the interface structures between Al and GaAs(2×4) or GaAs(4×6) are shown in Fig. 1(a) and (b). The star marks show the fundamental lattice points. The radius of the filled circles were defined as being proportional to the observed structure factor with corrections made for polarization, the Lorentz factor and the variation of the active sample area. At the reciprocal lattice points indicated by the open circles, no diffraction peaks were observed. These figures show that the interface between the Al and

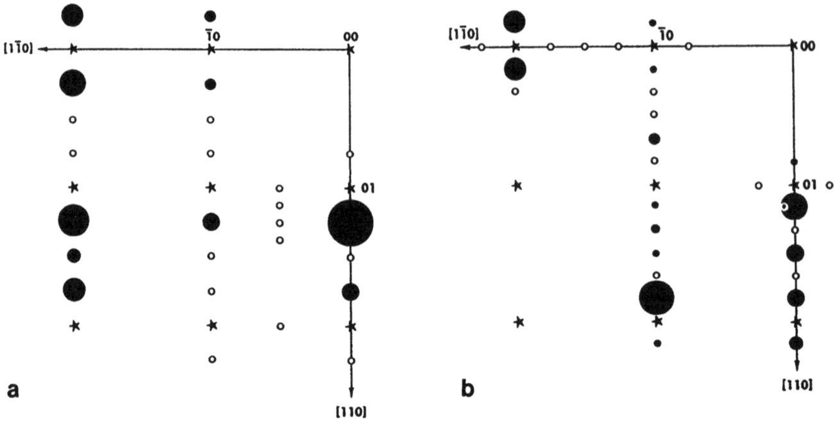

Figure 2. Observed structure factors from, (a) the 1×4 superstructure at the interface between Sb and GaAs(2×4) and (b) the 1×6 superstructure at the interface between Sb and GaAs(4×6). The radius of each filled circle is proportional to the observed structure factor. At the reciprocal-lattice points indicated by open circles, no diffraction peaks were observed. Star marks are the fundamental lattice points.

Table I. Observed interfacial superstructures and Schottky-barrier heights(SBH) for metal/GaAs interfaces. SBH(I-V) and SBH(C-V) were determined by current-voltage and capacitance-voltage methods, respectively.

metal	metal / GaAs (2 x 4)			metal / GaAs (4 x 6)		
	Super-structure	SBH (I-V) (eV)	SBH (C-V) (eV)	Super-structure	SBH (I-V) (eV)	SBH (C-V) (eV)
Al	no	0.83	0.82	4 x 6	0.82	0.79
Sb	1 x 4	0.71	0.68	1 x 6	0.78	0.80

4. Discussion

First, note that there are interfacial superstructures at the metal/GaAs interfaces with chemical reactions and/or diffusion, as well as at the ideal metal/Si interfaces. It is interesting to find that the chemical reaction and/or diffusion does not completely destroy the ordering at the metal/GaAs interfaces. Hence, the metal/GaAs interfaces are found to be homogeneous enough for us to clarify the relationship between the atomic structures and SBH. Furthermore, the observed interfacial superstructures have a part of or all of the symmetry which exists in the original GaAs surface structures. This indicates that the GaAs surface superstructures were modified according to the specific chemical reaction between the contact metal and the original GaAs surfaces.

We will now discuss our results on the SBH in connection with the interfacial atomic structure. A remarkable result is the relation between the SBH and the interfacial superstructure for the Sb/GaAs contacts. The SBH of the Sb/GaAs contact with the (1×6) interfacial superstructure was found to be larger than that of the Sb/GaAs contact with the (1×4) interfacial superstructure by as much as 0.1 eV. This result indicates the importance of the local electronic structure for

SB formation at the Sb/GaAs interfaces. On the other hand, the SBHs for Al/GaAs interfaces with and without the interfacial superstructure were found to be almost the same, revealing that the SBH is determined irrespective of the details of the interfacial atomic structures at the Al/GaAs interfaces.

It is interesting to compare our experimental results for the Sb case with the widely accepted SB models, namely the metal-induced gap states (MIGS) model[10] and the Advanced Unified Defect Model (AUDM)[11]. On the one hand, the observed SBH dependence on the interfacial atomic structure is found to be clearly inconsistent with the MIGS model, which emphasizes that the SBH is determined by semiconductor bulk property such as the charge neutrality level. On the other hand, our results can be explained by the AUDM which insists that the SBH is determined by interfacial defect states with associated energy variations.

Indeed, the observed SBH difference of 0.1 eV between the Sb/GaAs contacts is exactly the same as the reported difference in the surface Fermi pinning level [12] between these GaAs surfaces. Thereby, it can be concluded that the SBHs were determined by the defect states on the original GaAs surface when the interfacial superstructures existed after the metal deposition. However, since the observed interfacial structure was not the same in symmetry as those found on the original GaAs surface, we can not rule out the possibility that the SBHs are determined by interfacial states newly created during the interface evolution. Hence, a study should be carried out on the evolution of the interface states during the Sb deposition to clarify the mechanism of the observed SBH difference for Sb/GaAs contacts.

5. Conclusion

We have found ordered interfacial atomic structures, or interfacial superstructures, at metal(Al,Sb)/GaAs interfaces, by using the grazing incident x-ray diffraction (GID) method. The observed interfacial superstructures have a part or all of the symmetry which exists in the original GaAs surface structure, which indicates that the GaAs surface superstructures were modified according to the specific chemical reaction between the contact metal and the original GaAs surface. The SBHs of the Al/GaAs contacts with and without the interfacial superstructure are almost the same. However, we also found a close relationship between the SBH and the interfacial superstructure for the Sb/GaAs contacts, with the SBHs of the Sb/GaAs differing by as much as 0.1 eV according to the variations in interfacial superstructure. This result strongly suggests the importance of the local electronic structure for SB formation at these metal/GaAs interfaces.

Acknowledgement

The authors would like to thank Y. Miyagi for assistance in the experiment. Y. Matsumoto and M. Ogawa are also acknowledged for their encouragement.

References

[1] R.T. Tung, Phys.Rev.Lett.52,461(1984).
[2] D.R. Heslinga, H.H. Weitering, D.P. van der Werf, T.M.Klapwijk, and T. Hibma, Phys.Rev.Lett.64,1585(1990).

[3] R.T. Tung, K.K. Ng, J.M. Gibson, and A.F.J. Levi, Phys. Rev. B33, 7077(1986).
[4] J. Mizuki, K. Akimoto, I. Hirosawa, K. Hirose, T. Mizutani, and J. Matsui, J.Vac.Sci.Technol.B6,31(1988).
[5] K. Hirose, K. Akimoto, I. Hirosawa, J. Mizuki, T. Mizutani, and J. Matsui, Phys.Rev.B39,8037(1989).
[6] K. Akimoto, J. Mizuki, T. Tatsumi, N. Aizaki, and J. Matsui, Surf. Sci.183,L297(1987).
[7] K. Hirose, H. Tsuda and T. Mizutani, J.Appl.Phys.64,6575(1988).
[8] P. Drathen, W. Ranke and K. Jacobi, Surf.Sci.77,L162(1978).
[9] W. Schottky, Z. Phys.118,539(1942).
[10] J. Tersoff, Phys.Rev.Lett.52,465(1984).
[11] W.E. Spicer, Z. Liliental-Weber, E. Weber, N. Newman, T. Kendelewicz, R. Cao, C. McCants, P. Mahowald, K. Miyano, and I. Lindau, J. Vac. Sci. Technol.B6,1245(1988).
[12] S.P. Svensson, J. Kanski, T.G. Andersson, and P.-O.Nilsson, J. Vac.Sci.Technol.B2,235(1984).

7×7 Superstructure at the Amorphous Si/Si(111) Interface and Its Modification During Solid Phase Epitaxial Growth

A. Sakai[1], *T. Tatsumi*[2], *and K. Ishida*[1]

[1]Fundamental Research Laboratories, NEC Corporation,
34 Miyukigaoka, Tsukuba, Ibaraki 305, Japan
[2]Microelectronics Research Laboratories, NEC Corporation,
4-1-1 Miyazaki, Miyamae-ku, Kawasaki, Kanagawa 213, Japan

Abstract. The 7×7 superstructure buried at the amorphous Si (a-Si)/Si(111) interface was directly observed by transmission electron microscopy (TEM). High-resolution (HR) images of the ⟨110⟩ cross-sectional observation showed the atomic arrangement of the projected interface superstructure, which had a periodicity of 7 times the 1/3{224} periodicity along the interface. The atomic structure, which consists of a dimer-stacking fault layer and epitaxial layers extended over the vacant corner sites and the dimer-chains, was revealed by using computer image simulation. Structural modification of the interface 7×7 superstructure during solid phase epitaxial growth (SPE) of the a-Si overlayer was also investigated by TEM. Plan-view transmission electron diffraction observations of samples annealed at various temperatures indicated that a modification of the dimer structure occurred during the transformation from the 7×7 to the 1×1 structure (7×7→1×1). It was found by ⟨110⟩ cross-sectional HRTEM that 7×7→1×1 proceeded laterally within the stacking fault layer prior to the vertical SPE of the a-Si. Furthermore, it was confirmed that the unfaulting of the reversed stacking region, as well as the faulting of the normal stacking region, occurred simultaneously during 7×7→1×1. The mechanism of 7×7→1×1 through the dimer modification was also discussed.

1. Introduction

Surface and interface superstructure in Si materials have been widely studied by a large number of techniques. Transmission electron microscopy (TEM) is one of the most effective methods for analyzing surface and interface atomic structures of various kinds of materials. Compared with the other techniques, for example scanning tunneling microscopy [1,2], low energy electron diffraction [3] and grazing incidence X-ray diffraction [4], TEM provides us with both reciprocal and real space information from the same area of the samples. A transmission electron diffraction (TED) pattern represents the obtainable reciprocal space information and analyses of the TED pattern reveal the average atomic structures. High resolution (HR) TEM is one of the visualization techniques in real space and allows us to deduce the atomic configuration of the superstructures directly. In particular, cross-sectional HRTEM observation provides us with a projection of a side view of the superstructure and thus reveals the atomic arrangement perpendicular to the superstructure plane [5].

It has been confirmed using the TED method that a 7×7 periodicity on a Si(111) surface was preserved even after deposition of amorphous Si (a-Si) under ultra-high vacuum (UHV) conditions and that such a metastable 7×7 periodicity turned into the 1×1 periodicity owing to solid phase epitaxial growth (SPE) of the a-Si overlayer during annealing [6]. This finding provides us with two scientific interests concerning the interface 7×7 superstructure. One is the structural difference of the interface 7×7 superstructure from the surface reconstructed structure. The other is the transformation from the interface 7×7 superstructure into the 1×1 structure (7×7→1×1) owing to the crystallographic alignment with the substrate during annealing.

In this paper, first, we will demonstrate the direct imaging of the 7×7 superstructure buried at the a-Si/Si(111) interface by cross-sectional HRTEM and predict the interface atomic structure by using an image matching technique between the experimental and the simulated images. Then, we will investigate the structural modification of the interface superstructure during SPE of the a-Si overlayer and deduce the mechanism of 7×7→1×1.

2. Experimental

Since the projected image of the superstructure is obtained by cross-sectional observation, the image may be unclear unless only one domain of the superstructure is projected. Therefore, in order to form a large domain of a 7×7 superstructure layer, a thick Si buffer layer of 3000 Å was epitaxially grown on a Si(111) substrate at 750°C after surface cleaning in a molecular-beam-epitaxy chamber with base pressure 1.2×10^{-10} Torr. The thick buffer layer produces an atomically flat surface [7]. After the confirmation of a clean surface showing a 7×7 reconstructed structure by reflection high energy electron diffraction, a-Si was subsequently deposited by electron-beam evaporation to a thickness of 50 Å at room temperature. Annealing was performed at various temperatures from 400 to 600°C for 10 to 30 min in a N_2 atmosphere. Two types of TEM specimens were prepared. One was for plan-view observation and chemically thinned from the substrate side. The other is for cross-sectional observation along the <110> direction. During the preparation process, samples were not exposed to temperatures higher than 150°C and were Ar-ion thinned at the low energy of 3 kV. TEM was performed with AKASHI EM 002B operating at 200 kV in both plan-view and cross-sectional observations.

3. Results and Discussion

3.1. Structure of As-grown Interface Superstructure

3.1.1. TED Observation

Figure 1 shows a TED pattern taken from the sample in the [111] plan-view observation. Bragg reflection spots from the buffer layer denoted by the hexagonal unit cell index (11) (which is the cubic index {220} of a bulk crystals), diffuse rings from the deposited a-Si layer, and two-dimensional fundamental reflections denoted by (01) and (10)

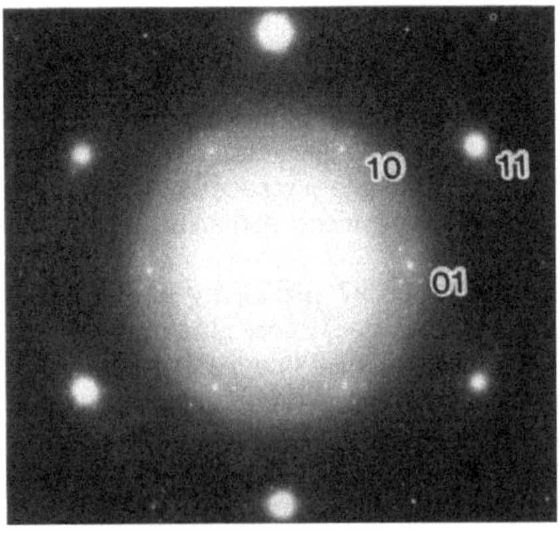

Fig. 1. A transmission electron diffraction pattern of the [111] plan-view observation taken from the area including the a-Si/Si interface.

(1/3{224}) are seen. Furthermore, extra spots of fractional indices (*h*/7, *k*/7) which show the existence of the 7×7 superstructure are clearly observed. Since the superstructure spots were always accompanied by the diffuse rings, {11} and {01} spots, they are proved to come from the a-Si/Si interface.

3.1.2. HRTEM Observation

Figure 2 shows a typical cross-sectional HRTEM image of the a-Si/Si interface region taken under the Scherzer focus. The specimen thickness of the observed region was thin enough to prevent appearance of the half-spacing contrast. The average terrace width at the interface was estimated to be about 500 Å by counting the number of the bi-layer steps in the observed region. Since the specimen thickness (< 100 Å) is sufficiently less than the terrace width, only one domain of the 7×7 superstructure is projected on the image.

The following distinct image variation can be noticed in the interface regions. First, as indicated by arrow heads, comparatively large bright spots are clearly observed between layers denoted by A and B at intervals of 23 Å. This periodicity corresponds to 7 times the 1/3{224} periodicity of 3.3 Å. Second, the region below the C layer shows a Si bulk crystal image, while in the B layer, elongated dark spots are gradually rotated relative to their orientations in the bulk crystal with a period of 23 Å. Such contrast variation is also recognized in the A layer.

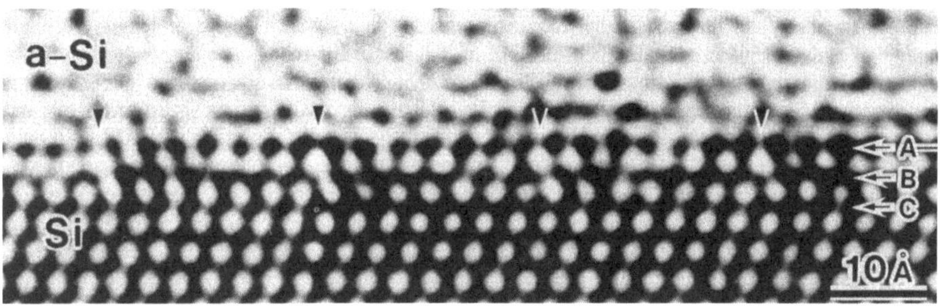

Fig. 2. A cross-sectional HRTEM image of the a-Si/Si interface showing the atomic arrangement of the buried 7×7 superstructure. Large bright spots between A and B layers can be seen at intervals of 23 Å (indicated by arrow heads). Image variation of elongated dark spots can also seen in A and B layers along the interface. Si <110> bulk crystal image is seen below C layer.

3.1.3. Identification of Structure

From the previous analysis of the TED pattern [6], it was suggested that the buried 7×7 superstructure at the a-Si/Si interface is different from the dimer-adatom-stacking fault structure which is a widely accepted model for the surface 7×7 superstructure [8]. Robinson *et al.* have also analyzed the partial ordering extending into the amorphous region on the dimer-stacking fault layer using X-ray diffraction [9]. Therefore, in order to reproduce the observed image, the model proposed by Robinson *et al.*, which consists of two epitaxial layers having normal and reversed stacking on the dimer-stacking fault layer, was employed as the starting position for the multi-slice image simulation. However, the simulation images of the <110> projection of Robinson's model could not explain the details of our experimental images [10].

We therefore propose a new model shown in Fig. 3(a). This model consists of a dimer-stacking fault layer and two epitaxial layers having normal and reversed stacking

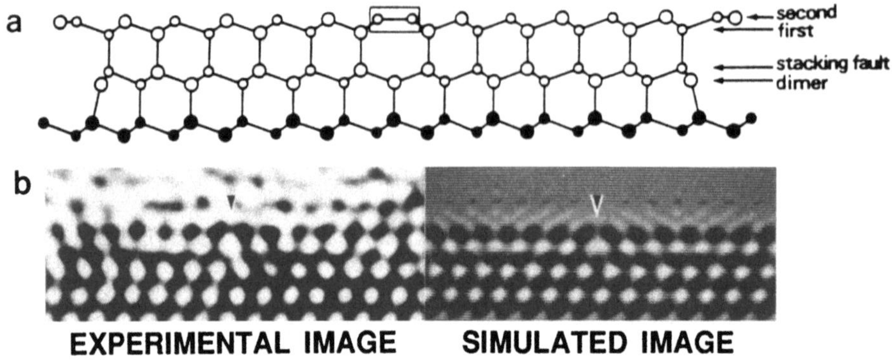

Fig. 3. (a) A model of the buried 7×7 superstructure at the a-Si/Si interface. The atomic bridge above the dimer-chain is surrounded by a rectangle. (b) A comparison between the experimental and the simulated image for the model of (a). The occupancy factor of the atoms in the first and second epitaxial layers are assumed to be 0.9 and 0.63, respectively.

regions in which the second epitaxial layer is extending over the vacant corner sites and the dimer-chains. Thus the atomic tunnels along the <110> direction are formed between the atomic bridge (surrounded by a rectangle) and the dimer-chains. Figure 3(b) shows a comparison between the part of the experimental image shown in Fig. 2 and the simulated image of the present model. In the simulation, it is assumed that an objective lens defocus value and a specimen thickness of -320 Å and 75 Å, respectively, are the experimental conditions and atoms in respective epitaxial layers have partial occupancies in order to vary the ordering gradually to the amorphous layer along the growth direction of [111]. We find good matching between the experimental and simulated images. Namely the large bright spots as well as the rotation of the elongated dark spots are clearly reproduced in the simulated image. Therefore, the existence of the atomic bridge in the second epitaxial layer can be identified. Such an atomic bridge indicates a possibility for the existence of new dimer structures which align perpendicularly to the dimer-chain underneath, although the detailed arrangements on (111) plane cannot be determined from our <110> projection.

3.2. Structural Modification due to SPE

3.2.1. TED Observation

Figure 4 shows a series of plan-view TED patterns taken from the annealed samples. In the TED patterns, extra spots of fractional indices ($h/7,k/7$) and the diffuse ring observed in the as-grown sample of Fig. 4(a) disappear in the sample annealed at 600°C for 10 min of Fig. 4(c), showing 7×7→1×1 due to SPE. No changes in the pattern feature were detected in the samples annealed at temperatures below 400°C. The variations in intensity distribution of the diffraction spots are mainly observed in Fig. 4(a) and (b), which show the intermediate state of 7×7→1×1. It is notable that the fractional spots on the line connecting {10}, {20}, and {11} spots fade with an increase in the annealing temperature but not equally, i.e., the spots except for those adjacent to the {10} spots fade faster. This is clearly seen in Fig. 4(b). The above pattern variations which were always accompanied with the amorphous diffuse ring, allow us to interpret that 7×7→1×1 occurred at the a-Si/Si interface. Furthermore, the distribution of the diffraction intensity among the fractional order spots is calculated kinematically for several structure models which varied from the original interface 7×7 superstructure. The contribution of the dimer structure modification to the variation of the spot intensity distribution was mainly tested. Comparison between

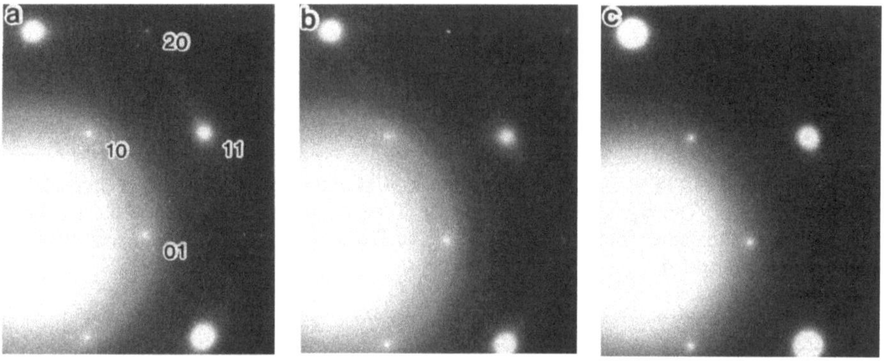

Fig. 4. A series of plan-view transmission electron diffraction patterns of the interface 7×7 superstructures taken from the samples annealed at; (a) 460°C for 10 min; (b) 500°C for 10 min; (c) 600°C for 10 min.

the observed and the calculated TED patterns suggested that the dimer structures were first modified through $7\times7\rightarrow1\times1$ during the annealing.

3.2.2. HRTEM Observation

Before reporting the HRTEM observations of the structural modification due to SPE, we will pay attention to the two characteristic features of the HRTEM image of the original interface 7×7 superstructure shown in Fig. 3(b). One is the dimer-chain under the atomic bridge indicated by arrow heads (hereafter we call the image feature Type A). The other is the dimer-stacking fault layer, where the elongated dark spots are observed to be gradually rotated with a half-periodicity of the unit cell (Type B).

Next, the results of cross-sectional HRTEM observations for the samples annealed under the above conditions will be reported. In the samples which were annealed at the temperatures above 480°C, SPE of the a-Si overlayer was observed. The HRTEM images of the interfaces no longer showed the original features as shown in Fig. 2 and the SPE growth front simply showed the 1×1 structure. The most characteristic images of the intermediate state of $7\times7\rightarrow1\times1$ were obtained for the sample annealed at 460°C for 10 min. Figure 5 is a typical HRTEM image showing an a-Si/Si interface region, taken under almost the same conditions as Fig. 2. In the interface region, the image feature similar to Type A can be identified at the points indicated by arrow heads. The variation of Type B is also recognized in the layer denoted by DS. We therefore believe that the DS layer corresponds to the dimer-stacking fault layer which varied from the original structure. Namely, Fig. 5 represents the very initial stage of SPE of the a-Si overlayer on the buried 7×7 superstructure.

It should be noticed that the image of the DS layer consists of about four types of half-unit cell images. The results are shown at the bottom of Fig. 5. The image α is similar to a half of the Type B image, i.e., the elongated dark spots are gradually rotated. In the image β, all elongated dark spots have the same orientation as that of the bulk crystal. On the other hand, the image γ shows only three or four elongated dark spots on the right-hand side having the same orientation as that of the bulk. Such orientation is shown to be reversed in the image δ. The present observation of the <110> direction is sensitive to the modification of the dimer structure since one of the dimer-chains aligned along the three <110> directions is projected as one atomic column in the Type A image. Therefore, invisibility of Type A directly means the modification of the dimer structure aligned at least along the incident electron beam direction. On the other hand, Type B is essentially due to the projection of both the normal and reversed stacking regions at the

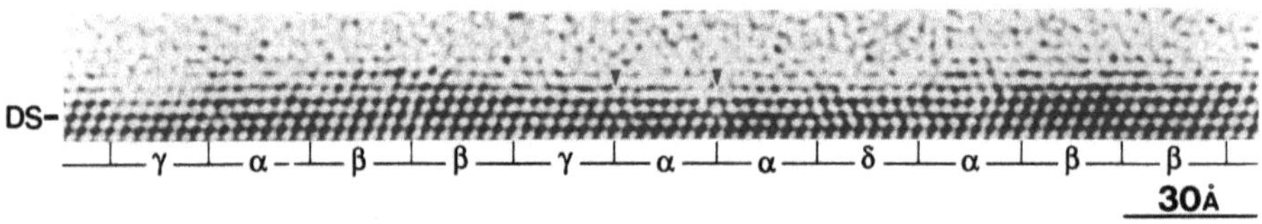

Fig. 5. A typical HRTEM image showing the a-Si/Si interface of the sample annealed at 460°C for 10 min. The arrow heads indicate the Type A (see text) image and the DS layer corresponds to the dimer-stacking fault layer. The DS layer is classified into four types of images; α, β, γ, and δ.

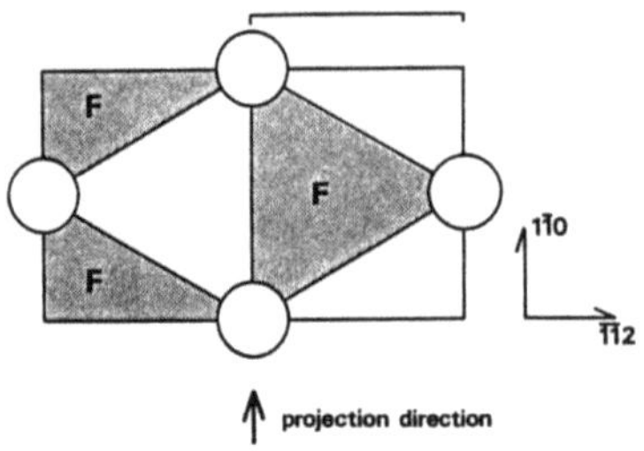

Fig. 6. Observation geometry of a unit cell of the 7×7 superstructure and its relation to the vacant corner sites and the reversed stacking regions labeled F.

same position. As shown in Fig. 6, in the half-unit cell, the normal stacking regions are projected predominantly on the right-hand side, whereas the reversed stacking regions are on the left. Therefore, the variation of the Type B image can be interpreted as the variation of the ratio of the domains sizes of the normal and reversed stacking regions within the thickness of the TEM specimen.

3.2.3. Modeling and Comparison

In order to explain the above image variations, we considered possible models in which the size of either the normal or the reversed stacking region was varied in several ways. In the modeling, the structural modification of only the dimer-stacking fault layer was taken into account for the sake of simplicity. Furthermore, each model described below represents the mean structure through the TEM specimen thickness since about three unit cells of the interface 7×7 superstructure are projected on the HRTEM image. Figures 7(a)-(f) show schematic illustrations of a half-unit cell of the stacking fault layer in which the open and filled circles represent the normal and the reversed stacking atoms, respectively. Taking the TED results into consideration, the unfaulting processes where the dimer structure modification occurs, are described in Figs. 7(a), (b) and (c). Figure 7(a) shows that only half of the reversed stacking region is unfaulted. Figure 7(b) shows that a portion of the dimer structures is modified, so that the reversed stacking atoms along the dimer-chain are partially transformed into the normal stacking atoms. The modification of all dimer structures is described in Fig. 7(c). On the other hand, in Fig. 7(d), the region excepting the dimer structures is unfaulted, i.e., the dimer structure is preserved. Figure 7(e) shows the 1×1 structure which is completely unfaulted. In the same manner, the faulting process within the normal stacking region is also possible. Figure 7(f) shows the 1×1 structure but it consists of only the reversed stacking region, i.e., the twin structure is present at the interface.

Employing these models, we carried out image simulation using the multi-slice method, where an objective lens defocus value and a specimen thickness of -340 Å and 75 Å, respectively, are assumed to be the experimental conditions. The resulting images of the computer simulations are shown in Figs. 8(a)-(f), which correspond to the models shown in Figs. 7(a)-(f), respectively. The simulated images show the variations in the

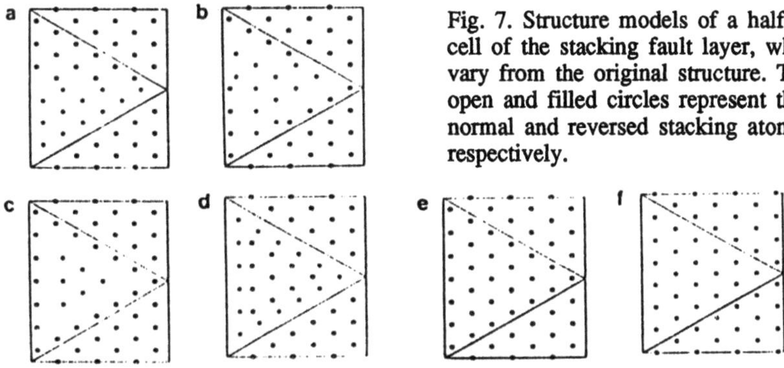

Fig. 7. Structure models of a half-unit cell of the stacking fault layer, which vary from the original structure. The open and filled circles represent the normal and reversed stacking atoms, respectively.

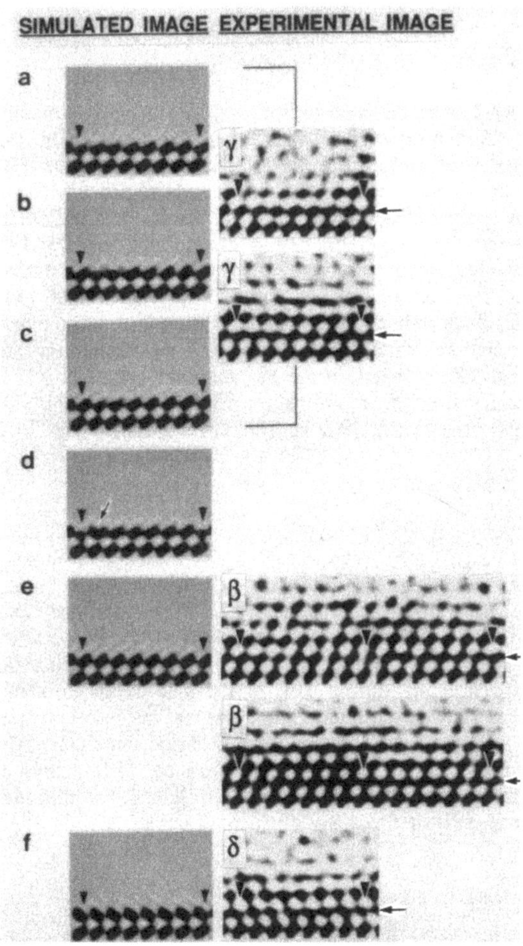

Fig. 8. Comparisons between the experimental and the simulated images for the models shown in Fig. 7. The arrows indicate the DS layers shown in Fig. 5. The arrow heads indicate the corresponding positions between the experimental and the simulated images.

features of the elongated dark spots. It is not easy to detect the differences in the image features between Figs. 8(a), (b) and (c) but the image of Fig. 8(d) is obviously distinguished from the former ones at the point indicated by the arrow, where the reversed elongated dark spot can be seen. Figure 8 also shows the comparison between the experimental images shown in Fig. 5 and the corresponding simulated images. We find that the simulated images account for the variations in the elongated dark spot images in the DS layer (indicated by arrows), except for the simulated image of Fig. 8(d). Therefore, the image γ is interpreted as the projections of the atomic configurations formed by the unfaulting process during 7×7→1×1, in which the dimer structure within the reversed stacking region was modified to a greater or lesser degree. On the contrary, in the region projected on the image δ, the dimer structures within the normal stacking regions are considered to be modified. Namely, the faulting process also occurs, forming the twin structure, during 7×7→1×1, although the unfaulting process is dominant. It is also notable in Fig. 8 that the epitaxial growth above the DS layer scarcely occurred, except in the region showing the 1×1 structure of the image β. This result indicates that 7×7→1×1 proceeds laterally in the DS layer prior to the vertical SPE.

3.2.4. Mechanism of 7×7→1×1

Now the mechanism of 7×7→1×1 can be deduced as follows. The original structure of the interface 7×7 superstructure is composed of the corner hole, the dimer-chain, the stacking fault, and the epitaxial layer structures. Since the strain in the interface 7×7 superstructure is attributed mainly to the dimer structure, the corner hole region where six adjacent dimers exist, has the highest strain fields, followed by the dimer-chain region. If we postulate that the metastable interface 7×7 superstructure is transformed into the 1×1 structure, which is energetically favorable, accompanied by the strain relaxation during the annealing, then the modification first occurs in the corner hole region and next propagates into the dimer-chain region. Such dimer structure modification introduces either an expansion or a shrinkage of the normal stacking region leading to a transformation into the complete 1×1 structure. The vertical SPE proceeds at the 1×1 interface after 7×7→1×1. Therefore, it is concluded that the dimer structure modification prior to the vertical SPE is essential to 7×7→1×1, which is supported by the TED and the cross-sectional HRTEM.

4. Summary

We have demonstrated the direct imaging of the interface 7×7 superstructure buried at the a-Si/Si(111) interface by <110> cross-sectional HRTEM. The interface 7×7 superstructure consists of the dimer-stacking fault layer and the epitaxial layers extended over the vacant corner sites and the dimer-chains. We also investigated structural modification of the interface 7×7 superstructure during SPE of the a-Si overlayer using TED and HRTEM. The HRTEM images showed the intermediate state of 7×7→1×1. It was found that the unfaulting process as well as the faulting process occurred simultaneously through the dimer structure modification, which was also supported by TED analyses, during 7×7→1×1. It was also confirmed that 7×7→1×1 proceeded laterally within the stacking fault layer prior to the vertical SPE of the a-Si.

Acknowledgement

The authors would like to thank T.Niino for her helpful assistance with the experiments , I.Hirosawa, N.Ikarashi, and H.Ono for their valuable discussions, and Y.Matsumoto, M.Nakamae, M.Ogawa, H.Watanabe, and F.Saito for their encouragement.

References

[1] G.Binnig, H.Rohrer, Ch.Gerber, and E.Weibel, Phys.Rev.Lett. 50, 120(1983).
[2] R.J.Hamers, R.M.Tromp, and J.E.Demuth, Surf.Sci.181, 346(1987).
[3] R.E.Schlier and H.E.Farnsworth, J.Chem.Phys.30, 917(1959).
[4] K.Akimoto, J.Mizuki, T.Tatsumi, N.Aizaki, and J.Matsui, Surf.Sci.183, L297(1987).
[5] J.M.Gibson, M.McDonald, and F.C.Unterwald, Phys.Rev.Lett.55, 1765(1985).
[6] J.M.Gibson, H.-J.Gossmann, J.C.Bean, R.T.Tung, and L.C.Feldman, Phys.Rev.Lett.56, 355(1986).
[7] A.Sakai, T.Tatsumi, T.Niino, H.Hirayama, and K.Ishida, Appl.Phys.Lett.55, 2500(1989).
[8] K.Takayanagi, Y.Tanishiro, M.Takahashi, and S.Takahashi, J.Vac.Sci.Technol.A3, 1502(1985).
[9] I.K.Robinson, W.K.Waskiewicz, R.T.Tung, and J.Bohr, Phys.Rev.Lett.57, 2714(1986).
[10] A.Sakai, T.Tatsumi, and K.Ishida, Surf.Sci.224, L956(1989).

Electronic Structure Calculations of Complex Surfaces and Interfaces

S. Ohnishi

Fundamental Research Laboratories, NEC Corporation,
34 Miyukigaoka, Tsukuba, Ibaraki 305, Japan

Abstract. Electronic structures for the metal-semiconductor systems are investigated by the cluster model of several tens of atoms embedded in a model potential for a bulk crystal based on the linear combination of atomic orbitals scheme using norm-conserving pseudopotentials. Active orbitals of Pt and Ir clusters in the scanning tunneling microscopy and Schottky-barrier height changes induced by the formation of reconstructed interfaces in Pb/Si(1$\bar{1}$1) and Au/Si(111) are discussed.

1. Introduction

Recent developments of scanning tunneling microscopy (STM) and microfabrication technologies by single-atom manipulation have introduced revolutionary innovations in material science. A growing number of applications of various kinds of interactions between metal tip and surface in the STM to atomic scale fabrications have presented new possibilities for creating artificial materials on surfaces. Several new material phases, also, are now realized by the ordering formation of monolayer metals and molecules on surfaces and reconstructed metal-metal and metal-semiconductor interfaces, which are expected to provide us with new unknown physical properties. It is of great interest to explore electronic structures of new material phases with the aid of computers. Geometrical structures of these systems of tip-surface and hybridized materials, however, are not always simple enough to be simulated by the present standard technology of the first-principles energy band calculations: the tip-surface interaction essentially localizes at the metal tip protrusion without a two-dimensional periodicity, and the reconstruction at surfaces and interfaces induces the formation of superstructures with a very large unit cell of several tens of atoms or incommensurate non-periodic systems as a whole. For investigating the local electronic structures of such complex systems, the finite atom cluster approach based on the linear combination of atomic orbitals (LCAO) scheme is flexible enough to model any complex structure by extracting relevant atoms from surfaces and interfaces and it is then easy to understand what happens to the electrons in these atoms.

The cluster method, however, has not been extensively applied to these complex systems because of the difficulties caused by the finite size effects associated with problems of (1) number of basis functions, (2) charge neutrality, and (3) boundary effects. As for (1) and (2), the application of the norm-conserving pseudopotential[1] to the LCAO method permits us to deal with large clusters of several tens of atoms of any element by using only valence wave functions effectively. If we focus on the atomic region in the Wigner-Seitz cell at the center of symmetry of such large clusters, both potential field and charge distribution are found to be determined effectively by the self-consistent calculation within the local density functional scheme [2]. The termination problem, (3), is the most difficult. To afford the complete boundary condition requires all the information on the bulk system represented by the on Green's function, which makes calulations intractable. In this paper, in order to avoid trouble caused by the boundary effects, a new model potential, V_M, which describes the local Hamiltonian is introduced so as to constitute an environment for the cluster extracted from a bulk crystal. Attaching V_M's to clusters enables us to discuss electronic structures of the metal tips of Pt and Ir by the model clusters of $Pt_{10}V_{M25}/Si_{21}V_{M25}$, $Ir_{10}V_{M25}/Si_{21}V_{M25}$ for the STM and metal-semiconductor interfaces of Pb/Si(111) and Au/Si(111) by $Pb_{13,10}/Si_{16}V_{M18}$, and $Au_{13,10}/Si_{16}V_{M18}$.

2. Method

The nodeless feature of atomic radial wave functions given by the norm-conserving pseudopotential in the LCAO scheme made it feasible to perform numerical integrations for Hamiltonian and overlap matrix elements and expansions for the augmentation of wave function tail and charge density. After getting a good convergence for the total energy, the model potential, V_M, of non-local separable type is constructed by the local orbitals Ψ_i for eigenstate i and potential V_{loc} as

$$V_M = \sum_i \frac{V_{loc}|\Psi_i\rangle\langle\Psi_i|V_{loc}}{\langle\Psi_i|V_{loc}|\Psi_i\rangle} ,$$

which works to force dangling bonds around the cluster to bulk like bonding orbitals. Figure 1 shows the structures of the $Si_{26}V_{M30}$ cluster and the Wigner-Seitz cell of the Si atom at the center of it. The local density of states (LDOS) is shown in Figure 2, where the s and p band widths agree very well with those by the standard

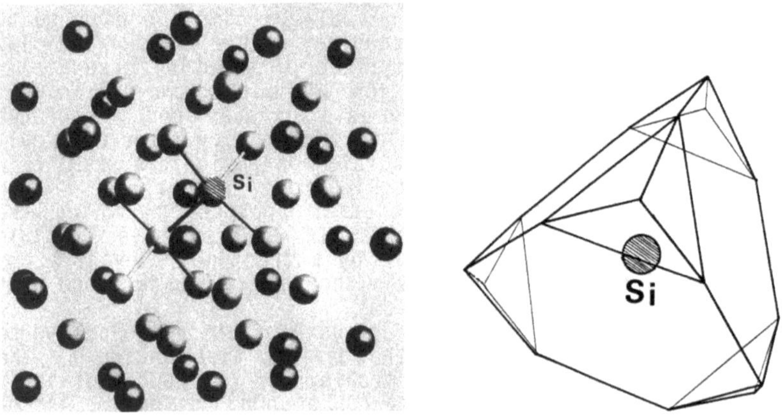

Figure 1: Si bulk model cluster of $Si_{26}V_{M30}$ and the Wigner-Seitz cell. White and black spheres represent Si and V_M, respectively.

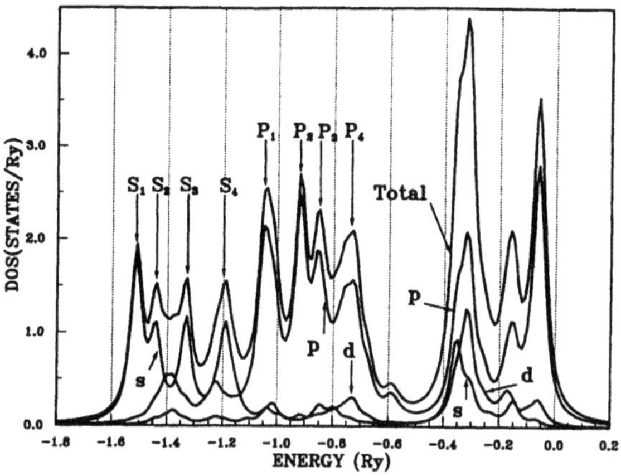

Figure 2: The LDOS of Si atom at the center of $Si_{26}V_{M30}$.

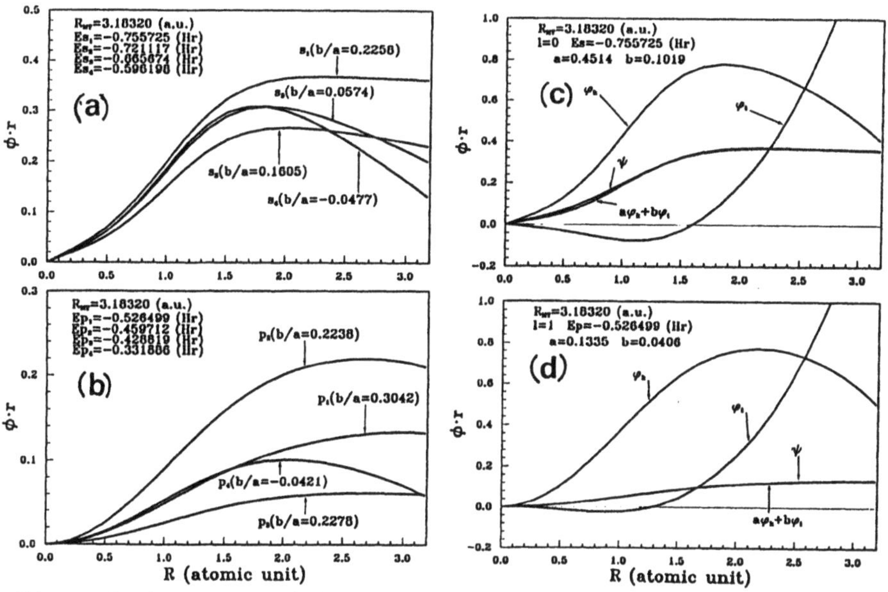

Figure 3: Average radial wave functions of s and p states, (a) and (b), respectively. S_{1-4} and P_{1-4} correspond to those in Figure 2. Two LMTO's and their linear combination to fit radial wave functions are shown in (c) and (d). Expansion parameters a and b are listed in (a) and (b).

method. There are four distinct LDOS peaks. Figure 3 shows their average radial wave functions. Each orbital at these energies is found to be expanded by only two orbitals of the linear muffin-tin orbital (LMTO), $\phi_\ell^k(r)$ and $\phi_\ell^i(r)$, which are given by the linear combination of so called phi and phidot connected to the modified spherical Bessel functions $k_\ell(r)$ and $i_\ell(r)$, respectively. In the present calculations, these four atomic-type local orbitals are applied to construct V_M for Si clusters. The total charge in the

Wigner-Seitz cell of Si given by the sum of the LDOS in Figure 2 is found to be almost neutral, 4.0 ± 0.02 within an accuracy of 5 meV of the highest occupied molecular orbital (HOMO) state.

3. Applications

3.1 Pt and Ir tip in the STM

The metal tip used in the STM is a typical example of a complex surface of which the electronic structures have not been analyzed intensively. For an understanding of the microscopic mechanism of the tunneling current between the tip and surface, atomic and electronic structures of the apex atom at the metal tip protrusion have been recognized to be very important [3]. In order to find the active orbital in the tip, tunneling currents from the Si dimer of Si(100) surface to Pt and Ir tips are calculated by applying model clusters of $Si_{21}V_{M26}$, $Pt_{10}V_{M25}$, and $Ir_{10}V_{M25}$ of which the geometrical structures are shown in Figure 4 together with the LDOS of the Si dimer atoms. V_M's for metals are given by the muffin-tin potential determined by superimposing atomic charge densities. The global feature for the surface LDOS agrees very well with the surface energy band calculations. The bonding state lies just below the Fermi level, E_F. Tunneling currents from this bonding state to various eigenstates of Pt cluster are shown in Figure 5 together with the LDOS of metal tips and, for the case of Ir, in Figure 6. The most active orbitals in both Pt and Ir clusters are found to comprise mainly $6p_z$, hybridized with $5d_{zz}$ and 6s orbitals localized at the apex atom of the cluster.

3.2 Pb/Si(111) and Au/Si(111) interfaces

The relation between atomic structure at the metal-semiconductor interface and Schottky barrier height has attracted considerable attention for a microscopic understanding of the Schottky barrier formation:especially the Si-silicide interfaces have been intensively studied both experimentally and theoretically because of their well-defined abruptness [4]. Electronic structures of a nonreacting metal-semiconductor interface, however, are less well-known because of their complexity. The recent report on Schottky barrier heights at epitaxial Pb/Si(111) interfaces of two types of contacts, 0.1 eV and 0.3 eV higher than the nonreacting case in Si(111)(7×7)-Pb and Si(111)($\sqrt{3}\times\sqrt{3}$)R30°-Pb, respectively, indicated that the reconstruction at the interface may play an important role in determing the Schottky barrier [5].

Since the nature of bonding of the Si dangling bond state at the interface is a key to the understanding of the mechanism for determining the Fermi level, we focus on electronic structures of Si atoms at the interface by the LCAO cluster model calculation. There are two types of Si atoms at the interface of Si(111)($\sqrt{3}\times\sqrt{3}$)R30°-Pb contact:Si with on-top site Pb and that without Pb, denoted by Si^B and Si^D, respectively, modelled by the clusters of $Si_{16}V_{M25}Pb_{13}{}^R$ and $Si_{16}V_{M25}Pb_{10}$ as shown in Figure 7. The LDOS's of Si^B and Pb shown in Figure 8 are similar to the LDOS of the bulk model shown in Figure 2 except for the enhancement of the p state peak due to the local bonding between Si^B and Pb. Table 1 shows total valence electron charges in the Wigner-Seitz cell shown in Figure 1 obtained by the sum of the LDOS at the Si site. Si^B has slightly excess

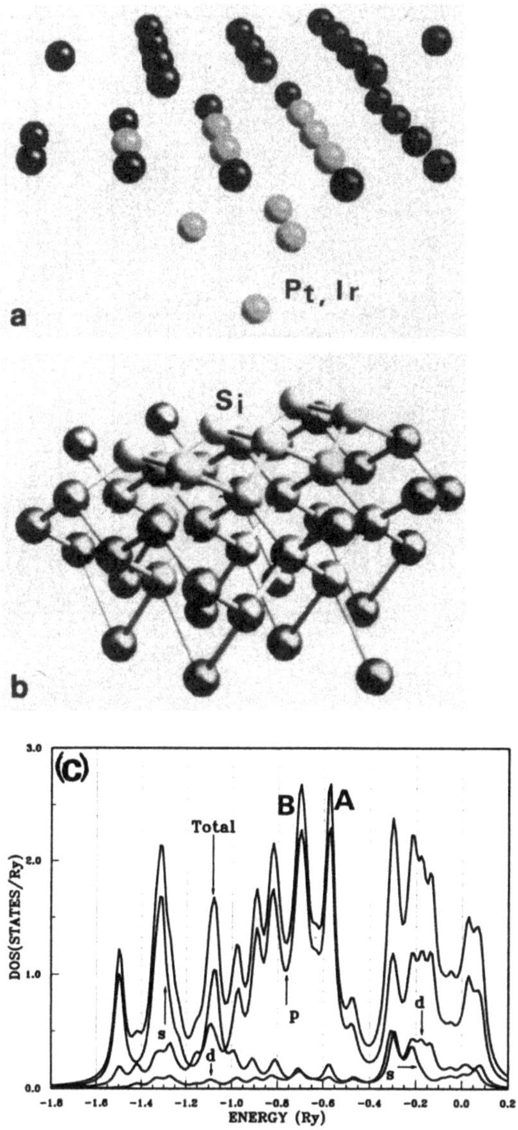

Figure 4: Atomic structures of the metal tip of $Pt(Ir)_{10}V_{M25}$, (a), and Si(100) surface dimer model of $Si_{21}V_{M26}$, (b). The LDOS of Si(100) dimer is shown in (c). B and A represent the bonding and anti-bonding state, respectively.

electrons. The dangling bond state of Si^D lies 0.4 eV below E_F without forming any bond with Pb. Si^D becomes negative at the interface. The cluster $Si_{16}V_{M25}Pb_{13}{}^A$ models an abrupt contact. The Pb cluster on Si(111) being the same as the Pb(111) layers is shifted by x=2, y=2 (atomic unit) parallel to the Si(111) surface so as not to form a Si-Pb bond. The Fermi level is 0.30 eV higher than that of the reconstructed model of $Si_{16}V_{M25}Pb_{13}{}^R$. It is consistent

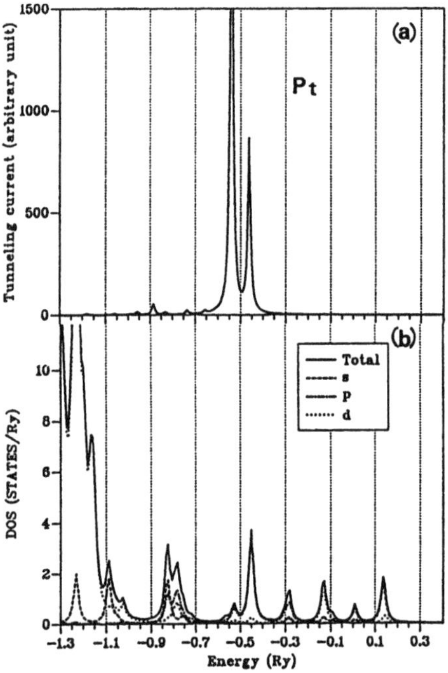

Figure 5

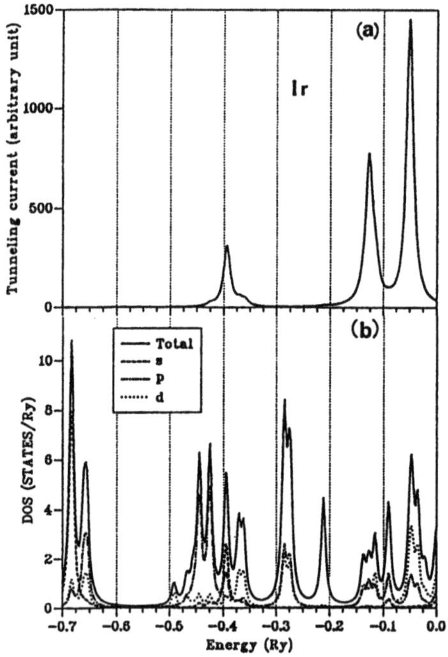

Figure 6

Figure 5: Tunneling current from the bonding state of Si(100) dimer to Pt tip, (a), and the LDOS of the apex atom of $Pt_{10}V_{M25}$, (b).

Figure 6: Tunneling current from the bonding state of Si(100) dimer to Ir tip, (a), and the LDOS of the apex atom of $Ir_{10}V_{M25}$, (b).

Table 1. Electron charges of Si atoms in the Wigner-Seitz cell.

Valence electrons in the Wigner-Seitz cell shown in Figure 1 are calculated by the sum of the LDOS. Si^B and Si^D in the first layer of Si(111) are shown in Figure 7. $Si^{(2)}$, being in the second layer, is the nearest neighbour of $Si^{B,D}$.

	Si^B	Si^D	$Si^{(2)}$
$Pb_{13}Si_{16}V_{M18}^{R}$	4.08		4.01
$Pb_{13}Si_{16}V_{M18}^{A}$	4.00		3.85
$Pb_{10}Si_{16}V_{M18}$		4.30	4.05
$Au_{13}Si_{16}V_{M18}$	3.62		4.02
$Au_{10}Si_{16}V_{M18}$		3.72	3.98
$Si_{26}V_{M30}$	4.02		

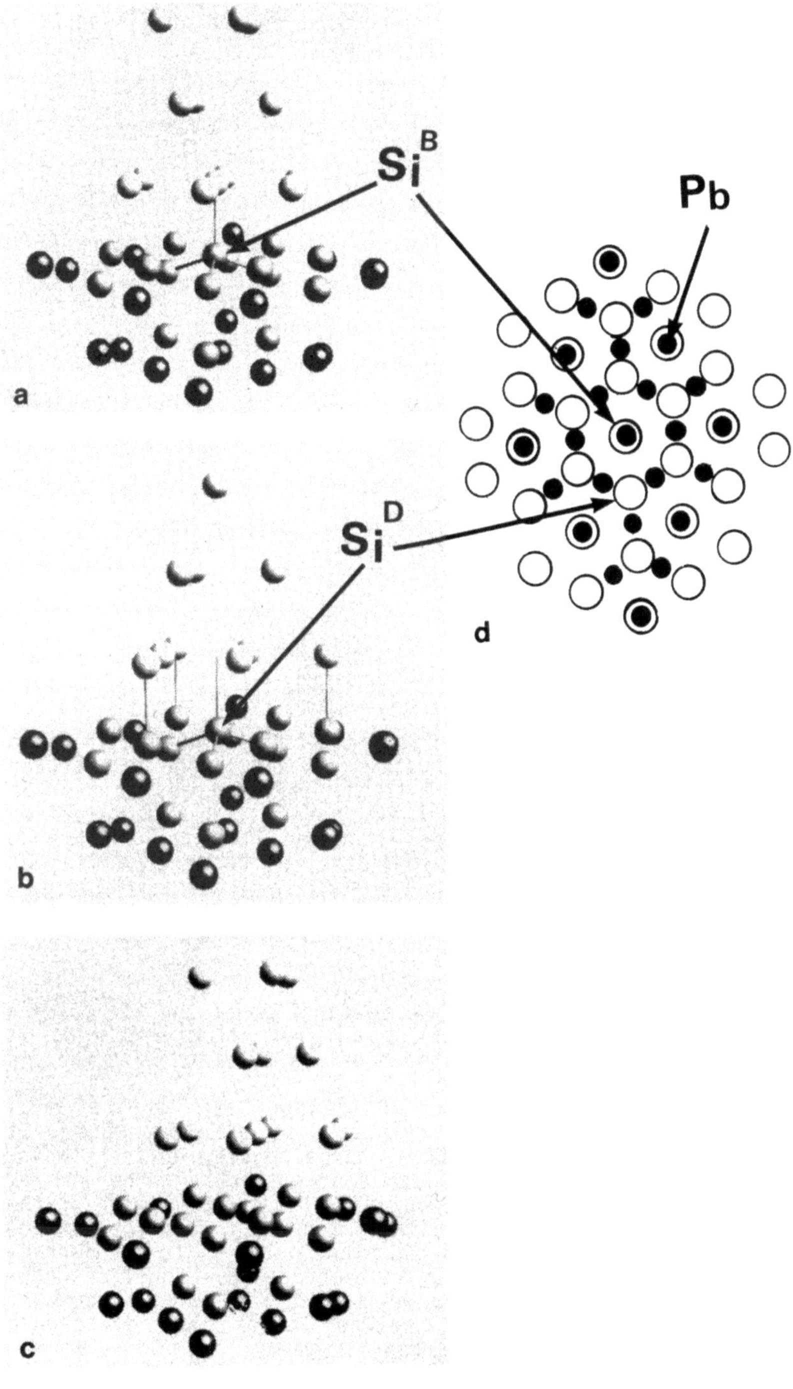

Figure 7: Atomic structures of $Si_{16}V_{M25}Pb_{13}{}^{R}$, (a), $Si_{16}V_{M25}Pb_{10}$, (b), and $Si_{16}V_{M25}Pb_{13}{}^{A}$, (c). (d) shows the top view of Si(111)($\sqrt{3}\times\sqrt{3}$)R30-Pb structure.

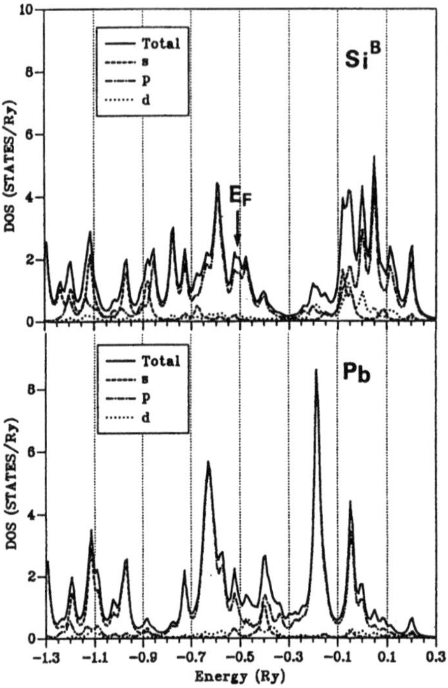

Figure 8: The LDOS of Si and Pb in $Si_{16}V_{M25}Pb_{13}{}^R$.

with the experimental result on the Schottky barrier height provided that the band gap of the Si bulk crystal does not change in the case of Si-Pb contact. Higher E_F is due to the charge transfer from Si to the Pb cluster. In the case of Au contact, Si loses electron charges at the interface, which was expected from the difference between Si and Au work function. The dangling bond state lies at E_F.

4. Summary

The effective model potential for the termination of the finite size cluster extracted from the system of complex structures made it tractable to perform the LCAO cluster model calculation with the aid of the norm-conserving pseudopotential.

The local electronic structures of the apex atom of the metal tip were found to be strongly correlated with the STM images. It should be noted that the single state charge density of the bonding state of the Si(100) dimer at the apex atom was found to be similar to that obtained by the present calculation for the STM image of the constant distance mode. The $6p_z$ orbital extending toward the surface is almost spherically symmetric on the plane parallel to the surface so that the convolution between tip and surface wave functions does not change the symmetry properties of the surface wave functions. Further studies on the relation between atomic and electronic structures of the tip will be reported elsewhere.

The anomalous Schottky barrier height in the reconstructed interface of Si(111)($\sqrt{3}\times\sqrt{3}$)R30-Pb is considered to be due to the fact that the number of dangling bond states is reduced by the ordering formation at the interface and the strong Si-Pb covalent bond works to pull down the Fermi level. The metal induced gap state in this case can be the dangling bond state of Si^D with excess electrons, which however does not pin the Fermi level. The eigenstate at E_F in $Si_{16}V_{M25}Pb_{13}{}^R$ comprises p_z and $p_{x,y}$ orbitals of Si^B and Si^D, respectively, and extended s and p orbitals of Pb with the bond charge in between, while it changes into separate orbitals of Si and Pb clusters in $Si_{16}V_{M25}Pb_{13}{}^A$. As for the Au/Si(111) system, the effect of the Si-Au distance on electronic structures might be more important.

Acknowledgments

The author would like to thank Dr. K. Hirose in NEC Fundamental Research Labs. for fruitful discussions and Mr. T. Ikeda and Mrs. N. Watari in NEC Scientific Information Systems Development LTD for collaborating in the development of the program for the LCAO cluster method.

References

[1] D.R. Hamann, M. Schlüter, and C. Chiang, Phys.Rev.Lett.**43**, 1494(1979);G.B.Bachelet, D.R. Hamann, and M. Schlüter, Phys. Rev.**B26**,4199(1982).
[2] S. Ohnishi,T. Ikeda, and N. Watari, to be published in "Proceedings of the International Conference on Computer Applications to Materials Science and Engineering" Elsevier (1991).
[3] S. Ohnishi and M. Tsukada, Solid State Comm. **71**,391(1989); J.Vac.Sci.Technol **A8**,174(1990);M. Tsukada, K. Kobayashi, and S. Ohnishi, J.Vac.Sci.Technol **A8**,160(1990).
[4] R.T. Tung, Phys.Rev.Lett. **52**,461(1984);M. Liehr, P.E. Schmid, F.K. LeGoues, and P.S. Ho, Phys.Rev.Lett. **54**, 2139(1985); D.R. Hamann and L.F. Mattheiss, Phys.Rev.Lett. **54**, 2517(1985); D.R. Hamann, Phys.Rev.Lett. **60**, 313(1988);H. Fujitani and S. Asano, J.Phys.Soc.Japan **57**, 2253(1988).
[5] D.R. Heslinga, H.H. Weitering, D.P. van der Werf, T.M. Klapwijk, and T. Hibma, Phys. Rev.Lett. **64**, 1589(1990).

Natural Superlattice in $Ga_{0.5}In_{0.5}P$ as a Result of Surface Reconstruction

T. Suzuki[1], *A. Gomyo*[1], and *S. Iijima*[2]

[1]Opto-Electronics Research Laboratories, NEC Corporation, 1-1 Miyazaki 4 chome, Miyamae-ku, Kawasaki-shi 213, Japan

[2]Fundamental Research Laboratories, NEC Corporation, 34 Miyukigaoka, Tsukuba, Ibaraki 305, Japan

Abstract. The formation mechanism of the CuPt-type natural superlattice (NSL), observed in $Ga_{0.5}In_{0.5}P$ grown on (001) GaAs by metalorganic vapor phase epitaxy (MOVPE), is discussed, based mainly on GaAs substrate surface orientation effects on the formation. Reconstruction of steps, extending in the [110] direction, and terraces at the phosphorus rich (001) surface plays an essential role in this model. The structure reconstruction with double periodicity in the [$\bar{1}$10] direction, observed on the flat (001) As-rich GaAs surface, is assumed to exist as a stable structure even in the (001) P-rich $Ga_{0.5}In_{0.5}P$ surface grown by MOVPE, and to persist still in the presence of steps (step-terrace-reconstruction (STR) model). The STR model, together with the assumptions of preferential sticking of one of the cations (probably In atoms) at step-edges due to a steric effect well explains the NSL formation. An effort for justification of these hypotheses is described. The formation mechanism of CuPt type NSL for III-V-V type alloys is also briefly discussed.

1. Introduction

Natural superlattice (NSL), i.e. spontaneously ordered atomic structure on a sublattice, formed during crystal growth, has now been widely reported for various III-V alloy semiconductors obtained by various kinds of growth methods, such as molecular beam epitaxy (MBE), metalorganic vapor phase epitaxy (MOVPE), and chloride transport vapor phase epitaxy (Cl-VPE) [1-19]. There is also one report describing a NSL in a crystal grown by liquid phase epitaxy (LPE) [20].

Among the several types of reported NSLs, CuPt-type NSL is most commonly observed [2-19] and, regarding the NSL, $Ga_{0.5}In_{0.5}P$ grown on GaAs has been a most intensively studied material. The semiconductor plays an important role as an "active layer" material in 600 nm-band red-light emitting visible laser diodes. The existence of the CuPt type NSL, whose structure model is shown in Fig.1, substantially influences the band-gap energy and, in turn, the wavelength of the laser diodes. The band-gap energy of $Ga_{0.5}In_{0.5}P$ has been found to be related to the degree

Eds.: A. Yoshimori · T. Shinjo · H. Watanabe

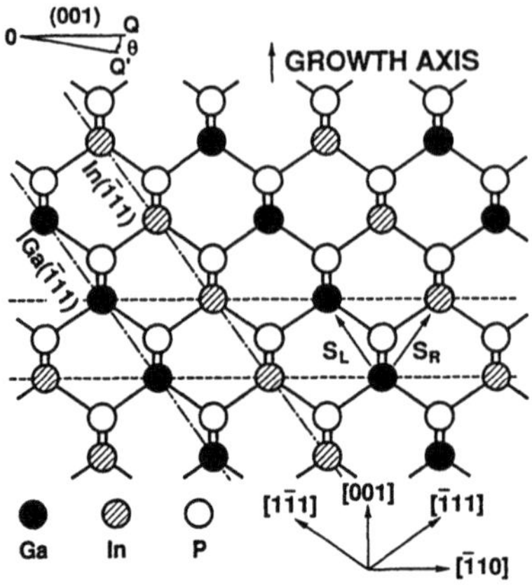

Fig.1 Structure model of a perfect CuPt-type natural superlattice (NSL) in $Ga_{0.5}In_{0.5}P$, projected onto the (110) plane. Ga-planes and In-planes are alternately aligned in the [1̄11] direction, forming a monolayer superlattice. The other NSL is possible in the [11̄1] direction. The period of NSL is 6.5Å. The meaning of $\alpha_1, \alpha_2, \cdots$, or S_L and S_R, or θ will be explained in Section 4.1.

of ordering of the NSL. The relation also has been demonstrated theoretically by Kurimoto and Hamada [21]. The relation has been found to be the case also in the AlGaInP quaternary system and has been used to achieve shorter wavelength (632.7 nm) visible laser diodes [22] and novel "window-structure" high-power visible laser diodes [23]. Thus, understanding and control of the NSL formation is of essential importance for developing and manufacturing visible laser diodes, while the NSL also deserves entire studies in its own right.

This paper discusses the formation mechanism of the CuPt-type NSL, by taking the NSL in $Ga_{0.5}In_{0.5}P$ as an illustrative example, based on experimental results mainly on the effects of GaAs substrate orientations on NSL formation. This paper partly revises the formation mechanism that is discribed in Ref. 24.

2. Experimental

$Ga_{0.5}In_{0.5}P$ epitaxial layers, which are lattice-matched to GaAs substrates, were grown by low-pressure (70 Torr) MOVPE in a hydrogen atomosphere, using triethylgallium (TEGa), trimethylindium (TMIn) for the group III atom (cation) sources, and phosphine (PH_3) for the group V atom (anion) source. Detailed growth conditions are described in Refs. 25, 26 and those referred to in Table 1.

Table 1

Effects of substrate orientation on appearance of NSL variants and on NSL TED intensity ratio for $Ga_{0.5}In_{0.5}P$

Sample No.	Substrate				Appearance of NSL variants				NSL TED intensity ratio
					Pair 1		Pair 2		
	Type [d]	Basic orientation	Off-angle [e] (deg)	Towards	V_1 $[\bar{1}11]$	V_2 $[1\bar{1}1]$	V_3 $[11\bar{1}]$	V_4 $[\bar{1}\bar{1}\bar{1}]$	$I(V_1)/I(V_2)$
1[a]	SUB-E	(001)	0	(exact)	yes	yes	no	no	~1
2[b]	SUB-AB2	(001)	2	[011]	yes	yes	no	no	>1
3[b]	SUB-A2	(001)	2	[110]	yes	yes	no	no	~1
4[b]	SUB-B2	(001)	2	$[\bar{1}10]$	yes	yes	no	no	$\gg 1$
5[a]	SUB-A6	(001)	6	[110]	yes	yes	—[g]	—	~1
6[a]	SUB-B6	(001)	6	$[\bar{1}10]$	yes	yes[f]	—	—	~∞
7[c]	SUB-B	$(\bar{1}\bar{1}\bar{1})$	0.5±0.1	$[01\bar{1}]$	no	no	no	no	
8[c]	SUB-A90	(110)	0	(exact)	no	no	no	no	

a) Ref. 11, b) Ref. 13, c) Ref. 9, d) Way of naming is given in Section 2, e) Orientation accuracies not indicated are all ±0.5°, f) Intensity is very faint, g) Observations were not carried out.

GaAs substrates with various orientations were used. SUB-Aθ, and SUB-Bθ represent the (001) surface substrates with misorientation angle θ (in degrees) towards the [110] and $[\bar{1}10]$ directions, respectively. For example, SUB-B6 has the (001) surface with a misorientation of 6° towards the $[\bar{1}10]$ direction. SUB-ABθ represents the (001) substrates with misorientation angle θ towards the [011] direction. SUB-E and SUB-B mean exactly (001) and $\{\bar{1}\bar{1}\bar{1}\}$ oriented substrates, respectively.

SUB-Aθ or SUB-Bθ will also be used to express surface orientations, even if the surface is a growing surface of an epitaxial layer.

For crystallographic structure studies, transmission electron microscopy (TEM) was used. Cross-section-view samples were prepared by cleaving and Ar ion etching, combined with mechanical polishing.

3. Effects of GaAs Substrate on NSL Formation

3.1 Substrate Orientation Effects on NSL Variant Appearance

This section summarizes the previous results on the substrate orientation effects [3, 9, 11].

In zinc-blende structure, there are four equivalent $\langle\bar{1}\bar{1}\bar{1}\rangle$ directions: $[\bar{1}11]$, $[1\bar{1}1]$, $[11\bar{1}]$, and $[\bar{1}\bar{1}\bar{1}]$. The NSL variants with these orientations, as well as these directions themselves, will hereafter often be referred to as v_1,v_2,v_3, and v_4, respectively. If the NSL formation is governed only by the $Ga_{0.5}In_{0.5}P$ bulk-crystal thermodynamics, the four variants would be observed equally in the crystal [3]. A remarkable asymmetry in the appearance of the NSL variants, however, was observed [3]. It was further found that $Ga_{0.5}In_{0.5}P$ grown on GaAs substrates with various kinds of orientations showed different NSL variant appearances [3, 9, 11]. The results are summarized in Table 1.

The orientation dependence means the following. (1) GaAs substrates only with the (001) or the vicinal (001) surfaces can produce the NSL. (2) Neither v_3 nor v_4 variants are observed. (3) The v_1 variant formation is enhanced, compared with the v_2 variant, for the vicinal (001) surfaces with an off-angle component towards the [$\bar{1}$10] direction (Samples 2,4, and 6 in Table 1).(4) The trend stated in the last item (3) becomes pronounced as the off-angle increases at least up to 6°. (5) The substrate with an off-angle towards the [110] direction does not produce the v_3 or v_4 variants. It rather tends to even inhibit the v_1 and v_2 NSL formation or introduce disorder when the off-angle magnitude increases.

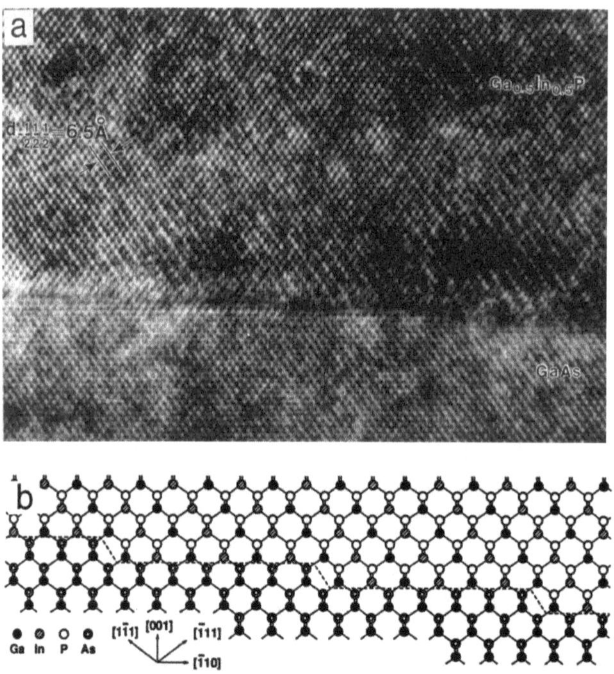

Fig.2 (a) [110]-zone-axis high resolution TEM image for epi-sub interface area for $Ga_{0.5}In_{0.5}P$ grown on SUB-B2 GaAs. NSL v_1 variant and one-molecular-layer (ML) high steps are visible at the interface. NSL is formed from the interface. (b) Structure model of one-ML-high B(+) steps. Although, in this figure, the Ga atom is drawn at each step edge, see Fig. 3 and Sections 4.3 and 4.4.

Considering the above results, together with no NSL appearances for the $\{\bar{1}\bar{1}\bar{1}\}$ and (110) surfaces, it is concluded that the substrate surfaces play a vital role in the NSL formation.

Items (3) to (5) indicate that step arrays on the (001) surface, which are extending in the [110] direction and descending towards the [$\bar{1}$10] direction, play an essential role in the v_1 NSL formation. We call this kind of step B(+) step. By symmetry consideration, the v_2 variant should be related to the step arrays extending in the [110] direction and descending towards the [$\bar{1}$10] direction. These steps are called B(−) steps. B-steps will be used as a general term for B(+) and/or B(−) steps. While the TED patterns for substrates SUB-AB2 and SUB-B2 still showed the existence of the v_2 variant, substrate SUB-B6 virtually did not show the v_2 variant. This is because the high B(+) step density due to the large off-angle for SUB-B6 substrate virtually suppressed the occurrence of B(−) steps.

Figure 2(a) shows a high resolution TEM image of the interface region between $Ga_{0.5}In_{0.5}P$ and SUB-B2 GaAs. An array of one-molecular layer high (2.83Å) B(+) steps is visible at the interface. Hereafter we assume one-molecular-layer (ML) high steps for B-steps. The (001) terrace surfaces are observed between the nearest steps. The v_1 NSL variant is seen in $Ga_{0.5}In_{0.5}P$ layer from the interface. These features are consistently observed in similar samples. Figure 2(b) diagrammatically illustrates an atomic arrangement around the interface region in Fig.2(a).

3.2 Size of Single NSL Domain

Another fact which should be mentioned here is the size of a single coherent domain for the v_1 NSL. In a dark field image for the v_1 NSL in $Ga_{0.5}In_{0.5}P$ grown on SUB-B6 GaAs, very coherent growth of the v_1 NSL was observed and defects, such as anti-phase boundaries (APBs), were scarcely visible [24], in accordance with the fact that this substrate had produced only v_1 variant NSL TED spots, whose intensity was strong and whose shape was spotty. The observation of larger areas for the same specimen revealed that the average lateral size of the domain is about 400Å. A 6° misorientation substrate should have an average terrace width of 27Å, which can accommodate only 3 NSL periods, indicating that, on average, 15 B(+) steps are contained in a single domain, without introducing any defect. *It is quite remarkable that so many steps scarcely introduce any defects for the v_1 NSL in the epitaxial layer on the substrate.* The domain size increased as the off-angle increased at least up to 6°.

4. NSL Formation Mechanism

4.1 Hypotheses for Intra-Plane and Inter-Plane Ordering

Figure 1 illustrates the v_1 variant NSL structure model which is projected onto the (110) plane. OQ' indicates the substrate surface with a misorientation θ. In order to obtain the v_1 variant, which is composed of alternating Ga-planes and In-planes in the $[\bar{1}11]$ direction, two kinds of ordering mechanism will be necessary. They are (1) intra-plane ordering in each (001) plane and (2) inter-plane ordering among the (001) planes in the growth direction, if *layer-by-layer* growth has been occurring on the (001) plane. The first mechanism should assure that Ga-lines and In-lines are arranged alternately in every (001) plane (α_1, α_2, ... in Fig.1). The second mechanism should assure that a definite phase-relation for the nearest two Ga-lines is retained. That is, as shown in Fig.1, the displacement vector S_L for Ga-lines is always taken for the successive (001) growth planes.The following considers the NSL formation process for a *phosphorus-rich (001)* surface. We do not ask whether or not the NSL is already formed beneath the phosphorus rich surface at issue; both cases are allowed.

If the next set of four hypotheses is to hold true, the intra- and inter-plane orderings will actually occur during growth. All the appearance symmetries and asymmetries for the $v_1 \sim v_4$ NSLs described in Section 3.1 are then naturally understood. *Consider the case of $Ga_{0.5}In_{0.5}P$ growth on vicinal (001) GaAs substrate with a misorientation of several degrees (e.g. 6°) towards the $[\bar{1}10]$ direction (i.e. SUB-B6)*; in Fig. 1 this means $\theta = 6°$.

(H1): One-molecular-layer-high steps (B(+) steps) move towards the $[\bar{1}10]$ direction.

(H2): In atoms preferentially stick to each step-edge.

(H3): Ga-line and In-line alternately align in the (001) plane, by the bond-length difference and the bonding-energy difference between Ga-P and In-P, and by a steric effect arising from the wide P-P distance at the step-edge.

(H4): The number of sites with which each terrace can accommodate Ga- or In-lines is all even, if the phosphorus-rich terraced (001) surface is stable. Hereafter the number is referred to as terrace-width. "Terrace width" without a hyphen is also used in the usual manner.

The details and the validity of these hypotheses are discussed in the following.

4.2 H1: One-Molecular-Layer(ML)-High Steps on Growing Surface

As shown in Fig.2(a), the interface between the $Ga_{0.5}In_{0.5}P$ epitaxial layer and the GaAs substrate with the vicinal surface having an off-angle towards the [110] direction shows an array of one ML high (2.83Å) steps

(B(+) steps). TEM observations, which are not described here in detail, revealed that, when a GaAs layer was grown on a $Ga_{0.5}In_{0.5}P$ layer on a similar vicinal (001) GaAs substrate, this upper interface also showed an array of B(+) steps. This fact suggests that the growing surface of $Ga_{0.5}In_{0.5}P$ is also exhibiting an array of B(+) steps, because the feature of the upper interface is considered to represent a general feature of every surface of $Ga_{0.5}In_{0.5}P$ during the layer-by-layer growth.

The hypothesis in Item H1, regarding the movement of steps, seems natural because the step-edges will act as adsorption sites for supplied Ga and In atoms. Lateral growth is also suggested by recent experiments, such as the MOVPE growth of fractional-layer superlattices of AlAs/GaAs, reported by Fukui et al. [27] and the Cl-VPE growth of supersteps of InP/GaInAs on vicinal wafers, reported by Cox et al.[28].

The previous report [10] assumed existence of $<\bar{1}11>$ microfacets at the growing surface and step movements towards the direction perpendicular to the microfacets. Instead of these previous assumptions, a more natural step feature and movement are presumed.

4.3 H2 and H3: Preferential Sticking at Step-Edges and Alternate Arrangement of Ga-Line and In-Line

The preferential sticking of Ga atoms to individual bottom step-edges, as shown in Fig.2(b), apparently seems natural, considering the large bonding energy difference between Ga-P and In-P, which can be supposed from the formation enthalpy difference [29]. Preferential In sticking, *however*, is more plausible, as discussed in Section 4.4. The alternate arrangement mechanism is also discussed there.

4.4 H4: Even Terrace-Width by Step-Terrace-Reconstruction (STR)

If growth progresses, via Figs. 3(b) and (c), obeying H1 ~ H2, the NSL formation results, as shown in Figs.3(d) and (e). Notice that, in Fig.3(e), the terrace-widths are all even.

Large Single Domain Compared with Terrace Width. As mentioned earlier (Section 3.2), the 6° misorientation substrate should have an average terrace width of 27Å, indicating that 15 B(+) steps, on average, are contained in a single domain, where only a very small density of defects is introduced [24]. If terrace-width for these 15 B(+) steps is taken randomly, the terraces with odd terrace-width will generate very many APBs. These, however, are rarely found. This fact strongly suggests that terrace-widths for the surface are all *even* (see the Fig. 3 caption).

2× Structure. As is well known, especially by the recent STM (scanning tunneling microscope) observations for MBE-grown As-stabilized (001) GaAs surfaces [31], As-dimers are formed to show 2×4 reconstruction, in a certain range of growth conditions. The 2× structure is in the $[\bar{1}10]$

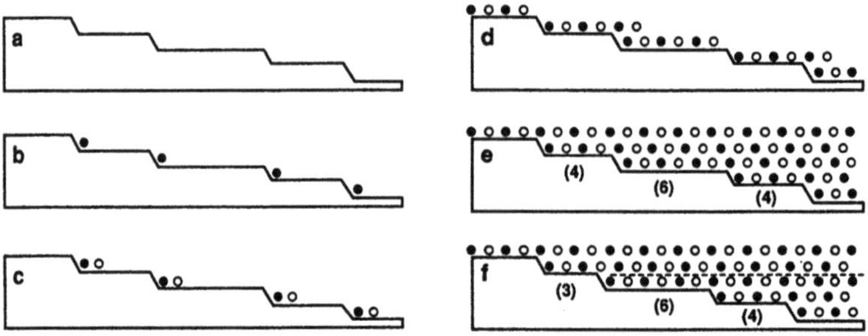

Fig. 3 Growth process of NSL in $Ga_{0.5}In_{0.5}P$ on a terraced (001) plane (projected onto (110) plane). Closed circles are In atoms, and open circles are Ga atoms. P atoms are not drawn in these figures. Numbers in the parentheses are those for cations on the terraces: "terrace-widths". (a) P-rich surface. (b) In atoms stick to each bottom step-edge. (c) Ga atoms stick to the sites next to the In atoms. (d) The process continues. (e) "Terrace-widths" are all even. (f) One odd terrace-width is included; then, an anti-phase boundary (APB), indicated as a dotted line, is introduced, if In atoms stick to the step-edges still in this case. However, as described in the Text (Section 4.4), the stablest structure of the vicinal terraced (001) surface may show even terrace-widths as a result of "step-terrace-reconstruction". Moreover, even if some terrace-widths become momentarily odd, the APBs do not arise, because of the site selectivity of Ga and In atoms on the reconstructed surface. (See the discussion about the STR model in Section 4.4.)

direction. A similar reconstruction will occur also on the P-rich (001) $Ga_{0.5}In_{0.5}P$ surface.

Step-Terrace-Reconstruction (STR) Model. *If the 2× structure persists still on every (001) terrace for a vicinal surface with many B(+) steps*, a structure shown in Fig.4 is expected on a *stable* surface. In this case, the top (s_6,t_6...) of each step-edges ends with a dimer-row in the [110] direction. Every bottom (s_1,t_1,...) of the step-edge starts with dangling-bond rows in the [110] direction. This model for a vicinal (001) surface with B(+) steps is referred to as a *step-terrace-reconstruction (STR) model* [24]. The 2× nature of each terrace on the P-rich surface assures the even terrace-width.

If a cation sticks to, e.g., site t_1 in Fig. 4, the terrace-width *momentarily becomes odd.* However, as this state is considered to be unstable, even terrace-width nature will soon be restored by catching another cation at site t_2 and by forming a P-P dimer on these cations, as is discussed in the following.

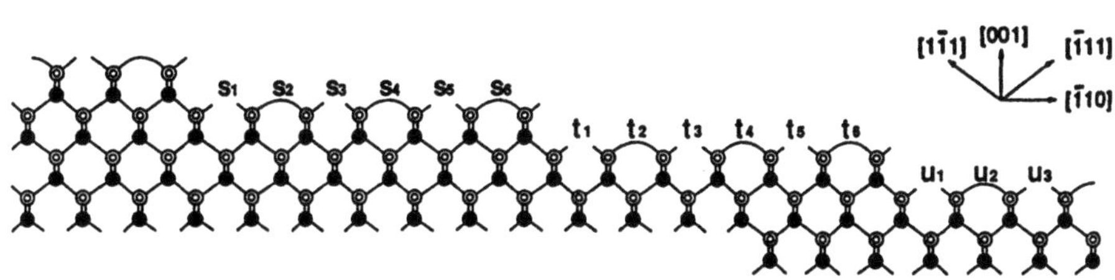

Fig.4 Structure model for anion-rich terraced (001) surface. There are three B(+) steps. Closed circles here represent cations (Ga, In) and double circles indicate anions (P). Due to the "step-terrace-reconstruction (STR)", in a stable state, every terrace shows even parity nature, preserving 2× structure. The present "bond-stick" drawing has not explicitly expressed the narrow spacing of anion atoms of dimers. However, in reality, the spacing may be signifcantly *narrower* (~35%) than that for the dangling bond pair.

Mechanism for Intra-Plane Ordering and Preferential Sticking at Step-Edges. The As-As spacing for an As-dimer is considered to be ~2.6Å [32], which is quite short, compared with the corresponding spacing of 4Å in the bulk GaAs. A similar narrow spacing (~2.5Å) is also reported for a Se-Se dimer in ZnSe [33]. The P-P spacing for a dimer of $Ga_{0.5}In_{0.5}P$ is, by analogy, expected to be considerably shorter than the spacing for the corresponding P-P spacing in the bulk crystal by an amount of at least ~35%. The P-P spacing for sites s_1, t_1, ··· in Fig. 4 will be wider than 4Å by at least ~17%, because the lower-left P atoms for these sites are nearly at the "right" positions. The lower-left P atoms of step-edge bottom sites of s_1, t_1, ··· have stuck to cations by three bonds. The dangling bond from each of these P atoms may be more reactive than the other dangling bonds such as those for sites s_3, s_5, t_3, t_5, ··· and may act as an adsorption bond for step-flow growth. Sticking of In atoms, can occur at step-edges s_1, t_1, ··· due to the relatively easy size-matching between the long In-P bond-length and the wider P-P distance at these sites. On the other hand, Ga sticking may be relatively difficult due to the shorter Ga-P bond-length (Ga-P is shorter than In-P by 7.1 % [29]) to "bridge" across the P-P distance for sites s_1, t_1, ···, in spite of the larger bonding energy of Ga-P.

Thus, by these *steric effects*, the occurrence of preferential In sticking at the step-edges seems plausible. P atoms will soon stick on the In atoms. The next sites s_2, t_2, ··· will be stuck by Ga atoms, in the following way. The dimers at sites s_2, t_2, ··· may break after the In-sticking, regenerating reactive dangling bonds for the next adsorption, while sites s_3, t_3, s_4, t_4, ··· may remain unaffected. This state of the surface will be quite unstable. Then, Ga will stick at these sites, because of the larger bonding energy than that for In-P [29]. The P-P distance at these sites after the breaking is shorter than 4Å by ~4% and thus also preferable for Ga sticking from a steric point of view. The Ga sticking after the In sticking is preferable also from the viewpoint of the stress minimum principle [10, 30], though the strain energy effect is smaller than the bonding-energy difference effect [34]. After that, on the Ga atoms, P atoms will stick quickly. The P-atoms will then form dimers to restore the initial step-edge reconstruction form as $s_6 \sim t_1$. These processes will be repeated to establish the intra-plane ordering. The inter-plane ordering, "phase-locking" [10], is also naturally realized, given the even-nature of terrace-widths due to the STR.

If the bonding-energy difference were predominant over the steric effect, preferential Ga-sticking would occur at sites s_1, t_1, ⋯ . Even in this case, sites s_2, t_2, ⋯ would be occupied by Ga atoms, due to the larger bonding-energy of Ga-P and a smaller contribution of the strain energy effect for In-sticking [34]. This arrangement of Ga-Ga is not preferable for the intra-plane ordering. This also means that preferential In sticking at step-edges due to the steric effect is more plausible than the Ga sticking at these step-edge sites.

In the above disscussion, the step-edges in the [110] direction are assumed to be straight. In reality, these step-edges would have many *kinks*, resulting in local variations in the terrace-widths. However, if the kinks occur in units of 2× structure with the local step-terrace-reconstructions, which are similar to that shown in Fig. 4, the even-nature of the local terrace-widths still holds. Thus, NSL is formed without introducing defects.

4.5 Other III-V Alloys

$Al_xGa_{1-x}As$. Bond-length difference and bonding-energy difference between constituent binaries play important roles in the present model. The roles will become clearer by examining the $Al_xGa_{1-x}As$ case. For $Al_xGa_{1-x}As$, CuPt-type NSL is not reported [1]. Although the formation enthalpy difference, and, in turn, the bonding-energy difference between the constituent binaries is quite large in this alloy system, compared with $Ga_{0.5}In_{0.5}P$ [29], a difference between GaInP and AlGaAs seems rather to lie in the bond-length (b) relation; b(Ga-P) is shorter than b(In-P) by 7%, and b(AlAs) is almost equal to b(GaAs) [29]. Indeed, every III-III-V type alloy which shows the CuPt-type NSL consists of two binaries with large bond-length difference [29]; examples include $Al_{0.5}In_{0.5}P$ [8], $(Al_xGa_{1-x})_{0.5}In_{0.5}P$ [8], $Ga_{0.47}In_{0.53}As$ [5], and $Al_{0.48}In_{0.52}As$ [35]. In these alloys, bonding-energy differences as well as bond-length differences between constituent binaries are significantly large. The fact that the Al-containing alloys mentioned here, in which Al-P and Al-As bonds have quite different bond-lengths from those of the remaining bonds in the respective alloys [29], also show the CuPt-type NSL, indicates that Al has, in itself, no special characteristic in the non-formation of CuPt NSL in $Al_xGa_{1-x}As$; the reason rather seems merely to be the very small bond-length difference (0.14%) between b(Al-As) and b(Ga-As).

Non-existence of the CuPt NSL in $Al_xGa_{1-x}As$, along with the existence of the NSL in the above mentioned alloys, is a strong *disproof* against the possibility that the bonding-energy difference alone is responsible for the alternate alignment of Ga-line and In-line. At the same time, these facts indicate that bond-length difference between constituent binaries is one of the essential factors in the alternate alignment, as discussed in Section 4.4. In this respect, studies on AlGaP and AlGaSb are desirable.

III-V-V Type Alloys. The application of the present formation mechanism to III-V-V type alloys, e.g. GaAsSb [16] and GaPAs [36], is not straightforward, but needs further consideration. However, it is possible to do it by taking the following into consideration : (1) Start with Fig. 4. Let the cation be Ga, and let anions be As and Sb. (2) Ga atoms stick to sites s_1, t_1, ⋯ . (There are only Ga atoms for the cation.) (3) As atoms will preferentially stick onto these Ga atoms, due to the larger As-Ga bonding-energy than that for Sb-Ga [29]. (4) Ga atoms stick to sites s_2, t_2, ⋯ . (5) Sb atoms will preferentially stick onto these Ga atoms to form As-Sb dimers. This will arise from a steric effect, or from the stress minimum condition, or from both. Both come from atom size difference between Sb and As. By steric effect we mean that Sb with the larger atomic radius reacts more easily with the left side As atom to form a dimer. The strain energy for As-Sb dimer formation, getting stored in the four bonds which connect the dimer to the Ga atoms beneath, will be lower than that for the As-As dimer, because the As-Sb spacing of the dimer will be wider than that for As-As (but still narrower than the average anion-anion spacing). (6) The initial step-edge structure is now restored. The process ((2)~(5)) repeats itself to form the v_1 NSL, which was unexpected from the previous model [10].

5. Conclusion

The CuPt-type natural superlattice formation for $Ga_{0.5}In_{0.5}P$ can be explained under the main hypothesis: the step-terrace-reconstruction (STR) model, combined with several other hypotheses, including preferential sticking of In at the step-edges. One of the consequences of the STR model is that each terrace-width in the $[\bar{1}10]$ direction is, in a stable state, even, i.e. the number of sites with which each (001) terrace can accommodate Ga- or In-lines is all even. In the present model, the large bond-length difference still plays a vital role, where the steric effect seems important. The present formation mechanism seems to have been considerably well justified from several aspects. An extension of the model for the NSL formation explains the NSL formation in III-V-V type alloys.

Acknowledgement

The authors express many thanks to I. Hino, K. Kobayashi, S. Kawata, H. Hotta, H. Fujii, and Y. Ueno for their discussions and growing many crystals. They also thank N. Hamada and T. Kurimoto for helpful discussions. They would like to thank M. Sakaguchi for his encouragement, and T. Kawamura for his cooperation at the early stage of this work.

References

[1] T. S. Kuan, T. F. Kuech, W. I. Wang, and E. L. Wilkie, Phys. Rev. Lett. 54(1985)201.
[2] I. J. Murgatroyd, A. G. Norman, G. R. Booker and T. M. Kerr, in: Proc. 11th Intern. Congr. on Electron Microscopy, Kyoto, 1986, Eds. T. Imura, et al. (Japan. Soc. Electron Microscopy, Tokyo,1986).
[3] A. Gomyo, T. Suzuki and S. Iijima, Phys. Rev. Lett. 60 (1988)2645.
[4] A. Gomyo, T. Suzuki, K. Kobayashi, S. Kawata, I. Hino and T. Yuasa, Appl. Phys. Lett. 50(1987)673.
[5] M. A. Shahid, S. Mahajan and D. E. Laughlin, Phys. Rev. Lett. 58(1987)2567.
[6] Y. E. Ihm, N. Ohtsuka, J. Klem and H. Morkoc, Appl. Phys. Lett. 51(1987)3013.
[7] A. Gomyo, K. Kobayashi, S. Kawata, I. Hino, T. Suzuki and T. Yuasa, J. Crystal Growth 77(1986)367.
[8] T. Suzuki, A. Gomyo, S. Iijima, K. Kobayashi, S. Kawata, I. Hino and T. Yuasa, Japan. J. Appl. Phys. 27(1988)2098.
[9] A. Gomyo, T. Suzuki, S. Iijima, H. Hotta,H. Fujii, S. Kawata, K. Kobayashi, Y. Ueno and I. Hino, Japan. J. Appl.Phys. 27(1988)L2370.
[10] T. Suzuki, A. Gomyo and S. Iijima, J. Crystal Growth 93(1988)396. The NSL formation mechanism presented in this reference has been largely revised in Ref. 26 and further revised in the present paper.
[11] A. Gomyo, S. Kawata, T. Suzuki, S. Iijima and I. Hino, Jpn. J. Appl. Phys 28(1989) L1728.
[12] A. Gomyo, H. Hotta, I. Hino, S. Kawata, K. Kobayashi and T. Suzuki, Jpn. J. Appl. Phys. 28(1989) L1330.
[13] T. Suzuki, A. Gomyo, and S. Iijima, J. Cryst. Growth, 99(1990)60.
[14] O. Ueda, M. Takikawa, J. Komeno and I. Umebu, Japan. J. Appl. Phy. 26(1987)L1824.
[15] P. Bellon, J. P. Chevallier, G. P. Martin, E. Dupont-Nivet, C. Thievaut and J. P. Ahdre, Appl. Phys. Lett. 52(1988)567.
[16] I. J. Murgatroyd, A. G. Norman and G. R. Bokker, J. Appl. Phys. 67(1990) 2310.
[17] E. Morita, M. Ikeda, O. Kumagai and K. Kaneko, Appl. Phys. Lett. 53(1988) 2164.

[18] S. Yasuami, C. Nozaki and Y. Ohba, Appl. Phys. Lett. 52(1988) 2031.
[19] S. Minagawa and M. Kondow, Electron. Lett. 25(1989)758.
[20] H. Nakayama and H. Fujita, Inst. Phys. Conf. Ser. No. 79, 1986, Adam Hilger, Bristol, p.289.
[21] T. Kurimoto, and N. Hamada, Phys. Rev. B 40(1989)3889
[22] K. Kobayashi, Y. Ueno, H. Hotta, A. Gomyo, K. Tada, K. Hara, and T. Yuasa, Jpn. J. Appl. Phys. 29 (1990) L1669.
[23] Y. Ueno, H. Fujii, K. Kobayashi, K. Endo, A. Gomyo, K. Hara, S. Kawata, T. Yuasa, and T. Suzuki, Jpn. J. Appl. Phys. 29 (1990) L1666.
[24] T. Suzuki, and A. Gomyo, Proceedings of the 6th International Conference on Molecular Beam Epitaxy, J. Cryst. Growth 111 (1991) 353. The STR model was presented at the conference, which was held in August, 1990, at San Diego, USA. The content in this reference has been partly revised in the present paper.
[25] K. Kobayashi, I. Hino, A. Gomyo, S. Kawata and T. Suzuki, IEEE J. Quantum Electron. QE-23 (1987)704.
[26] T. Suzuki, I. Hino, K. Kobayashi, A. Gomyo, and S. Kawata, OPTOELECTRONICS-Devices and Technologies 4(1989)317.
[27] T. Fukui and H. Saito, Appl. Phys. Lett. 50(1987)824
[28] H. M. Cox, P. S. Lin, A. Yi-Yan, K. Kash, M. Seto, and Bastos, Appl. Phys. Lett. 55(1989)72.
[29] Formation enthalpy (kcal/mole) and lattice constant (-A) values for several binary III-V alloys are cited in this order in the following, for convenience. AlP(39.8, 5.451), GaP(24.4, 5.45), InP(21.2, 5.869), AlAs(27.8, 5.661), GaAs(17, 5.653), InAs(14.0, 6.058), AlSb(20*, 6.136), GaSb(10, 6.096), InSb(7.3, 6.479), where the formation enthalpy value indicated by a mark * *for AlSb* is an extrapolated one. Formation enthalpy (exp.), from: J. C. Philips, Bonds and Bands in Semiconductors, Academic Press, New York, 1973, and lattice constant, from: H. C. Casey,-Jr., and M. B. Panish, *Heterostructure Lasers (B)*, Academic Press, New York, 1978.
[30] J. C. Mikkelsen, Jr. and J. B. Boyce, Phys. Rev. Lett. 49(1982) 1412.
[31] M. D. Pashley, K. W. Haberern, W. Friday, J. M. Woodall, and P. D. Kirchner, Phys. Rev. Lett. 21(1988) 2176.
[32] R.I.G. Uhrberg, R.D. Bringans, R.Z. Bachrach, and J.E. Northrup, Phys. Rev. Lett. 56(1986) 520.
[33] P. H. Fuoss, D. W. Kisker, G. Renaud, K. L. Tokuda, S. Brennan, and J. L. Kahn, Phys. Rev. Lett. 63 (1989)2389.
[34] N. Hamada, and T. Kurimoto, 20th International Conf. on Physics of Semiconductors, Thessaloniki, Greece, August 1990, TuP-129.
[35] O. Ueda, T. Fujii, Y. Nakada, H. Yamada, and I. Umebu, J. Cryst. Growth. 95(1989)38.
[36] T. Suzuki and A. Gomyo, Abstracts for Sixth International Conference on Molecular Beam Epitaxy, San Diego, USA, August 27-31, 1990, XII-3.

Index of Contributors

The Third NEC Symposium October 7—11. 1990 Hakone

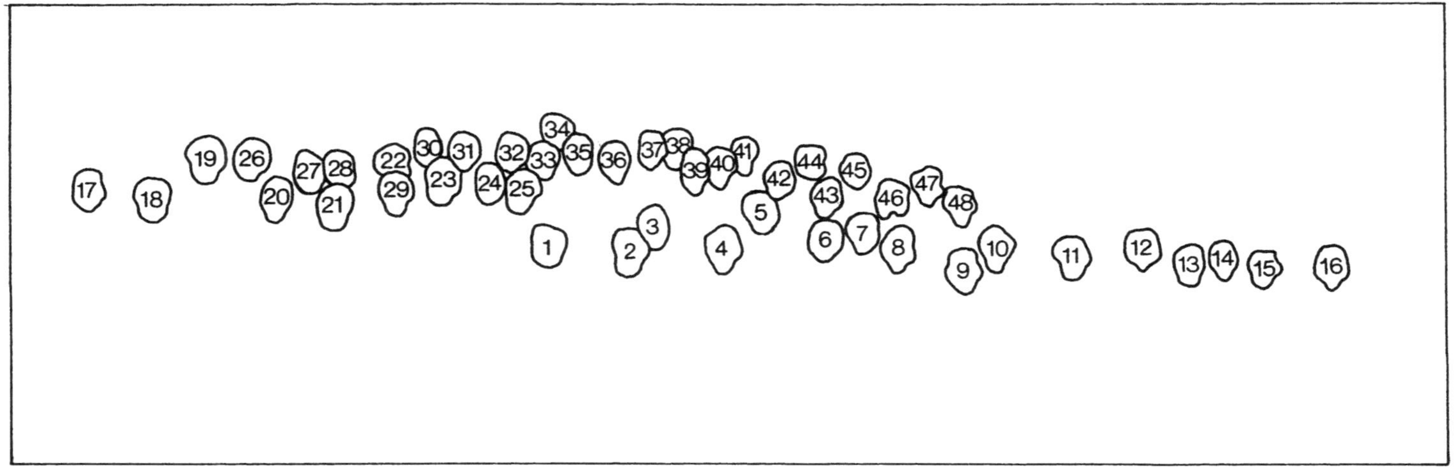

1. A. Ichimiya
2. Tien T. Tsong
3. C. Oshima
4. M. Weinert
5. J. Kanamori
6. Charles M. Falco
7. O. Nishikawa
8. H. Watanabe
9. H. Kamimura
10. D. Dijkkamp
11. S. Sugano
12. T. Mizutani
13. T. Shinjo
14. N. Hamada
15. O. Sugino
16. S. Iijima
17. H. Yasunaga
18. Y. Nishina
19. D. R. Hamann
20. D. R. Halpert
21. K. Tamaru
22. S. Ohnishi
23. J. F. van der Veen
24. K. W. Jacobsen
25. K. Takayanagi
26. H. Tsuda
27. N. Ikarashi
28. M. C. Payne
29. C. J. Bryant
30. H. Hasegawa
31. K. Yagi
32. Y. Murata
33. M. Tsukada
34. A. Sakai
35. S. Kono
36. P. G. Soukiassian
37. J. Mizuki
38. T. Aruga
39. A. Yoshimori
40. K. Hirose
41. C. F. Quate
42. A. Okiji
43. S. Okiji
44. S. Sawada
45. K. Cho
46. S. Cho
47. K. Terakura
48. K. Ishida

MIX
Papier aus verantwortungsvollen Quellen
Paper from responsible sources
FSC® C105338

If you have any concerns about our products,
you can contact us on
ProductSafety@springernature.com

In case Publisher is established outside the EU,
the EU authorized representative is:
Springer Nature Customer Service Center GmbH
Europaplatz 3, 69115 Heidelberg, Germany

Printed by Libri Plureos GmbH
in Hamburg, Germany